T0134444

Intelligent Systems Reference Library

Volume 146

The aim of this series is to publish a Reference Library, including novel advances and developments in all aspects of Intelligent Systems in an easily accessible and well structured form. The series includes reference works, handbooks, compendia, textbooks, well-structured monographs, dictionaries, and encyclopedias. It contains well integrated knowledge and current information in the field of Intelligent Systems. The series covers the theory, applications, and design methods of Intelligent Systems. Virtually all disciplines such as engineering, computer science, avionics, business, e-commerce, environment, healthcare, physics and life science are included. The list of topics spans all the areas of modern intelligent systems such as: Ambient intelligence, Computational intelligence, Social intelligence, Computational neuroscience, Artificial life, Virtual society, Cognitive systems, DNA and immunity-based systems, e-Learning and teaching, Human-centred computing and Machine ethics, Intelligent control, Intelligent data analysis, Knowledge-based paradigms, Knowledge management, Intelligent agents, Intelligent decision making, Intelligent network security, Interactive entertainment, Learning paradigms, Recommender systems, Robotics and Mechatronics including human-machine teaming, Self-organizing and adaptive systems, Soft computing including Neural systems, Fuzzy systems, Evolutionary computing and the Fusion of these paradigms, Perception and Vision, Web intelligence and Multimedia.

More information about this series at http://www.springer.com/series/8578

Hassan AbouEisha · Talha Amin
Igor Chikalov · Shahid Hussain
Mikhail Moshkov

Extensions of Dynamic Programming for Combinatorial Optimization and Data Mining

Hassan AbouEisha
Computer, Electrical and Mathematical Sciences and Engineering Division
King Abdullah University of Science and Technology
Thuwal
Saudi Arabia

Talha Amin
Computer, Electrical and Mathematical Sciences and Engineering Division
King Abdullah University of Science and Technology
Thuwal
Saudi Arabia

Igor Chikalov
Computer, Electrical and Mathematical Sciences and Engineering Division
King Abdullah University of Science and Technology
Thuwal
Saudi Arabia

Shahid Hussain
Computer, Electrical and Mathematical Sciences and Engineering Division
King Abdullah University of Science and Technology
Thuwal
Saudi Arabia

Mikhail Moshkov
Computer, Electrical and Mathematical Sciences and Engineering Division
King Abdullah University of Science and Technology
Thuwal
Saudi Arabia

ISSN 1868-4394 ISSN 1868-4408 (electronic)
Intelligent Systems Reference Library
ISBN 978-3-030-06309-2 ISBN 978-3-319-91839-6 (eBook)
https://doi.org/10.1007/978-3-319-91839-6

Printed on acid-free paper

This Springer imprint is published by the registered company Springer International Publishing AG part of Springer Nature
The registered company address is: Gewerbestrasse 11, 6330 Cham, Switzerland

To our families

Preface

Dynamic programming is an efficient technique to solve optimization problems. It is based on decomposing the initial problem into simpler ones and solving these subproblems beginning from the simplest ones. The aim of a dynamic programming algorithm is to find an optimal object from a given set of objects.

We develop extensions of dynamic programming which allow us to describe the set of objects under consideration, to perform a multi-stage optimization of objects relative to different criteria, to count the number of optimal objects, to find the set of Pareto optimal points for bi-criteria optimization problem, and to study relationships between two criteria.

We present different applications of this technique in the areas of (i) optimization of decision trees, (ii) optimization of decision rules and systems of decision rules, (iii) optimization of element partition trees which are used in finite element methods for solving PDEs, and (iv) study of combinatorial optimization problems.

The applications include optimization of decision trees and decision rule systems as algorithms for problem solving, as ways for knowledge representation, and as classifiers. In addition, we study optimal element partition trees for rectangular meshes, and create the multi-stage optimization approach for such classic combinatorial optimization problems as matrix chain multiplication, binary search trees, global sequence alignment, and shortest paths.

The results presented in this book can be useful for researchers in combinatorial optimization, data mining, knowledge discovery, machine learning, and finite element methods, especially for those who are working in rough set theory, test theory, logical analysis of data, and PDE solvers. The book can be used for the creation of courses for graduate students.

Thuwal, Saudi Arabia
December 2017

Hassan AbouEisha
Talha Amin
Igor Chikalov
Shahid Hussain
Mikhail Moshkov

Acknowledgements

This book is an outcome of our research and teaching work at King Abdullah University of Science and Technology. We are thankful to the administration of KAUST and to our university colleagues for bringing together a great community of people inspired by science that has become a true home for our research group for many years.

We are grateful to our coauthors in papers devoted to the creation of extensions of dynamic programming: Jewahir AbuBekr, Mohammed Al Farhan, Abdulaziz Alkhalid, Maram Alnafie, Saad Alrawaf, Fawaz Alsolami, Mohammad Azad, Monther Busbait, Victor Calo, Pawel Gepner, Damian Goik, Piotr Gurgul, Konrad Jopek, Jacek Kitowski, Krzysztof Kuznik, Andrew Lenharth, Bartlomiej Medygral, Donald Nguyen, Szymon Nosek, Enas Odat, Anna Paszynska, Maciej Paszynski, Keshav Pingali, Marcin Skotniczny, Maciej Wozniak, and Beata Zielosko.

We are thankful to Prof. Andrzej Skowron for stimulating discussions.

We extend an expression of gratitude to Prof. Janusz Kacprzyk, to Dr. Thomas Ditzinger, and to the Series Intelligent Systems Reference Library staff at Springer for their support in making this book possible.

Contents

Part II Decision Trees

Chapter 1
Introduction

This book is devoted to the development of extensions of dynamic programming. The term dynamic programming was proposed by Richard Bellman (see [5]). The idea of dynamic programming can be described as follows: for a given problem, we define the notion of a subproblem and an ordering of subproblems from "smallest" to "largest". If (i) the number of subproblems is polynomial, and (ii) the solution of each subproblem can be computed in polynomial time from the solutions of smaller subproblems then we can design a polynomial algorithm for the initial problem. Even if we cannot guarantee that the number of subproblems is polynomial, the dynamic programming approach continues to be applicable.

The aim of conventional dynamic programming is to find an optimal object (solution) from a finite set of objects (set of possible solutions). We consider extensions of dynamic programming which allow us to

- describe the set of all objects;
- perform multi-stage optimization relative to different criteria;
- describe the set of objects after optimization and to count the number of such objects;
- construct the set of Pareto optimal points for two criteria;
- describe relationships between two criteria.

The areas of applications for extensions of dynamic programming include (but are not limited to) combinatorial optimization, finite element method, fault diagnosis, complexity of algorithms, machine learning, data mining, and knowledge representation.

H. AbouEisha et al., *Extensions of Dynamic Programming for Combinatorial Optimization and Data Mining*, Intelligent Systems Reference Library 146,
https://doi.org/10.1007/978-3-319-91839-6_1

1.1 Multi-stage and Bi-criteria Optimization

The considered algorithms operate with a directed acyclic graph (DAG) which describes the structure of subproblems of the initial problem. This DAG represents the set of objects (the set of possible solutions) under consideration. Each subproblem (node of DAG) is a subset of this set.

We create procedures of optimization relative to different cost functions which start at the terminal nodes and finish in the "topmost" node corresponding to the initial problem. These procedures associate with each node (subproblem) the minimum cost of an object from the subproblem and remove from the initial DAG some edges. The obtained graph represents the set of objects which minimize the considered cost function for the so-called strictly increasing cost functions or a subset of this set for the so-called increasing cost functions. The depth and the number of nodes in a decision tree are examples of increasing and strictly increasing cost functions, respectively. The procedures of optimization can be applied to the DAG sequentially relative to a number of cost functions. We can count the number of objects after each step of this *multi-stage optimization* procedure.

Special attention is given to the study of *bi-criteria optimization*. We correspond to each object a point whose coordinates are values of the two considered criteria for the object. We study the problem of minimization of these criteria. Our aim is to construct the set of all Pareto optimal points (nondominated points) for this problem. We design algorithms which construct the set of Pareto optimal points for each node (subproblem) and use the sets attached to child nodes to construct the set of Pareto optimal points for the node. We also show how the constructed set can be transformed into the graphs of functions which describe the relationships between the considered criteria.

The two approaches we propose are complementary. In contrast to the bi-criteria optimization, the multi-stage optimization relative to two cost functions allows us to construct only one Pareto optimal point corresponding to objects with minimum value of the first cost function. However, it is possible to make multi-stage optimization relative to more than two cost functions, to describe the set of optimal objects for strictly increasing cost functions, and to count the number of such objects.

1.2 Directions of Study and Complexity of Algorithms

The number of nodes in the DAG and the number of Pareto optimal points have crucial impact on the complexity of the considered algorithms.

The main directions of our study are the following:

1. Optimization of decision trees for decision tables;
2. Optimization of decision rules and systems of decision rules for decision tables;

3. Optimization of element partition trees controlling the LU factorization of systems of linear equations which result from the finite element method discretization over two-dimensional meshes with rectangular elements;
4. Classic combinatorial optimization problems including matrix chain multiplication, binary search trees, global sequence alignment, and shortest paths.

For the first two directions, the number of nodes in DAG (the number of so-called separable subtables of the initial decision table described by conditions "attribute = value") can grow exponentially with the growth of the size of the input decision table. However, the number of Pareto optimal points is bounded from above by a polynomial on the size of the input decision table for various cost functions: for the depth and number of nodes of decision trees, for the length and coverage of decision rules, etc. As a result, the considered algorithms are polynomial depending on the number of nodes in the DAG and the size of the decision table. Earlier, we have studied the class of so-called infinite restricted information systems [27, 28]. For decision tables over an infinite restricted information system, the size of tables and the number of nodes in DAGs are bounded from above by polynomials on the number of attributes in the decision table. For such tables, the time complexity of the considered algorithms is polynomial depending on the number of attributes in the input decision table.

For the third direction, both the number of nodes in DAG (the number of submeshes of the considered mesh) and the number of Pareto optimal points for various cost functions are bounded from above by polynomials on the size of the mesh. As a result, the considered algorithms have polynomial time complexity depending on the size of the input mesh.

And finally for the fourth direction, the number of nodes in DAG (the number of subproblems in each of the considered four problems) is polynomial depending on the input length. Corresponding multi-stage optimization procedures (we do not consider here bi-criteria optimization) have polynomial time complexity depending on the length of input.

1.3 Applications

For the decision trees (the first direction), the applications of multi-stage optimization approach include the study of totally optimal (simultaneously optimal relative to a number of cost functions) decision trees for Boolean functions, improvements on the bounds on the depth of decision trees for diagnosis of constant faults in iteration-free circuits over monotone basis, computation of minimum average depth for a decision tree for sorting eight elements (this problem was open since 1968 [9]), study of optimal tests and decision trees for modified majority problem, and designing an algorithm for the problem of reduct optimization. The applications of bi-criteria optimization approach include the comparison of different greedy algorithms for construction of decision trees, analysis of trade-offs for decision trees for corner point detection (used in computer vision [34, 35]), study of derivation of decision rules from decision trees, and a new technique called multi-pruning of decision trees which is used for data mining, knowledge representation and classification. The classifiers constructed by multi-pruning process often have better accuracy than the classifiers constructed by CART [8].

For the decision rules and systems of rules (the second direction), the applications of multi-stage optimization approach include the study of decision rules that are totally optimal relative to the length and coverage (have minimum length and maximum coverage simultaneously), investigation of a simulation of a greedy algorithm for the construction of relatively small sets of decision rules that cover almost all objects (rows), and comparison minimum depth of deterministic and nondeterministic decision trees for total Boolean functions. The set of Pareto optimal points is used to measure the quality of multiple heuristic methods for the construction of systems of decision rules.

For the element partition trees (the third direction), the applications include study of meshes based on multi-stage and bi-criteria optimization approaches, in particular, the investigation of totally optimal element partition trees that are optimal relative to a number of criteria simultaneously. The study of the structure of optimal element partition trees allows the creation of new heuristics for element partition tree optimization [1, 31].

For the classic combinatorial optimization problems (the fourth direction), the study of four known problems (matrix chain multiplication, binary search trees, global sequence alignment, and shortest paths) shows that conventional dynamic programming algorithms can be easily extended to multi-stage optimization procedures.

All the applications considered in this work are related to combinatorial optimization. Most of the tools and an essential part of applications for decision trees, rules and systems of rules is connected with data mining: we consider algorithms for discovery of knowledge in the form of decision trees and rules, and for optimization of trees and rules for better knowledge representation.

1.4 Some Remarks

We have used extensions of dynamic programming for the comparison of heuristics for construction of decision trees and rules. In the case of single-criterion optimization, we have compared trees and rules constructed by heuristics with optimal ones constructed by dynamic programming algorithms. In the case of bi-criteria optimization, we have compared values of two criteria for trees and rules constructed by heuristics with Pareto optimal points.

The existence of, in many cases, totally optimal (optimal relative to a number of criteria simultaneously) decision trees, decision rules, decision rule systems, and element partition trees is one of the most interesting results of our study. We have used two tools for the investigation of this phenomenon: multi-stage optimization (totally optimal objects exist if the result of sequential optimization relative to a number of criteria does not depend on the order of criteria) and bi-criteria optimization (totally optimal relative to two criteria objects exist if and only if there is exactly one Pareto optimal point for the corresponding bi-criteria optimization problem).

For problems of existence of totally optimal decision trees and decision rules, and for the modified majority problem, experimental results allowed us to formulate hypotheses which were proven later. In particular, we have proved that for almost all total Boolean functions, there exist totally optimal decision trees relative to the depth and number of nodes.

1.5 Comparison with Other Investigations

The conventional dynamic programming approach for decision tree optimization was created earlier by Garey [18] (see also [4, 26, 39]).

We began to work in this direction in 2000 [28]. In 2002 we published the first paper related to the multi-stage optimization of decision trees [29] (see also extended version of this paper [30]). Since 2011 we have published several papers about relationships between different parameters of decision trees (see, for example, [11, 13, 25]). The considered algorithms also construct the set of Pareto optimal points.

We are not aware of any previous results related to extensions of dynamic programming (multi-stage optimization and bi-criteria optimization) for the study of decision trees. To the best of our knowledge, there are no previous results connected with the dynamic programming approach for decision rule optimization and for element partition tree optimization.

The situation with extensions of dynamic programming for classic combinatorial optimization problems is different. We should mention, in particular, Algebraic Dynamic Programming (ADP) approach created by Robert Giegerich and his colleagues in 1998 in which the problem decomposition is described by a tree grammar and the optimization criterion is given by an evaluation algebra satisfying Bellman's principle [20–23, 38]. Applications of this approach are mainly related to bioinformatics. Initially, this approach included optimization relative to one criterion. In 2004 the authors added the multi-stage (lexicographic) optimization [24] (see also [42]), and in 2014 they added the construction of the set of Pareto optimal points for bi-criteria optimization problems [19, 36, 37].

1.6 Contents of Book

This book includes Introduction, five parts consistsing of 18 chapters, and an appendix.

1.6.1 Part I. Common Tools: Pareto Optimal Points and Decision Tables

In Part I, we consider tools that are common and can be considered prerequisites for some of the later parts of the book.

Note that the results presented in this part were obtained jointly by our research group in KAUST including Fawaz Alsolami and Mohammad Azad who are not the coauthors of the book.

In Chap. 2, we discuss tools (statements and algorithms) which are used for the study of Pareto optimal points for bi-criteria optimization problems related to decision trees (Part II), decision rules and decision rule systems (Part III), and element partition trees (Part IV). We correspond to each object (tree, rule, etc.) a point where the coordinates are values of the two considered criteria for the given object. We consider the problem of minimization of these criteria, and construct for this problem all Pareto optimal points (nondominated points). It is easy if the number of objects under consideration is reasonable. However, if the number of objects is huge (as for the case of decision trees, decision rules, decision rule systems, and element partition trees), we use special algorithms which are based on the tools proposed in this chapter.

In Chap. 3, we focus on tools connected with decision tables. These tools are used in Parts II and III. We describe the main notions related to decision tables, consider the structure of subtables of a given decision table represented as a directed acyclic graph (DAG), and discuss time complexity of algorithms on DAGs. We also consider classes of decision tables over so-called restricted information systems for which these algorithms have polynomial time complexity depending on the number of conditional attributes in the input decision table.

1.6.2 Part II. Decision Trees

Part II is devoted to the development of extensions of dynamic programming for the study of decision trees. The considered extensions include multi-stage optimization of decision trees relative to a sequence of cost functions, counting the number of optimal trees, and studying of relationships of cost versus cost and cost versus uncertainty for decision trees by construction the set of Pareto optimal points for the corresponding bi-criteria optimization problem.

In Chap. 4, we introduce different kinds of decision trees, including approximate decision trees and irredundant decision trees relative to two types of uncertainties. We describe the set of decision trees corresponding to a subgraph of a DAG constructed for a decision table, and show how to compute the cardinality of this set. We also discuss the notion of a cost function for decision trees.

In Chap. 5, we consider multi-stage optimization for decision trees and two of its applications: study of totally optimal decision trees for Boolean functions, and bounds on the depth of decision trees for diagnosis of constant faults in iteration-free circuits over monotone basis.

Chapter 6 contains more applications of multi-stage optimization for decision trees: computation of minimum average depth of a decision tree for sorting eight elements, study of optimal tests and decision trees for modified majority problem, and design of an algorithm for the problem of reduct optimization.

In Chap. 7, we consider an algorithm which constructs the sets of Pareto optimal points for bi-criteria optimization problems for decision trees relative to two cost functions. We describe how the constructed set of Pareto optimal points can be transformed into the graphs of functions which describe the relationships between the considered cost functions. We also show three applications of bi-criteria optimization for two cost functions: comparison of different greedy algorithms for construction of decision trees, analysis of trade-offs for decision trees for corner point detection, and study of derivation of decision rules from decision trees.

Finally, Chap. 8 is connected with the bi-criteria optimization problem for decision trees relative to a cost function and an uncertainty measure. We explain how to construct the set of Pareto optimal points and graphs of functions which describe the relationships between the considered cost function and uncertainty measure. We end this chapter with some illustrative examples and a discussion of multi-pruning of decision trees with applications from data mining, knowledge representation, and classification.

1.6.3 Part III. Decision Rules and Systems of Decision Rules

Part III is devoted to the development of extensions of dynamic programming for design and analysis of decision rules and systems of decision rules.

In Chap. 9, we discuss main notions related to decision rules and systems of decision rules including different kinds of rules and rule systems, and notions of uncertainty and cost function.

In Chap. 10, we consider multi-stage optimization of decision rules with regards to different criteria, describe how to count the number of optimal rules, and discuss three applications. We study from experimental and theoretical points of view decision rules that are totally optimal relative to the length and coverage, investigate simulation of a greedy algorithm for the construction of relatively small sets of decision rules that cover almost all rows, and compare minimum depth of deterministic and nondeterministic decision trees for total Boolean functions.

In Chap. 11, we discuss algorithms which construct the sets of Pareto optimal points for bi-criteria optimization problems for decision rules and decision rule systems relative to two cost functions. We also show how the constructed set of Pareto optimal points can be transformed into the graphs of functions which describe the relationships between the considered cost functions. Further, we use the set of Pareto optimal points to measure the viability of multiple heuristic methods for the construction of systems of decision rules.

Finally, in Chap. 12, we consider algorithms which construct the sets of Pareto optimal points for bi-criteria optimization problems for decision rules and decision rule systems relative to cost and uncertainty. We also present experiments displaying the existence of trade-offs in various decision tables.

1.6.4 Part IV. Element Partition Trees

Part IV is devoted to the optimization of element partition trees controlling the LU factorization of systems of linear equations. These systems result from the finite element method discretization over two-dimensional meshes with rectangular elements.

Chapter 13 starts by formally defining the class of meshes under study. We describe the notion of element partition tree and present an abstract way of defining optimization criteria of element partition trees in terms of cost functions. A definition of a cost function is provided in addition to a few examples of cost functions under study along with some of their properties.

Chapter 14 presents tools and applications of multi-stage optimization of element partition trees. It starts by describing how the set of element partition trees can be represented compactly in the form of a DAG. It then suggests algorithms to count the number of element partition trees represented by a DAG and to optimize the corresponding set of element partition trees with respect to some criterion. The latter algorithm can be used for multi-stage optimization of element partition trees relative to a sequence of cost functions and for the study of totally optimal element partition trees. Finally, we present results of experiments on three different classes of adaptive meshes.

In Chap. 15, we study bi-criteria optimization of element partition trees. This chapter presents an algorithm for finding Pareto optimal points corresponding to Pareto optimal element partition trees, and some applications of this algorithm.

1.6.5 Part V. Combinatorial Optimization Problems

Part V is devoted to the development of the multi-stage optimization procedures for four different classic combinatorial optimization problems which have simple dynamic programming solutions.

In Chap. 16, we study the matrix chain multiplication problem. For this problem, we consider different cost functions and present a multi-stage optimization procedure relative to a sequence of such functions.

In Chap. 17, we present a procedure of multi-stage optimization of binary search trees relative to the weighted average depth and weighted depth. The considered procedure is based on the representation of all possible binary search trees by a DAG.

In Chap. 18, we study the global sequence alignment problem based on the extension of dynamic programming related to multi-stage optimization. We provide a method to model this problem using a DAG which allows us to describe all optimal solutions specific to a certain cost function. Furthermore, we can optimize sequence alignments successively relative to different optimization criteria.

In Chap. 19, we present an algorithm for multi-stage optimization of paths in a directed graph relative to different cost functions. We assume graph edges are assigned with several sets of weights representing different criteria. We do optimization sequentially: at each stage we consider one set of weights and either minimize sum of weights along a path or maximize the minimum edge weight.

1.6.6 Appendix

The appendix describes the capabilities and architecture of Dagger, a software system developed in KAUST for research and educational purposes. The system implements several algorithms dealing with decision trees, systems of decision rules, and more general class of problems than can be solved by dynamic programming. The first section of the appendix describes features of the system, lists third-party libraries used and states necessary environment requirements for running Dagger. The second section highlights key design decisions.

Note that the algorithms for the optimization of element partition trees are implemented in a special software system.

1.7 Use of Book

The algorithms described in this book can work with medium-sized decision tables, meshes, and classic combinatorial optimization problems. Current versions of these tools have research goals and are not intended for industrial applications.

The created tools can be useful in rough set theory and its applications [32, 33, 40] where decision rules and decision trees are widely used, in logical analysis of data [6, 7, 15] which is based mainly on the use of patterns (decision rules), in test theory [10, 41, 43] which study both decision trees and decision rules, in finite element methods [16, 17] as a way to improve quality of solvers, and in the study of combinatorial optimization problems – the considered examples [2, 3, 12, 14] show that conventional dynamic programming algorithms can be extended often to multi-stage optimization dynamic programming algorithms.The book can be used also for the creation of courses for graduate students.

References

1. AbouEisha, H., Gurgul, P., Paszynska, A., Paszynski, M., Kuznik, K., Moshkov, M.: An automatic way of finding robust elimination trees for a multi-frontal sparse solver for radical 2D hierarchical meshes. In: Wyrzykowski, R., Dongarra, J., Karczewski, K., Wasniewski, J. (eds.) Parallel Processing and Applied Mathematics – 10th International Conference, PPAM 2013, Warsaw, Poland, September 8–11, 2013, Revised Selected Papers, Part II. Lecture Notes in Computer Science, vol. 8385, pp. 531–540. Springer, Berlin (2013)
2. AbuBekr, J., Chikalov, I., Hussain, S., Moshkov, M.: Sequential optimization of paths in directed graphs relative to different cost functions. In: Sato, M., Matsuoka, S., Sloot, P.M.A., van Albada, G.D., Dongarra, J. (eds.) International Conference on Computational Science, ICCS 2011, Nanyang Technological University, Singapore, June 1–3, 2011. Procedia Computer Science, vol. 4, pp. 1272–1277. Elsevier (2011)
3. Alnafie, M., Chikalov, I., Hussain, S., Moshkov, M.: Sequential optimization of binary search trees for multiple cost functions. In: Potanin, A., Viglas, T. (eds.) Seventeenth Computing: The Australasian Theory Symposium, CATS 2011, Perth, Australia, January 2011. CRPIT, vol. 119, pp. 41–44. Australian Computer Society (2011)
4. Bayes, A.J.: A dynamic programming algorithm to optimise decision table code. Aust. Comput. J. **5**(2), 77–79 (1973)
5. Bellman, R.: The theory of dynamic programming. Bull. Am. Math. Soc. **60**(6), 503–516 (1954)
6. Boros, E., Hammer, P.L., Ibaraki, T., Kogan, A.: Logical analysis of numerical data. Math. Program. **79**, 163–190 (1997)
7. Boros, E., Hammer, P.L., Ibaraki, T., Kogan, A., Mayoraz, E., Muchnik, I.: An implementation of logical analysis of data. IEEE Trans. Knowl. Data Eng. **12**, 292–306 (2000)
8. Breiman, L., Friedman, J.H., Olshen, R.A., Stone, C.J.: Classification and Regression Trees. Wadsworth and Brooks, Monterey (1984)
9. Césari, Y.: Questionnaire, codage et tris. Ph.D. thesis, Institut Blaise Pascal, Centre National de la Recherche (1968)
10. Chegis, I.A., Yablonskii, S.V.: Logical methods of control of work of electric schemes. Trudy Math. Inst. Steklov (in Russian) **51**, 270–360 (1958)
11. Chikalov, I., Hussain, S., Moshkov, M.: Relationships between depth and number of misclassifications for decision trees. In: Kuznetsov, S.O., Slezak, D., Hepting, D.H., Mirkin, B.G. (eds.) 13th International Conference Rough Sets, Fuzzy Sets, Data Mining and Granular Computing (RSFDGrC 2011), Moscow, Russia, June 25–27, 2011. Lecture Notes in Computer Science, vol. 6743, pp. 286–292. Springer, Berlin (2011)
12. Chikalov, I., Hussain, S., Moshkov, M.: Sequential optimization of matrix chain multiplication relative to different cost functions. In: Cerná, I., Gyimóthy, T., Hromkovic, J., Jeffery, K.G, Královic, R., Vukolic, M., Wolf, S. (eds.) SOFSEM 2011: Theory and Practice of Computer Science – 37th Conference on Current Trends in Theory and Practice of Computer Science, Nový Smokovec, Slovakia, January 22–28, 2011. Lecture Notes in Computer Science, vol. 6543, pp. 157–165. Springer, Berlin (2011)
13. Chikalov, I., Hussain, S., Moshkov, M.: Relationships between number of nodes and number of misclassifications for decision trees. In: Yao, J., Yang, Y., Slowinski, R., Greco, S., Li, H., Mitra, S., Polkowski, L. (eds.) 8th International Conference Rough Sets and Current Trends in Computing, RSCTC 2012, Chengdu, China, August 17–20, 2012. Lecture Notes in Computer Science, vol. 7413, pp. 212–218. Springer, Berlin (2012)
14. Chikalov, I., Hussain, S., Moshkov, M., Odat, E.: Sequential optimization of global sequence alignments relative to different cost functions. In: ACM International Conference on Convergence and Hybrid Information Technology, ICHIT 2010, Daejeon, Korea, August 26–28, 2010. ACM (2010)
15. Crama, Y., Hammer, P.L., Ibaraki, T.: Cause-effect relationships and partially defined Boolean functions. Ann. Oper. Res. **16**, 299–326 (1988)

16. Demkowicz, L.: Computing with hp-Adaptive Finite Elements, Volume 1: One and Two Dimensional Elliptic and Maxwell Problems. Chapman & Hall/CRC Applied Mathematics and Nonlinear Science Series. Chapman & Hall/CRC, Boca Raton, FL (2006)
17. Demkowicz, L., Kurtz, J., Pardo, D., Paszynski, M., Rachowicz, W., Zdunek, A.: Computing with hp-Adaptive Finite Elements, Volume 2: Frontiers: Three Dimensional Elliptic and Maxwell Problems with Applications. Chapman & Hall/CRC Applied Mathematics and Nonlinear Science Series. Chapman & Hall/CRC, Boca Raton (2007)
18. Garey, M.R.: Optimal binary identification procedures. SIAM J. Appl. Math. **23**, 173–186 (1972)
19. Gatter, T., Giegerich, R., Saule, C.: Integrating pareto optimization into dynamic programming. Algorithms **9**(1), 12 (2016). https://doi.org/10.3390/a9010012
20. Giegerich, R.: A declarative approach to the development of dynamic programming algorithms, applied to RNA folding. Report 98-02, Faculty of Technology, Bielefeld University (1998)
21. Giegerich, R.: A systematic approach to dynamic programming in bioinformatics. Bioinformatics **16**(8), 665–677 (2000)
22. Giegerich, R., Kurtz, S., Weiller, G.F.: An algebraic dynamic programming approach to the analysis of recombinant DNA sequences. In: Workshop on Algorithmic Ascpects of Advanced Programming Languages, WAAAPL'99, Paris, France, September 30, pp. 77–88 (1999)
23. Giegerich, R., Meyer, C., Steffen, P.: A discipline of dynamic programming over sequence data. Sci. Comput. Program. **51**(3), 215–263 (2004)
24. Giegerich, R., Steffen, P.: Pair evaluation algebras in dynamic programming. In: 21st Workshop of the GI-Fachgruppe Programming Languages and Computing Concepts, Bad Honnef, Germany, May 3–5, 2004, pp. 115–124 (2005). http://www.uni-kiel.de/journals/servlets/MCRFileNodeServlet/jportal_derivate_00001127/2004_tr10.pdf
25. Hussain, S.: Relationships among various parameters for decision tree optimization. In: Faucher, C., Jain, L.C. (eds.) Innovations in Intelligent Machines-4 – Recent Advances in Knowledge Engineering. Studies in Computational Intelligence, vol. 514, pp. 393–410. Springer, Berlin (2014)
26. Martelli, A., Montanari, U.: Optimizing decision trees through heuristically guided search. Commun. ACM **21**(12), 1025–1039 (1978)
27. Moshkov, M.: On the class of restricted linear information systems. Discret. Math. **307**(22), 2837–2844 (2007)
28. Moshkov, M., Chikalov, I.: On algorithm for constructing of decision trees with minimal depth. Fundam. Inform. **41**(3), 295–299 (2000)
29. Moshkov, M., Chikalov, I.: Sequential optimization of decision trees relatively different complexity measures. In: 6th International Conference Soft Computing and Distributed Processing. Rzeszòw, Poland, June 24–25, pp. 53–56 (2002)
30. Moshkov, M., Chikalov, I.: Consecutive optimization of decision trees concerning various complexity measures. Fundam. Inform. **61**(2), 87–96 (2004)
31. Paszynska, A., Paszynski, M., Jopek, K., Wozniak, M., Goik, D., Gurgul, P., AbouEisha, H., Moshkov, M., Calo, V.M., Lenharth, A., Nguyen, D., Pingali, K.: Quasi-optimal elimination trees for 2D grids with singularities. Sci. Program. **2015**, 303,024:1–303,024:18 (2015)
32. Pawlak, Z.: Rough Sets - Theoretical Aspect of Reasoning About Data. Kluwer Academic Publishers, Dordrecht (1991)
33. Pawlak, Z., Skowron, A.: Rudiments of rough sets. Inf. Sci. **177**(1), 3–27 (2007)
34. Rosten, E., Drummond, T.: Fusing points and lines for high performance tracking. In: 10th IEEE International Conference on Computer Vision, ICCV 2005, Beijing, China, October 17–20, 2005, pp. 1508–1515. IEEE Computer Society (2005)
35. Rosten, E., Drummond, T.: Machine learning for high-speed corner detection. In: Leonardis, A., Bischof, H, Pinz, A. (eds.) 9th European Conference on Computer Vision, ECCV 2006, Graz, Austria, May 7–13, 2006. Lecture Notes in Computer Science, vol. 3951, pp. 430–443. Springer (2006)
36. Saule, C., Giegerich, R.: Observations on the feasibility of exact pareto optimization. In: Jossinet, F., Ponty, Y., Waldispühl, J. (eds.) 1st Workshop on Computational Methods for

Structural RNAs, CMSR 2014, Strasbourg, France, September 7, 2014, pp. 43–56. McGill University (2014). https://doi.org/10.15455/CMSR.2014.0004
37. Saule, C., Giegerich, R.: Pareto optimization in algebraic dynamic programming. Algorithms Mol. Biol. **10**, 22 (2015). https://doi.org/10.1186/s13015-015-0051-7
38. Sauthoff, G., Möhl, M., Janssen, S., Giegerich, R.: Bellman's GAP - a language and compiler for dynamic programming in sequence analysis. Bioinformatics **29**(5), 551–560 (2013)
39. Schumacher, H., Sevcik, K.C.: The synthetic approach to decision table conversion. Commun. ACM **19**(6), 343–351 (1976)
40. Skowron, A., Rauszer, C.: The discernibility matrices and functions in information systems. In: Słowiński, R. (ed.) Intelligent Decision Support: Handbook of Applications and Advances of the Rough Sets Theory, pp. 331–362. Kluwer Academic Publishers, Dordrecht (1992)
41. Soloviev, N.A.: Tests (Theory, Construction, Applications). Nauka, Novosibirsk (1978). (in Russian)
42. Steffen, P., Giegerich, R.: Versatile and declarative dynamic programming using pair algebras. BMC Bioinform. **6**:224 (2005). https://doi.org/10.1186/1471-2105-6-224
43. Zhuravlev, J.I.: On a class of partial Boolean functions. Diskret. Analiz (in Russian) **2**, 23–27 (1964)

Part I
Common Tools: Pareto Optimal Points and Decision Tables

In this part, we consider tools that are common for two or more parts of the book. In Chap. 2, we discuss tools which are used for the study of Pareto optimal points for bi-criteria optimization problems in Parts II, III, and IV. In Chap. 3, we consider tools connected with decision tables. These tools are used in Parts II and III.

Chapter 2
Tools for Study of Pareto Optimal Points

In this chapter, we consider tools (statements and algorithms) which are used for the study of Pareto optimal points for bi-criteria optimization problems related to decision trees (Part II), decision rules and decision rule systems (Part III), and element partition trees (Part IV). The term Pareto optimal point is named after Vilfredo Pareto (1848–1923).

We correspond to each object (tree, rule, etc.) a point which coordinates are values of the two considered criteria for the given object. We study the problem of minimization of these criteria, and construct for this problem all Pareto optimal points (nondominated points). It is easy to do if the number of objects under consideration is reasonable. However, if the number of objects can be huge (as for the case of decision trees, decision rules, decision rule systems, and element partition trees), we should use special algorithms which are based on the tools considered in this chapter.

Let $\mathbb{R}^2$ be the set of pairs of real numbers (*points*). We consider a partial order $\leq$ on the set $\mathbb{R}^2$ (on the plane): $(c, d) \leq (a, b)$ if $c \leq a$ and $d \leq b$. Two points α and β are *comparable* if $\alpha \leq \beta$ or $\beta \leq \alpha$. A subset of $\mathbb{R}^2$ in which no two different points are comparable is called an *antichain*. We will write $\alpha < \beta$ if $\alpha \leq \beta$ and $\alpha \neq \beta$. If α and β are comparable then $\min(\alpha, \beta) = \alpha$ if $\alpha \leq \beta$ and $\min(\alpha, \beta) = \beta$ if $\alpha > \beta$.

Let A be a nonempty finite subset of $\mathbb{R}^2$. A point $\alpha \in A$ is called a *Pareto optimal point for* A if there is no a point $\beta \in A$ such that $\beta < \alpha$. We denote by $Par(A)$ the set of Pareto optimal points for A. It is clear that $Par(A)$ is an antichain. Note that the considered notations are not unique in the literature [1, 2].

Lemma 2.1 *Let A be a nonempty finite subset of the set $\mathbb{R}^2$. Then, for any point $\alpha \in A$, there is a point $\beta \in Par(A)$ such that $\beta \leq \alpha$.*

Proof Let $\beta = (a, b)$ be a point from A such that $(a, b) \leq \alpha$ and $a + b = \min\{c + d : (c, d) \in A, (c, d) \leq \alpha\}$. Then $(a, b) \in Par(A)$. □

H. AbouEisha et al., *Extensions of Dynamic Programming for Combinatorial Optimization and Data Mining*, Intelligent Systems Reference Library 146,
https://doi.org/10.1007/978-3-319-91839-6_2

Lemma 2.2 *Let A, B be nonempty finite subsets of the set $\mathbb{R}^2$, $A \subseteq B$, and, for any $\beta \in B$, there exists $\alpha \in A$ such that $\alpha \leq \beta$. Then $Par(A) = Par(B)$.*

Proof Let $\beta \in Par(B)$. Then there exists $\alpha \in A$ such that $\alpha \leq \beta$. By Lemma 2.1, there exists $\gamma \in Par(A)$ such that $\gamma \leq \alpha$. Therefore $\gamma \leq \beta$ and $\gamma = \beta$ since $\beta \in Par(B)$. Hence $Par(B) \subseteq Par(A)$.

Let $\alpha \in Par(A)$. By Lemma 2.1, there exists $\beta \in Par(B)$ such that $\beta \leq \alpha$. We know that there exists $\gamma \in A$ such that $\gamma \leq \beta$. Therefore $\gamma \leq \alpha$ and $\gamma = \alpha$ since $\alpha \in Par(A)$. As a result, we have $\beta = \alpha$ and $Par(A) \subseteq Par(B)$. Hence $Par(A) = Par(B)$. □

Lemma 2.3 *Let A be a nonempty finite subset of $\mathbb{R}^2$. Then*

$$|Par(A)| \leq \min\left(\left|A^{(1)}\right|, \left|A^{(2)}\right|\right)$$

where $A^{(1)} = \{a : (a, b) \in A\}$ and $A^{(2)} = \{b : (a, b) \in A\}$.

Proof Let $(a, b), (c, d) \in Par(A)$ and $(a, b) \neq (c, d)$. Then $a \neq c$ and $b \neq d$ (otherwise, (a, b) and (c, d) are comparable). Therefore $|Par(A)| \leq \min\left(\left|A^{(1)}\right|, \left|A^{(2)}\right|\right)$. □

Points from $Par(A)$ can be ordered in the following way: $(a_1, b_1), \ldots, (a_t, b_t)$ where $a_1 < \ldots < a_t$. Since points from $Par(A)$ are incomparable, $b_1 > \ldots > b_t$. We will refer to the sequence $(a_1, b_1), \ldots, (a_t, b_t)$ as the *normal representation of the set $Par(A)$*.

It is well known (see [3]) that there exists an algorithm which, for a given set A, constructs the set $Par(A)$ and makes $O(|A| \log |A|)$ comparisons.

For the sake of completeness, we describe an algorithm which, for a given nonempty finite subset A of the set $\mathbb{R}^2$, constructs the normal representation of the set $Par(A)$. We assume that A is a multiset containing, possibly, repeating elements. The cardinality $|A|$ of A is the total number of elements in A.

Algorithm $\mathcal{A}_1$ (construction of normal representation for the set of POPs).

Input: A nonempty finite subset A of the set $\mathbb{R}^2$ containing, possibly, repeating elements (multiset).

Output: Normal representation P of the set $Par(A)$ of Pareto optimal points for A.

1. Set P equal to the empty sequence.
2. Construct a sequence B of all points from A ordered according to the first coordinate in the ascending order.
3. If there is only one point in the sequence B, then add this point to the end of the sequence P, return P, and finish the work of the algorithm. Otherwise, choose the first $\alpha = (\alpha_1, \alpha_2)$ and the second $\beta = (\beta_1, \beta_2)$ points from B.
4. If α and β are comparable then remove α and β from B, add the point $\min(\alpha, \beta)$ to the beginning of B, and proceed to step 3.

5. If α and β are not comparable (in this case $\alpha_1 < \beta_1$ and $\alpha_2 > \beta_2$) then remove α from B, add the point α to the end of P, and proceed to step 3.

Proposition 2.1 *Let A be a nonempty finite subset of the set $\mathbb{R}^2$ containing, possibly, repeating elements (multiset). Then the algorithm $\mathscr{A}_1$ returns the normal representation of the set $Par(A)$ of Pareto optimal points for A and makes $O(|A| \log |A|)$ comparisons.*

Proof The step 2 of the algorithm requires $O(|A| \log |A|)$ comparisons. Each call to step 3 (with the exception of the last one) leads to two comparisons. The number of calls to step 3 is at most $|A|$. Therefore the algorithm $\mathscr{A}_1$ makes $O(|A| \log |A|)$ comparisons.

Let the output sequence P be equal to $(a_1, b_1), \ldots, (a_t, b_t)$ and let us set $Q = \{(a_1, b_1), \ldots, (a_t, b_t)\}$. It is clear that $a_1 < \ldots < a_t, b_1 > \ldots > b_t$ and, for any $\alpha \in A, \alpha \notin Q$, there exists $\beta \in Q$ such that $\beta < \alpha$. From here it follows that $Par(A) \subseteq Q$ and Q is an antichain. Let us assume that there exists $\gamma \in Q$ which does not belong to $Par(A)$. Then there exists $\alpha \in A$ such that $\alpha < \gamma$. Since Q is an antichain, $\alpha \notin Q$. We know that there exists $\beta \in Q$ such that $\beta \leq \alpha$. This results in two different points β and γ from Q being comparable, which is impossible. Therefore $Q = Par(A)$ and P is the normal representation of the set $Par(A)$. □

Remark 2.1 Let A be a nonempty finite subset of $\mathbb{R}^2$, $(a_1, b_1), \ldots, (a_t, b_t)$ be the normal representation of the set $Par(A)$, and $rev(A) = \{(b, a) : (a, b) \in A\}$. Then $Par(rev(A)) = rev(Par(A))$ and $(b_t, a_t), \ldots, (b_1, a_1)$ is the normal representation of the set $Par(rev(A))$.

Lemma 2.4 *Let A be a nonempty finite subset of $\mathbb{R}^2$, $B \subseteq A$, and $Par(A) \subseteq B$. Then $Par(B) = Par(A)$.*

Proof It is clear that $Par(A) \subseteq Par(B)$. Let us assume that, for some β, $\beta \in Par(B)$ and $\beta \notin Par(A)$. Then there exists $\alpha \in A$ such that $\alpha < \beta$. By Lemma 2.1, there exists $\gamma \in Par(A) \subseteq B$ such that $\gamma \leq \alpha$. Therefore $\gamma < \beta$ and $\beta \notin Par(B)$. Hence $Par(B) = Par(A)$. □

Lemma 2.5 *Let $A_1, \ldots, A_k$ be nonempty finite subsets of $\mathbb{R}^2$. Then $Par(A_1 \cup \ldots \cup A_k) \subseteq Par(A_1) \cup \ldots \cup Par(A_k)$.*

Proof Let $\alpha \in (A_1 \cup \ldots \cup A_k) \setminus (Par(A_1) \cup \ldots \cup Par(A_k))$. Then there is $i \in \{1, \ldots, k\}$ such that $\alpha \in A_i$ but $\alpha \notin Par(A_i)$. Therefore there is $\beta \in A_i$ such that $\beta < \alpha$. Hence $\alpha \notin Par(A_1 \cup \ldots \cup A_k)$, and $Par(A_1 \cup \ldots \cup A_k) \subseteq Par(A_1) \cup \ldots \cup Par(A_k)$. □

A function $f : \mathbb{R}^2 \to \mathbb{R}$ is called *increasing* if $f(x_1, y_1) \leq f(x_2, y_2)$ for any $x_1, x_2, y_1, y_2 \in \mathbb{R}$ such that $x_1 \leq x_2$ and $y_1 \leq y_2$. For example, $sum(x, y) = x + y$ and $\max(x, y)$ are increasing functions.

Let f, g be increasing functions from $\mathbb{R}^2$ to $\mathbb{R}$, and A, B be nonempty finite subsets of the set $\mathbb{R}^2$. We denote by $A\langle fg\rangle B$ the set $\{(f(a, c), g(b, d)) : (a, b) \in A, (c, d) \in B\}$.

Lemma 2.6 *Let A, B be nonempty finite subsets of $\mathbb{R}^2$, and f, g be increasing functions from $\mathbb{R}^2$ to $\mathbb{R}$. Then $Par(A \langle fg \rangle B) \subseteq Par(A) \langle fg \rangle Par(B)$.*

Proof Let $\beta \in Par(A \langle fg \rangle B)$ and $\beta = (f(a, c), g(b, d))$ where $(a, b) \in A$ and $(c, d) \in B$. Then, by Lemma 2.1, there exist $(a', b') \in Par(A)$ and $(c', d') \in Par(B)$ such that $(a', b') \leq (a, b)$ and $(c', d') \leq (c, d)$. It is clear that $\alpha = (f(a', c'), g(b', d')) \leq (f(a, c), g(b, d)) = \beta$, and $\alpha \in Par(A) \langle fg \rangle Par(B)$. Since $\beta \in Par(A \langle fg \rangle B)$, we have $\beta = \alpha$. Therefore $Par(A \langle fg \rangle B) \subseteq Par(A) \langle fg \rangle Par(B)$. □

Let f, g be increasing functions from $\mathbb{R}^2$ to $\mathbb{R}$, $P_1, \ldots, P_t$ be nonempty finite subsets of $\mathbb{R}^2$, $Q_1 = P_1$, and, for $i = 2, \ldots, t$, $Q_i = Q_{i-1} \langle fg \rangle P_i$. We assume that, for $i = 1, \ldots, t$, the sets $Par(P_1), \ldots, Par(P_t)$ are already constructed. We now describe an algorithm that constructs the sets $Par(Q_1), \ldots, Par(Q_t)$ and returns $Par(Q_t)$.

Algorithm $\mathscr{A}_2$ (fusion of sets of POPs)

Input: Increasing functions f, g from $\mathbb{R}^2$ to $\mathbb{R}$, and sets $Par(P_1), \ldots, Par(P_t)$ for some nonempty finite subsets $P_1, \ldots, P_t$ of $\mathbb{R}^2$.
Output: The set $Par(Q_t)$ where $Q_1 = P_1$, and, for $i = 2, \ldots, t$, $Q_i = Q_{i-1} \langle fg \rangle P_i$.

1. Set $B_1 = Par(P_1)$ and $i = 2$.
2. Construct the multiset

$$A_i = B_{i-1} \langle fg \rangle Par(P_i) = \{(f(a, c), g(b, d)) : (a, b) \in B_{i-1}, (c, d) \in Par(P_i)\}$$

   (we will not remove equal pairs from the constructed set).
3. Using algorithm $\mathscr{A}_1$, construct the set $B_i = Par(A_i)$.
4. If $i = t$ then return B_i and finish the work of the algorithm. Otherwise, set $i = i + 1$ and proceed to step 2.

Proposition 2.2 *Let f, g be increasing functions from $\mathbb{R}^2$ to $\mathbb{R}$, $P_1, \ldots, P_t$ be nonempty finite subsets of $\mathbb{R}^2$, $Q_1 = P_1$, and, for $i = 2, \ldots, t$, $Q_i = Q_{i-1} \langle fg \rangle P_i$. Then the algorithm $\mathscr{A}_2$ returns the set $Par(Q_t)$.*

Proof We will prove by induction on i that, for $i = 1, \ldots, t$, the set B_i (see the description of the algorithm $\mathscr{A}_2$) is equal to the set $Par(Q_i)$. Since $B_1 = Par(P_1)$ and $Q_1 = P_1$, we have $B_1 = Par(Q_1)$. Let for some $i - 1, 2 \leq i \leq t$, the considered statement hold, i.e., $B_{i-1} = Par(Q_{i-1})$. Then $B_i = Par(B_{i-1} \langle fg \rangle Par(P_i)) = Par(Par(Q_{i-1}) \langle fg \rangle Par(P_i))$. We know that $Q_i = Q_{i-1} \langle fg \rangle P_i$. By Lemma 2.6, we see that $Par(Q_i) \subseteq Par(Q_{i-1}) \langle fg \rangle Par(P_i)$. By Lemma 2.4,

$$Par(Q_i) = Par(Par(Q_{i-1}) \langle fg \rangle Par(P_i)) .$$

Therefore $B_i = Par(Q_i)$. So we have $B_t = Par(Q_t)$, and the algorithm $\mathscr{A}_2$ returns the set $Par(Q_t)$. □

Proposition 2.3 *Let f, g be increasing functions from $\mathbb{R}^2$ to $\mathbb{R}$,*

$$f \in \{x + y, \max(x, y)\},$$

$P_1, \ldots, P_t$ be nonempty finite subsets of $\mathbb{R}^2$, $Q_1 = P_1$, and, for $i = 2, \ldots, t$, $Q_i = Q_{i-1} \langle fg \rangle P_i$. Let $P_i^{(1)} = \{a : (a, b) \in P_i\}$ for $i = 1, \ldots, t$, $m \in \omega$, and

$$P_i^{(1)} \subseteq \{0, 1, \ldots, m\}$$

for $i = 1, \ldots, t$, or $P_i^{(1)} \subseteq \{0, -1, \ldots, -m\}$ for $i = 1, \ldots, t$. Then, during the construction of the set $Par(Q_t)$, the algorithm $\mathscr{A}_2$ makes

$$O(t^2 m^2 \log(tm))$$

elementary operations (computations of f, g and comparisons) if $f = x + y$, and

$$O(tm^2 \log m)$$

elementary operations (computations of f, g and comparisons) if $f = \max(x, y)$. If $f = x + y$ then $|Par(Q_t)| \leq tm + 1$, and if $f = \max(x, y)$ then $|Par(Q_t)| \leq m + 1$.

Proof For $i = 1, \ldots, t$, we denote $p_i = |Par(P_i)|$ and $q_i = |Par(Q_i)|$. Let $i \in \{2, \ldots, t\}$. To construct the multiset $A_i = Par(Q_{i-1}) \langle fg \rangle Par(P_i)$, we need to compute the values of f and g a number of times equal to $q_{i-1} p_i$. The cardinality of A_i is equal to $q_{i-1} p_i$. We apply to A_i the algorithm $\mathscr{A}_1$ which makes $O(q_{i-1} p_i \log(q_{i-1} p_i))$ comparisons. As a result, we find the set $Par(A_i) = Par(Q_i)$. To construct the sets $Par(Q_1), \ldots, Par(Q_t)$, the algorithm $\mathscr{A}_2$ makes $\sum_{i=2}^{t} q_{i-1} p_i$ computations of f, $\sum_{i=2}^{t} q_{i-1} p_i$ computations of g, and $O\left(\sum_{i=2}^{t} q_{i-1} p_i \log(q_{i-1} p_i)\right)$ comparisons.

We know that $P_i^{(1)} \subseteq \{0, 1, \ldots, m\}$ for $i = 1, \ldots, t$, or $P_i^{(1)} \subseteq \{0, -1, \ldots, -m\}$ for $i = 1, \ldots, t$. Then, by Lemma 2.3, $p_i \leq m + 1$ for $i = 1, \ldots, t$.

Let $f = x + y$. Then, for $i = 1, \ldots, t$, $Q_i^{(1)} = \{a : (a, b) \in Q_i\} \subseteq \{0, 1, \ldots, im\}$ or, for $i = 1, \ldots, t$, $Q_i^{(1)} \subseteq \{0, -1, \ldots, -im\}$ and, by Lemma 2.3, $q_i \leq im + 1$. In this case, to construct the sets $Par(Q_1), \ldots, Par(Q_t)$ the algorithm $\mathscr{A}_2$ makes $O(t^2 m^2)$ computations of f, $O(t^2 m^2)$ computations of g, and $O(t^2 m^2 \log(tm))$ comparisons, i.e.,

$$O(t^2 m^2 \log(tm))$$

elementary operations (computations of f, g, and comparisons). Since $q_t \leq tm + 1$, $|Par(Q_t)| \leq tm + 1$.

Let $f = \max(x, y)$. Then, for $i = 1, \ldots, t$, $Q_i^{(1)} = \{a : (a, b) \in Q_i\} \subseteq \{0, 1, \ldots, m\}$ or, for $i = 1, \ldots, t$, $Q_i^{(1)} \subseteq \{0, -1, \ldots, -m\}$ and, by Lemma 2.3, $q_i \leq m + 1$. In this case, to construct the sets $Par(Q_1), \ldots, Par(Q_t)$ the algorithm $\mathscr{A}_2$

makes $O(tm^2)$ computations of f, $O(tm^2)$ computations of g, and $O(tm^2 \log m)$ comparisons, i.e.,

$$O(tm^2 \log m)$$

elementary operations (computations of f, g, and comparisons). Since $q_t \leq m+1$, $|Par(Q_t)| \leq m+1$. □

Similar analysis can be done for the sets $P_i^{(2)} = \{b : (a,b) \in P_i\}$, $Q_i^{(2)} = \{b : (a,b) \in Q_i\}$, and the function g.

A function $p : \mathbb{R} \to \mathbb{R}$ is called *strictly increasing* if $p(x) < p(y)$ for any $x, y \in \mathbb{R}$ such that $x < y$. Let p and q be strictly increasing functions from $\mathbb{R}$ to $\mathbb{R}$. For $(a,b) \in \mathbb{R}^2$, we denote by $(a,b)^{pq}$ the pair $(p(a), q(b))$. For a nonempty finite subset A of $\mathbb{R}^2$, we denote $A^{pq} = \{(a,b)^{pq} : (a,b) \in A\}$.

Lemma 2.7 *Let A be a nonempty finite subset of $\mathbb{R}^2$, and p , q be strictly increasing functions from $\mathbb{R}$ to $\mathbb{R}$. Then $Par(A^{pq}) = Par(A)^{pq}$.*

Proof Let $(a,b), (c,d) \in A$. It is clear that $(c,d) = (a,b)$ if and only if $(c,d)^{pq} = (a,b)^{pq}$, and $(c,d) < (a,b)$ if and only if $(c,d)^{pq} < (a,b)^{pq}$. From here it follows that $Par(A^{pq}) = Par(A)^{pq}$. □

Let A be a nonempty finite subset of $\mathbb{R}^2$. We correspond to A a partial function $\mathscr{F}_A : \mathbb{R} \to \mathbb{R}$ defined in the following way: $\mathscr{F}_A(x) = \min\{b : (a,b) \in A, a \leq x\}$.

Lemma 2.8 *Let A be a nonempty finite subset of $\mathbb{R}^2$, and $(a_1, b_1), \ldots, (a_t, b_t)$ be the normal representation of the set $Par(A)$. Then, for any $x \in \mathbb{R}$, $\mathscr{F}_A(x) = \mathscr{F}(x)$ where*

$$\mathscr{F}(x) = \begin{cases} \text{undefined}, & x < a_1 \\ b_1, & a_1 \leq x < a_2 \\ \ldots & \ldots \\ b_{t-1}, & a_{t-1} \leq x < a_t \\ b_t, & a_t \leq x \end{cases} .$$

Proof One can show that $a_1 = \min\{a : (a,b) \in A\}$. Therefore the value $\mathscr{F}_A(x)$ is undefined if $x < a_1$. Let $x \geq a_1$. Then both values $\mathscr{F}(x)$ and $\mathscr{F}_A(x)$ are defined. It is easy to check that $\mathscr{F}(x) = \mathscr{F}_{Par(A)}(x)$. Since $Par(A) \subseteq A$, we have $\mathscr{F}_A(x) \leq \mathscr{F}(x)$. By Lemma 2.1, for any point $(a,b) \in A$, there is a point $(a_i, b_i) \in Par(A)$ such that $(a_i, b_i) \leq (a,b)$. Therefore $\mathscr{F}(x) \leq \mathscr{F}_A(x)$ and $\mathscr{F}_A(x) = \mathscr{F}(x)$. □

Remark 2.2 We can consider not only function $\mathscr{F}_A$ but also function $\mathscr{F}_{rev(A)} : \mathbb{R} \to \mathbb{R}$ defined in the following way:

$$\mathscr{F}_{rev(A)}(x) = \min\{a : (b,a) \in rev(A), b \leq x\} = \min\{a : (a,b) \in A, b \leq x\} .$$

From Remark 2.1 and Lemma 2.8 it follows that

$$\mathscr{F}_{rev(A)}(x) = \begin{cases} undefined, & x < b_t \\ a_t, & b_t \leq x < b_{t-1} \\ \dots & \dots \\ a_2, & b_2 \leq x < b_1 \\ a_1, & b_1 \leq x \end{cases}.$$

References

1. Arora, J.S.: Introduction to Optimum Design, 3rd edn. Elsevier, Waltham, MA (2012)
2. Ehrgott, M.: Multicriteria Optimization, 2nd edn. Springer, Heidelberg (2005)
3. Kung, H.T., Luccio, F., Preparata, F.P.: On finding the maxima of a set of vectors. J. ACM **22**(4), 469–476 (1975)

Chapter 3
Some Tools for Decision Tables

In Parts II and III, we develop extensions of dynamic programming for the study of decision trees, decision rules, and systems of decision rules for decision tables. In this chapter, we consider some tools that are common for these parts: we describe main notions related to decision tables, consider the structure of subtables of a given decision table represented as a directed acyclic graph (DAG), and discuss time complexity of algorithms on DAGs. We also consider classes of decision tables over restricted information systems for which these algorithms have polynomial time complexity depending on the number of conditional attributes in the input decision table.

The notion of a decision table is considered in Sect. 3.1. The notion of an uncertainty measure which is used under consideration of approximate decision trees and rules is discussed in Sect. 3.2. In Sect. 3.3, we study DAGs for decision tables which are used by algorithms optimizing decision trees and rules. Sections 3.4 and 3.5 are devoted to the discussion of restricted information systems and time complexity of algorithms on DAGs, respectively.

3.1 Decision Tables

A *decision table* is a rectangular table T with $n \geq 1$ columns filled with numbers from the set $\omega = \{0, 1, 2, \ldots\}$ of nonnegative integers. Columns of the table are labeled with *conditional* attributes $f_1, \ldots, f_n$. Rows of the table are pairwise different, and each row is labeled with a number from ω which is interpreted as a decision (a value of the *decision* attribute d). Rows of the table are interpreted as tuples of values of conditional attributes.

A decision table can be represented by a word over the alphabet $\{0, 1, ;, |\}$ in which numbers from ω are in binary representation (are represented by words over the alphabet $\{0, 1\}$), the symbol ";" is used to separate two numbers from ω, and the symbol "|" is used to separate two rows (we add to each row corresponding decision

H. AbouEisha et al., *Extensions of Dynamic Programming for Combinatorial Optimization and Data Mining*, Intelligent Systems Reference Library 146,
https://doi.org/10.1007/978-3-319-91839-6_3

as the last number in the row). The length of this word will be called the *size* of the decision table.

A decision table is called *empty* if it has no rows. We denote by $\mathcal{T}$ the set of all decision tables and by $\mathcal{T}^+$ – the set of nonempty decision tables. Let $T \in \mathcal{T}$. The table T is called *degenerate* if it is empty or all rows of T are labeled with the same decision. We denote by $\dim(T)$ the number of columns (conditional attributes) in T. Let $D(T)$ be the set of decisions attached to rows of T. We denote by $N(T)$ the number of rows in the table T and, for any $t \in \omega$, we denote by $N_t(T)$ the number of rows of T labeled with the decision t. By $mcd(T)$ we denote the *most common decision* for T which is the minimum decision t_0 from $D(T)$ such that $N_{t_0}(T) = \max\{N_t(T) : t \in D(T)\}$. If T is empty then $mcd(T) = 0$. A nonempty decision table is called *diagnostic* if rows of this table are labeled with pairwise different decisions.

For any conditional attribute $f_i \in \{f_1, \dots, f_n\}$, we denote by $E(T, f_i)$ the set of values of the attribute f_i in the table T. We denote by $E(T)$ the set of conditional attributes for which $|E(T, f_i)| \geq 2$. Let $range(T) = \max\{|E(T, f_i)| : i = 1, \dots, n\}$.

Let T be a nonempty decision table. A *subtable* of T is a table obtained from T by removal of some rows. Let $f_{i_1}, \dots, f_{i_m} \in \{f_1, \dots, f_n\}$ and $a_1, \dots, a_m \in \omega$. We denote by $T(f_{i_1}, a_1) \dots (f_{i_m}, a_m)$ the subtable of the table T containing the rows from T which at the intersection with the columns $f_{i_1}, \dots, f_{i_m}$ have numbers $a_1, \dots, a_m$, respectively. Such nonempty subtables, including the table T, are called *separable* subtables of T. We denote by $SEP(T)$ the set of separable subtables of the table T. Note that $N(T) \leq |SEP(T)|$. It is clear that the size of each subtable of T is at most the size of T.

3.2 Uncertainty Measures

Let $\mathbb{R}$ be the set of real numbers and $\mathbb{R}_+$ be the set of all nonnegative real numbers. An *uncertainty measure* is a function $U : \mathcal{T} \to \mathbb{R}$ such that $U(T) \geq 0$ for any $T \in \mathcal{T}$, and $U(T) = 0$ if and only if T is a degenerate table. One can show that the following functions (we assume that, for any empty table, the value of each of the considered functions is equal to 0) are uncertainty measures:

- *Misclassification error* $me(T) = N(T) - N_{mcd(T)}(T)$.
- *Relative misclassification error* $rme(T) = (N(T) - N_{mcd(T)}(T))/N(T)$.
- *Entropy* $ent(T) = -\sum_{t \in D(T)} (N_t(T)/N(T)) \log_2(N_t(T)/N(T))$.
- *Gini index* $gini(T) = 1 - \sum_{t \in D(T)} (N_t(T)/N(T))^2$.
- Function R where $R(T)$ is the number of unordered pairs of rows of T labeled with different decisions (note that $R(T) = N(T)^2 gini(T)/2$).

We assume that each of the following numerical operations (we call these operations *basic*) has time complexity $O(1)$: $\max(x, y)$, $x + y$, $x \times y$, $x \div y$, $\log_2 x$. This assumption is reasonable for computations with floating-point numbers. Under this assumption, each of the considered five uncertainty measures has polynomial time complexity depending on the size of decision tables.

An uncertainty measure U is called *increasing* if, for any nondegenerate decision table $T, f \in E(T)$, and $a \in E(T, f)$, the inequality $U(T) \geq U(T(f, a))$ holds. An uncertainty measure U is called *strictly increasing* if, for any nondegenerate decision table $T, f \in E(T)$, and $a \in E(T, f)$, the inequality $U(T) > U(T(f, a))$ holds. It is clear that each strictly increasing uncertainty measure is also increasing uncertainty measure.

Proposition 3.1 *The uncertainty measures R is strictly increasing. The uncertainty measure me is increasing but not strictly increasing. The uncertainty measures rme, ent, and gini are not increasing.*

Proof Let T be a nondegenerate decision table, $f \in E(T)$, $a \in E(T, f)$, and $T' = T(f, a)$.

Let B be the set of unordered pairs of rows of T labeled with different decisions, and B' be the set of unordered pairs of rows of T' labeled with different decisions. It is clear that $B' \subseteq B$. Since $f \in E(T)$, at least one row r from T does not belong to T'. Since T is nondegenerate, r belongs to at least one pair from B. This pair cannot belong to B'. Therefore $B' \subset B$, and $R(T') = |B'| < |B| = R(T)$. From here it follows that R is a strictly increasing uncertainty measure.

Let C be the set of rows of T labeled with decisions different from $mcd(T)$, and C' be the set of rows of T' labeled with decisions different from $mcd(T)$. It is clear that $C' \subseteq C$ and $N_{mcd(T')}(T') \geq N_{mcd(T)}(T')$. Therefore

$$\begin{aligned} me(T') &= N(T') - N_{mcd(T')}(T') \leq N(T') - N_{mcd(T)}(T') = |C'| \\ &\leq |C| = N(T) - N_{mcd(T)}(T) = me(T) . \end{aligned}$$

From here it follows that me is an increasing uncertainty measure.

Let

$$T_0 = \begin{array}{cc|c} f_1 & f_2 & \\ \hline 0 & 0 & 1 \\ 0 & 1 & 2 \\ 1 & 1 & 2 \end{array} \quad \text{and } T_0' = T_0(f_1, 0) = \begin{array}{cc|c} f_1 & f_2 & \\ \hline 0 & 0 & 1 \\ 0 & 1 & 2 \end{array}$$

It is clear that T_0 is a nondegenerate decision table, $f_1 \in E(T_0)$, and $0 \in E(T_0, f_1)$. We have $me(T_0) = 1 = me(T_0')$. Therefore me is not strictly increasing.

It is easy to see that $rme(T_0) = 1/3 < 1/2 = rme(T_0')$,

$$\begin{aligned} ent(T_0) &= \left(-(1/3)\log_2(1/3) - (2/3)\log_2(2/3)\right) < 0.92 < 1 \\ &= \left(-(1/2)\log_2(1/2) - (1/2)\log_2(1/2)\right) = ent(T_0') , \end{aligned}$$

and $gini(T_0) = 1 - (1/3)^2 - (2/3)^2 < 0.45 < 0.5 = 1 - (1/2)^2 - (1/2)^2 = gini(T_0')$. Therefore rme, ent, and $gini$ are not increasing. □

Remark 3.1 Let U_1, U_2 be uncertainty measures, and a, b be positive real numbers. Then $U_3 = U_1 \times U_2$ and $U_4 = aU_1 + bU_2$ are also uncertainty measures. If U_1 and U_2 are increasing then U_3 and U_4 are also increasing. If U_1 is increasing and U_2 is strictly increasing then U_3 and U_4 are strictly increasing. In particular, $me + R$ is a strictly increasing uncertainty measure.

An uncertainty measure U is called *sum-increasing* if, for any nondegenerate decision table T and any $f_i \in E(T)$, $U(T) \geq \sum_{j=1}^{t}(U(T(f_i, a_j))$ where $\{a_1, \ldots, a_t\} = E(T, f_i)$. An uncertainty measure U is called *strictly sum-increasing* if, for any nondegenerate decision table T and any $f_i \in E(T)$, $U(T) > \sum_{j=1}^{t}(U(T(f_i, a_j))$ where $\{a_1, \ldots, a_t\} = E(T, f_i)$.

It is clear that each strictly sum-increasing uncertainty measure is also sum-increasing uncertainty measure. If U is a sum-increasing uncertainty measure then U is an increasing uncertainty measure. If U is a strictly sum-increasing uncertainty measure then U is a strictly increasing uncertainty measure.

Proposition 3.2 *The uncertainty measure R is strictly sum-increasing. The uncertainty measure me is sum-increasing but not strictly sum-increasing. The uncertainty measures rme, ent, and gini are not sum-increasing.*

Proof By Proposition 3.1, the uncertainty measures *rme*, *ent*, and *gini* are not increasing. Therefore they are not sum-increasing.

By Proposition 3.1, the uncertainty measure *me* is not strictly increasing. Therefore *me* is not strictly sum-increasing. Let us show that *me* is sum-increasing. Let T be a nondegenerate decision table, $f_i \in E(T)$, and $E(T, f_i) = \{a_1, \ldots, a_t\}$. For $j = 1, \ldots, t$, let $T_j = T(f_i, a_j)$. Then

$$\begin{aligned}
me(T) &= N(T) - N_{mcd(T)}(T) \\
&= (N(T_1) - N_{mcd(T)}(T_1)) + \ldots + (N(T_t) - N_{mcd(T)}(T_t)) \\
&\geq (N(T_1) - N_{mcd(T_1)}(T_1)) + \ldots + (N(T_t) - N_{mcd(T_t)}(T_t)) \\
&= me(T_1) + \ldots + me(T_t) \ .
\end{aligned}$$

Therefore *me* is sum-increasing.

Let us show that R is a strictly sum-increasing uncertainty measure. We denote by B the set of unordered pairs of rows from T labeled with different decisions. For $j = 1, \ldots, t$, we denote by B_j the set of unordered pairs of rows from T_j labeled with different decisions. One can show that $B_j \subseteq B$ for $j = 1, \ldots, t$, and $B_{j_1} \cap B_{j_2} = \emptyset$ for any $j_1, j_2 \in \{1, \ldots, t\}$ such that $j_1 \neq j_2$. Therefore $R(T) \geq R(T_1) + \ldots + R(T_t)$. If we show that $B \neq B_1 \cup \ldots \cup B_t$ then we obtain $R(T) > R(T_1) + \ldots + R(T_t)$. Since T is nondegenerate, there are two rows r_1 and r_2 in T which are labeled with different decisions. It is clear that the pair (r_1, r_2) belongs to B. If there are two different $j_1, j_2 \in \{1, \ldots, t\}$ such that r_1 belongs to T_{j_1} and r_2 belongs to T_{j_2} then (r_1, r_2) does not belong to $B_1 \cup \ldots \cup B_t$. Let there exist $j \in \{1, \ldots, t\}$ such that r_1 and r_2 belong to T_j. Since $t > 1$, there is a row r_3 of T which does not belong to T_j. It is clear that (r_1, r_3) or (r_2, r_3) belongs to B and does not belong to $B_1 \cup \ldots \cup B_t$. Therefore R is strictly sum-increasing. □

Remark 3.2 Let U_1, U_2 be uncertainty measures, a, b be positive real numbers, $U_3 = U_1 \times U_2$, and $U_4 = aU_1 + bU_2$. If U_1 and U_2 are sum-increasing then U_3 and U_4 are sum-increasing. If U_1 is sum-increasing and U_2 is strictly sum-increasing then U_3 and U_4 are strictly sum-increasing. In particular, $me + R$ is a strictly sum-increasing uncertainty measure.

3.3 Directed Acyclic Graph $\Delta_{U,\alpha}(T)$

Let T be a nonempty decision table with n conditional attributes $f_1, \ldots, f_n$, U be an uncertainty measure, and $\alpha \in \mathbb{R}_+$. We now consider an algorithm $\mathscr{A}_3$ for the construction of a directed acyclic graph $\Delta_{U,\alpha}(T)$ which will be used for the description and study of decision rules and decision trees. Nodes of this graph are some separable subtables of the table T. During each iteration we process one node. We start with the graph that consists of one node T which is not processed and finish when all nodes of the graph are processed.

Algorithm $\mathscr{A}_3$ (construction of DAG $\Delta_{U,\alpha}(T)$).

Input: A nonempty decision table T with n conditional attributes $f_1, \ldots, f_n$, an uncertainty measure U, and a number $\alpha \in \mathbb{R}_+$.
Output: Directed acyclic graph $\Delta_{U,\alpha}(T)$.

1. Construct the graph that consists of one node T which is not marked as processed.
2. If all nodes of the graph are processed then the work of the algorithm is finished. Return the resulting graph as $\Delta_{U,\alpha}(T)$. Otherwise, choose a node (table) Θ that has not been processed yet.
3. If $U(\Theta) \leq \alpha$ mark the node Θ as processed and proceed to step 2.
4. If $U(\Theta) > \alpha$ then, for each $f_i \in E(\Theta)$, draw a bundle of edges from the node Θ (this bundle of edges will be called f_i*-bundle*). Let $E(\Theta, f_i) = \{a_1, \ldots, a_k\}$. Then draw k edges from Θ and label these edges with the pairs $(f_i, a_1), \ldots, (f_i, a_k)$. These edges enter nodes $\Theta(f_i, a_1), \ldots, \Theta(f_i, a_k)$, respectively. If some of the nodes $\Theta(f_i, a_1), \ldots, \Theta(f_i, a_k)$ are not present in the graph then add these nodes to the graph. Mark the node Θ as processed and return to step 2.

We now analyze time complexity of the algorithm $\mathscr{A}_3$. By $L(\Delta_{U,\alpha}(T))$ we denote the number of nodes in the graph $\Delta_{U,\alpha}(T)$.

Proposition 3.3 *Let the algorithm $\mathscr{A}_3$ use an uncertainty measure U which has polynomial time complexity depending on the size of the input decision table T. Then the time complexity of the algorithm $\mathscr{A}_3$ is bounded from above by a polynomial on the size of the input table T and the number $|SEP(T)|$ of different separable subtables of T.*

Proof Since the uncertainty measure U has polynomial time complexity depending on the size of decision tables, each step of the algorithm $\mathscr{A}_3$ has polynomial time complexity depending on the size of the table T and the number $L(\Delta_{U,\alpha}(T))$. The number of steps is $O(L(\Delta_{U,\alpha}(T)))$. Therefore the time complexity of the algorithm $\mathscr{A}_3$ is bounded from above by a polynomial on the size of the input table T and the number $L(\Delta_{U,\alpha}(T))$. The number $L(\Delta_{U,\alpha}(T))$ is bounded from above by the number $|SEP(T)|$ of different separable subtables of T. □

Remark 3.3 Note that $L(\Delta_{U,0}(T)) = |SEP(T)|$ for any uncertainty measure U and any diagnostic decision table T. Note also that, for any decision table T, the graph $\Delta_{U,0}(T)$ does not depend on the uncertainty measure U. We denote this graph $\Delta(T)$.

In Sect. 3.4, we describe classes of decision tables such that the number of separable subtables in tables from the class is bounded from above by a polynomial on the number of conditional attributes in the table, and the size of tables is bounded from above by a polynomial on the number of conditional attributes. For each such class, time complexity of the algorithm $\mathscr{A}_3$ is polynomial depending on the number of conditional attributes in decision tables.

A node of directed graph is called *terminal* if there are no edges starting in this node. A *proper subgraph* of the graph $\Delta_{U,\alpha}(T)$ is a graph G obtained from $\Delta_{U,\alpha}(T)$ by removal of some bundles of edges such that each nonterminal node of $\Delta_{U,\alpha}(T)$ keeps at least one bundle of edges starting in this node. By definition, $\Delta_{U,\alpha}(T)$ is a proper subgraph of $\Delta_{U,\alpha}(T)$. A node Θ of the graph G is terminal if and only if $U(\Theta) \leq \alpha$. We denote by $L(G)$ the number of nodes in the graph G.

3.4 Restricted Information Systems

In this section, we describe classes of decision tables for which algorithms that deal with the graphs $\Delta_{U,\alpha}(T)$ have polynomial time complexity depending on the number of conditional attributes in the input table T.

Let A be a nonempty set and F be a nonempty set of non-constant functions from A to $E_k = \{0, ..., k-1\}$, $k \geq 2$. Functions from F are called *attributes*, and the pair $\mathscr{U} = (A, F)$ is called a *k-valued information system*.

For arbitrary attributes $f_1, ..., f_n \in F$ and a mapping $\nu : E_k^n \to \{0, ..., k^n - 1\}$, we denote by $T_\nu(f_1, ..., f_n)$ the decision table with n conditional attributes $f_1, ..., f_n$ which contains the row $(\delta_1, ..., \delta_n) \in E_k^n$ if and only if the system of equations

$$\{f_1(x) = \delta_1, ..., f_n(x) = \delta_n\} \tag{3.1}$$

is *compatible* (has a solution from the set A). This row is labeled with the decision $\nu(\delta_1, ..., \delta_n)$. The table $T_\nu(f_1, ..., f_n)$ is called a *decision table over the information system* $\mathscr{U}$. We denote by $\mathscr{T}(\mathscr{U})$ the set of decision tables over $\mathscr{U}$.

Let us consider the function

$$SEP_{\mathscr{U}}(n) = \max\{|SEP(T)| : T \in \mathscr{T}(\mathscr{U}), \dim(T) \leq n\},$$

where $\dim(T)$ is the number of conditional attributes in T, which characterizes the maximum number of separable subtables depending on the number of conditional attributes in decision tables over $\mathscr{U}$.

A system of equations of the kind (3.1) is called a *system of equations over* $\mathscr{U}$. Two systems of equations are called *equivalent* if they have the same set of solutions from A. A compatible system of equations will be called *irreducible* if each of its proper subsystems is not equivalent to the system. Let r be a natural number. An information system $\mathscr{U}$ will be called *r-restricted* if each irreducible system of equations over $\mathscr{U}$ consists of at most r equations. An information system $\mathscr{U}$ will be called *restricted* if it is r-restricted for some natural r.

Theorem 3.1 [3] *Let $\mathscr{U} = (A, F)$ be a k-valued information system. Then the following statements hold:*

1. *If $\mathscr{U}$ is an r-restricted information system then $SEP_{\mathscr{U}}(n) \leq (nk)^r + 1$ for any natural n.*
2. *If $\mathscr{U}$ is not a restricted information system then $SEP_{\mathscr{U}}(n) \geq 2^n$ for any natural n.*

We now evaluate the time complexity of the algorithm $\mathscr{A}_3$ for decision tables over a restricted information system $\mathscr{U}$ under the assumption that the uncertainty measure U used by $\mathscr{A}_3$ has polynomial time complexity.

Lemma 3.1 *Let $\mathscr{U}$ be a restricted information system. Then, for decision tables from $\mathscr{T}(\mathscr{U})$, both the size and the number of separable subtables are bounded from above by polynomials on the number of conditional attributes.*

Proof Let $\mathscr{U}$ be a k-valued information system which is r-restricted. For any decision table $T \in \mathscr{T}(\mathscr{U})$, each value of each conditional attribute is at most k, each value of the decision attribute is at most $k^{\dim(T)}$, and, by Theorem 3.1, $N(T) \leq |SEP(T)| \leq (\dim(T)k)^r + 1$. From here it follows that the size of decision tables from $\mathscr{T}(\mathscr{U})$ is bounded from above by a polynomial on the number of conditional attributes in decision tables. By Theorem 3.1, the number of separable subtables for decision tables from $\mathscr{T}(\mathscr{U})$ is bounded from above by a polynomial on the number of conditional attributes in decision tables. □

Proposition 3.4 *Let $\mathscr{U}$ be a restricted information system, and the uncertainty measure U used by the algorithm $\mathscr{A}_3$ have polynomial time complexity depending on the size of decision tables. Then the algorithm $\mathscr{A}_3$ has polynomial time complexity for decision tables from $\mathscr{T}(\mathscr{U})$ depending on the number of conditional attributes.*

Proof By Proposition 3.3, the time complexity of the algorithm $\mathscr{A}_3$ is bounded from above by a polynomial on the size of the input table T and the number $|SEP(T)|$ of different separable subtables of T. From Lemma 3.1 it follows that, for decision tables from $\mathscr{T}(\mathscr{U})$, both the size and the number of separable subtables are bounded from above by polynomials on the number of conditional attributes. □

Remark 3.4 Let $\mathscr{U}$ be an information system which is not restricted. Using Remark 3.3 and Theorem 3.1 one can show that there is no an algorithm which constructs the graph $\Delta_{U,0}(T)$ for decision tables $T \in \mathscr{T}(\mathscr{U})$ and which time complexity is bounded from above by a polynomial on the number of conditional attributes in the considered decision tables.

Let us consider a family of restricted information systems. Let d and t be natural numbers, $f_1, \ldots, f_t$ be functions from $\mathbb{R}^d$ to $\mathbb{R}$, and s be a function from $\mathbb{R}$ to $\{0, 1\}$ such that $s(x) = 0$ if $x < 0$ and $s(x) = 1$ if $x \geq 0$. Then the 2-valued information system $U = (\mathbb{R}^d, F)$ where $F = \{s(f_i + c) : i = 1, \ldots, t, c \in \mathbb{R}\}$ is a $2t$-restricted information system.

If $f_1, \ldots, f_t$ are linear functions then we deal with attributes corresponding to t families of parallel hyperplanes in $\mathbb{R}^d$ which is usual for decision trees for datasets with t numerical attributes only [1].

We consider now a class of so-called linear information systems for which all restricted systems are known. Let P be the set of all points in the plane and l be a straight line (line in short) in the plane. This line divides the plane into two open half-planes H_1 and H_2, and the line l. Two attributes correspond to the line l. The first attribute takes value 0 on points from H_1, and value 1 on points from H_2 and l. The second one takes value 0 on points from H_2, and value 1 on points from H_1 and l. We denote by $\mathscr{L}$ the set of all attributes corresponding to lines in the plane. Information systems of the kind (P, F) where $F \subseteq \mathscr{L}$, will be called *linear* information systems. We describe all restricted linear information systems.

Let l be a line in the plane. Let us denote by $\mathscr{L}(l)$ the set of all attributes corresponding to lines which are parallel to l. Let p be a point in the plane. We denote by $\mathscr{L}(p)$ the set of all attributes corresponding to lines which pass through p. A set C of attributes from $\mathscr{L}$ is called a *clone* if $C \subseteq \mathscr{L}(l)$ for some line l or $C \subseteq \mathscr{L}(p)$ for some point p.

Theorem 3.2 [2] *A linear information system (P, F) is restricted if and only if F is the union of a finite number of clones.*

3.5 Time Complexity of Algorithms on $\Delta_{U,\alpha}(T)$

In this book, we consider a number of algorithms which deal with the graph $\Delta_{U,\alpha}(T)$. To evaluate time complexity of these algorithms, we will count the number of elementary operations made by the algorithms. These operations can either be basic numerical operations or computations of numerical parameters of decision tables. We assume, as we already mentioned, that each basic numerical operation ($\max(x, y)$, $x + y, x - y, x \times y, x \div y, \log_2 x$) has time complexity $O(1)$. This assumption is reasonable for computations with floating-point numbers. Furthermore, computing the considered parameters of decision tables usually has polynomial time complexity depending on the size of the decision table.

Proposition 3.5 *Let, for some algorithm $\mathscr{A}$ working with decision tables, the number of elementary operations (basic numerical operations and computations of numerical parameters of decision tables) be polynomial depending on the size of the input table T and on the number of separable subtables of T, and the computations of parameters of decision tables used by the algorithm $\mathscr{A}$ have polynomial time complexity depending on the size of decision tables. Then, for any restricted information system $\mathscr{U}$, the algorithm $\mathscr{A}$ has polynomial time complexity for decision tables from $\mathscr{T}(\mathscr{U})$ depending on the number of conditional attributes.*

Proof Let $\mathscr{U}$ be a restricted information system. From Lemma 3.1 it follows that the size and the number of separable subtables for decision tables from $\mathscr{T}(\mathscr{U})$ are bounded from above by polynomials on the number of conditional attributes in the tables. From here it follows that the algorithm $\mathscr{A}$ has polynomial time complexity for decision tables from $\mathscr{T}(\mathscr{U})$ depending on the number of conditional attributes. □

References

1. Breiman, L., Friedman, J.H., Olshen, R.A., Stone, C.J.: Classification and Regression Trees. Wadsworth and Brooks, Monterey (1984)
2. Moshkov, M.: On the class of restricted linear information systems. Discret. Math. **307**(22), 2837–2844 (2007)
3. Moshkov, M., Chikalov, I.: On algorithm for constructing of decision trees with minimal depth. Fundam. Inform. **41**(3), 295–299 (2000)

Part II
Decision Trees

This part is devoted to the development of extensions of dynamic programming to the study of decision trees. The considered extensions allow us to make multi-stage optimization of decision trees relative to a sequence of cost functions, to count the number of optimal trees, and to study relationships: cost versus cost and cost versus uncertainty for decision trees by the construction of the set of Pareto-optimal points for the corresponding bi-criteria optimization problem.

The applications include the study of totally optimal (simultaneously optimal relative to a number of cost functions) decision trees for Boolean functions, improvement of bounds on complexity of decision trees for diagnosis of circuits, computation of minimum average depth for a decision tree for sorting eight elements, study of optimal tests and decision trees for modified majority problem, design of an algorithm for the problem of reduct optimization, study of time and memory trade-off for corner point detection, study of decision rules derived from decision trees, creation of new procedure (multi-pruning) for construction of classifiers, and comparison of heuristics for decision tree construction.

In Chap. 4, we consider different kinds of decision trees. In Chaps. 5 and 6, we discuss multi-stage optimization of decision trees relative to different cost functions and its applications. We study bi-criteria optimization of decision trees and its applications in Chap. 7 for two cost functions, and in Chap. 8 for cost and uncertainty.

Chapter 4
Different Kinds of Decision Trees

A *decision tree* is a finite directed tree with the root in which terminal nodes are labeled with *decisions*, nonterminal nodes with *attributes*, and edges are labeled with *values of attributes*.

Decision trees are widely used as predictors [6, 16, 26] where a decision tree is considered as a model of a data and is used to predict value of the decision attribute for a new object given by values of conditional attributes. There are two other well known areas of applications of decision trees where decision trees are considered as algorithms for problem solving (see, for example, [25]) and as tools for data mining and knowledge representation (see, for example, [29]).

When decision trees are considered as algorithms, we are interested in minimizing the depth of decision trees (worst-case time complexity), average depth of decision trees (average-case time complexity), and number of nodes in decision trees (space complexity). In such situations, it is natural to study both single- and bi-criteria optimization of decision trees.

When decision trees are considered as tools for knowledge representation the goal is to have understandable trees. The standard way is to try to construct a decision tree with minimum number of nodes. However, the depth of decision trees is also important since we need to understand conjunctions of conditions corresponding to paths in decision trees from the root to terminal nodes. Another example of decision tree usage, where bi-criteria optimization is required, is connected with induction of decision rules from the decision trees [26]: the number of rules is equal to the number of terminal nodes in the tree and the maximum length of a rule is equal to the depth of the tree.

Most optimization problems connected with decision trees are NP-hard (see, for example, [13, 19, 25]). However, several exact algorithms for decision tree optimization are known including brute-force algorithms [28], algorithms based on dynamic programming [15, 21, 30], and algorithms using branch-and-bound technique [7]. Similarly, different algorithms and techniques for approximate optimization of deci-

H. AbouEisha et al., *Extensions of Dynamic Programming for Combinatorial Optimization and Data Mining*, Intelligent Systems Reference Library 146,
https://doi.org/10.1007/978-3-319-91839-6_4

sion trees have been extensively studied by researchers in the field, for example, using genetic algorithms [8], simulated annealing [17], and ant colony [5].

Most approximation algorithms for decision trees are greedy, in nature. Generally, these algorithms employ a top-down approach and at each step minimize some impurity. Different impurity criteria are known in literature [6, 24, 25, 27]. See [1–3, 14, 20, 22, 23, 27] for comparison of such criteria.

Dynamic programming for optimization of decision trees has been studied before by Garey [15] and others [21, 30], however not in the context of multi-objective optimization.

Previously, we studied multi-stage optimization techniques [4] but did not distinguished optimal and strictly optimal decision trees for which each subtree is optimal for the corresponding subtable of the decision table. However, the considered procedures of optimization allow us to describe the whole set of optimal trees only for such cost functions as number of nodes or average depth. For the depth, we can only obtain the set of strictly optimal decision trees which is a subset of the set of optimal trees.

Earlier in [11, 18], we considered relationships cost versus cost for different pairs of cost functions for decision trees separately with individual algorithm for each pair. The current work is a result of understanding the fact that, instead of considering relationships, we can consider constructing the set of Pareto optimal points for corresponding bi-criteria optimization problem (the set of Pareto optimal points can be easily derived from the graph describing the relationship, and this graph can be easily constructed if we know the set of Pareto optimal points). It gives us possibility to design a universal approach and simplify the algorithms. Similar situation is with relationships cost versus uncertainty (see our previous works in [9, 10, 12]).

In the following chapters, we study algorithms for multi-stage optimization of decision trees and for construction of the set of Pareto optimal points for bi-criteria optimization problems related to decision trees, and their applications.

In this chapter, we consider different kinds of decision trees, including approximate decision trees and decision trees irredundant relative to two types of decision tree uncertainty. We describe the set of decision trees corresponding to a subgraph of a directed acyclic graph constructed for a decision table, and show how to compute the cardinality of this set. We discuss also the notion of a cost function for decision trees.

4.1 Decision Trees for T

Let T be a decision table with n conditional attributes $f_1, \ldots, f_n$.

A decision tree over T is a finite directed tree with root in which nonterminal nodes are labeled with attributes from the set $\{f_1, \ldots, f_n\}$, terminal nodes are labeled with numbers from $\omega = \{0, 1, 2, \ldots\}$, and, for each nonterminal node, edges starting from this node are labeled with pairwise different numbers from ω.

Let Γ be a decision tree over T and v be a node of Γ. We denote by $\Gamma(v)$ the subtree of Γ for which v is the root. We define now a subtable $T(v) = T_\Gamma(v)$ of the table T. If v is the root of Γ then $T(v) = T$. Let v be not the root of Γ and $v_1, e_1, \ldots, v_m, e_m, v_{m+1} = v$ be the directed path from the root of Γ to v in which nodes $v_1, \ldots, v_m$ are labeled with attributes $f_{i_1}, \ldots, f_{i_m}$ and edges $e_1, \ldots, e_m$ are labeled with numbers $a_1, \ldots, a_m$, respectively. Then $T(v) = T(f_{i_1}, a_1) \ldots (f_{i_m}, a_m)$.

A decision tree Γ over T is called a *decision tree for* T if, for any node v of Γ,

- If $T(v)$ is a degenerate table then v is a terminal node labeled with $mcd(T(v))$.
- If $T(v)$ is not degenerate then either v is a terminal node labeled with $mcd(T(v))$, or v is a nonterminal node which is labeled with an attribute $f_i \in E(T(v))$ and, if $E(T(v), f_i) = \{a_1, \ldots, a_t\}$, then t edges start from the node v that are labeled with $a_1, \ldots, a_t$, respectively.

We denote by $DT(T)$ the set of decision trees for T.

For $b \in \omega$, we denote by $tree(b)$ the decision tree that contains only one (terminal) node labeled with b.

Let $f_i \in \{f_1, \ldots, f_n\}$, $a_1, \ldots, a_t$ be pairwise different numbers from ω, and $\Gamma_1, \ldots, \Gamma_t$ be decision trees over T. We denote by $tree(f_i, a_1, \ldots, a_t, \Gamma_1, \ldots, \Gamma_t)$ the following decision tree over T: the root of the tree is labeled with f_i, and t edges start from the root which are labeled with $a_1, \ldots, a_t$ and enter the roots of decision trees $\Gamma_1, \ldots, \Gamma_t$, respectively.

Let $f_i \in E(T)$ and $E(T, f_i) = \{a_1, \ldots, a_t\}$. We denote

$$DT(T, f_i) = \{tree(f_i, a_1, \ldots, a_t, \Gamma_1, \ldots, \Gamma_t) : \Gamma_j \in DT(T(f_i, a_j)), j = 1, \ldots, t\}\ .$$

Proposition 4.1 *Let T be a decision table. Then $DT(T) = \{tree(mcd(T))\}$ if T is degenerate, and $DT(T) = \{tree(mcd(T))\} \cup \bigcup_{f_i \in E(T)} DT(T, f_i)$ if T is nondegenerate.*

Proof Let T be a degenerate decision table. From the definition of a decision tree for T it follows that $tree(mcd(T))$ is the only decision tree for T.

Let T be a nondegenerate decision table. It is clear that $tree(mcd(T))$ is the only decision tree for T with one node.

Let $\Gamma \in DT(T)$ and Γ have more that one node. Then, by definition of the set $DT(T)$, $\Gamma = tree(f_i, a_1, \ldots, a_t, \Gamma_1, \ldots, \Gamma_t)$ where $f_i \in E(T)$ and $\{a_1, \ldots, a_t\} = E(T, f_i)$. Using the fact that $\Gamma \in DT(T)$ it is not difficult to show that $\Gamma_j \in DT(T(f_i, a_j))$ for $j = 1, \ldots, t$. From here it follows that $\Gamma \in DT(T, f_i)$ for $f_i \in E(T)$. Therefore $DT(T) \subseteq \{tree(mcd(T))\} \cup \bigcup_{f_i \in E(T)} DT(T, f_i)$.

Let, for some $f_i \in E(T)$, $\Gamma \in DT(T, f_i)$. Then

$$\Gamma = tree(f_i, a_1, \ldots, a_t, \Gamma_1, \ldots, \Gamma_t)$$

where $\{a_1, \ldots, a_t\} = E(T, f_i)$ and $\Gamma_j \in DT(T(f_i, a_j))$ for $j = 1, \ldots, t$. Using these facts it is not difficult to show that $\Gamma \in DT(T)$. Therefore

$$\{tree(mcd(T))\} \cup \bigcup_{f_i \in E(T)} DT(T, f_i) \subseteq DT(T) .$$

□

For each node Θ of the directed acyclic graph $\Delta(T)$, define the set

$$Tree^*(\Delta(T), \Theta)$$

of decision trees in the following way. If Θ is a terminal node of $\Delta(T)$ (in this case Θ is degenerate) then $Tree^*(\Delta(T), \Theta) = \{tree(mcd(\Theta))\}$. Let Θ be a nonterminal node of $\Delta(T)$ (in this case Θ is nondegenerate), $f_i \in E(T)$, and $E(\Theta, f_i) = \{a_1, \ldots, a_t\}$. We denote by $Tree^*(\Delta(T), \Theta, f_i)$ the set of decision trees $\{tree(f_i, a_1, \ldots, a_t, \Gamma_1, \ldots, \Gamma_t) : \Gamma_j \in Tree^*(\Delta(T), \Theta(f_i, a_j)), j = 1, \ldots, t\}$. Then

$$Tree^*(\Delta(T), \Theta) = \{tree(mcd(\Theta))\} \cup \bigcup_{f_i \in E(\Theta)} Tree^*(\Delta(T), \Theta, f_i) .$$

Proposition 4.2 *For any decision table T and any node Θ of the graph $\Delta(T)$, the equality $Tree^*(\Delta(T), \Theta) = DT(\Theta)$ holds.*

Proof We prove this statement by induction on nodes of $\Delta(T)$. Let Θ be a terminal node of $\Delta(T)$. Then $Tree^*(\Delta(T), \Theta) = \{tree(mcd(\Theta))\} = DT(\Theta)$. Let now Θ be a nonterminal node of $\Delta(T)$, and let us assume that $Tree^*(\Delta(T), \Theta(f_i, a_j)) = DT(\Theta(f_i, a_j))$ for any $f_i \in E(\Theta)$ and $a_j \in E(\Theta, f_i)$. Then, for any $f_i \in E(\Theta)$, we have $Tree^*(\Delta(T), \Theta, f_i) = DT(\Theta, f_i)$. By Proposition 4.1, $Tree^*(\Delta(T), \Theta) = DT(\Theta)$. □

4.2 (U, α)-Decision Trees for T

Let U be an uncertainty measure and $\alpha \in \mathbb{R}_+$.

A decision tree Γ over T is called a *(U, α)-decision tree for T* if, for any node v of Γ,

- If $U(T(v)) \leq \alpha$ then v is a terminal node which is labeled with $mcd(T(v))$.
- If $U(T(v)) > \alpha$ then v is a nonterminal node labeled with an attribute $f_i \in E(T(v))$, and if $E(T(v), f_i) = \{a_1, \ldots, a_t\}$ then t edges start from the node v which are labeled with $a_1, \ldots, a_t$, respectively.

We denote by $DT_{U,\alpha}(T)$ the set of (U, α)-decision trees for T. It is easy to show that $DT_{U,\alpha}(T) \subseteq DT(T)$.

Let $f_i \in E(T)$ and $E(T, f_i) = \{a_1, \ldots, a_t\}$. We denote

$$DT_{U,\alpha}(T, f_i) = \{tree(f_i, a_1, \ldots, a_t, \Gamma_1, \ldots, \Gamma_t) : \Gamma_j \in DT_{U,\alpha}(T(f_i, a_j)), j = 1, \ldots, t\} .$$

Proposition 4.3 *Let U be an uncertainty measure, $\alpha \in \mathbb{R}_+$, and T be a decision table. Then $DT_{U,\alpha}(T) = \{tree(mcd(T))\}$ if $U(T) \leq \alpha$, and $DT_{U,\alpha}(T) = \bigcup_{f_i \in E(T)} DT_{U,\alpha}(T, f_i)$ if $U(T) > \alpha$.*

Proof Let $U(T) \leq \alpha$. Then $tree(mcd(T))$ is the only (U, α) -decision tree for T.

Let $U(T) > \alpha$ and $\Gamma \in DT_{U,\alpha}(T)$. Then, by definition,

$$\Gamma = tree(f_i, a_1, \ldots, a_t, \Gamma_1, \ldots, \Gamma_t)$$

where $f_i \in E(T)$ and $\{a_1, \ldots, a_t\} = E(T, f_i)$. Using the fact that $\Gamma \in DT_{U,\alpha}(T)$ it is not difficult to show that $\Gamma_j \in DT_{U,\alpha}(T(f_i, a_j))$ for $j = 1, \ldots, t$. From here it follows that $\Gamma \in DT_{U,\alpha}(T, f_i)$ for $f_i \in E(T)$. Therefore $DT_{U,\alpha}(T) \subseteq \bigcup_{f_i \in E(T)} DT_{U,\alpha}(T, f_i)$.

Let, for some $f_i \in E(T)$, $\Gamma \in DT_{U,\alpha}(T, f_i)$. Then $\Gamma = tree(f_i, a_1, \ldots, a_t, \Gamma_1, \ldots, \Gamma_t)$ where $\{a_1, \ldots, a_t\} = E(T, f_i)$, and $\Gamma_j \in DT_{U,\alpha}(T(f_i, a_j))$ for $j = 1, \ldots, t$. Using these facts it is not difficult to show that $\Gamma \in DT_{U,\alpha}(T)$. Therefore $\bigcup_{f_i \in E(T)} DT_{U,\alpha}(T, f_i) \subseteq DT_{U,\alpha}(T)$. □

Let G be a proper subgraph of the graph $\Delta_{U,\alpha}(T)$. For each nonterminal node Θ of the graph G, we denote by $E_G(\Theta)$ the set of attributes f_i from $E(\Theta)$ such that f_i-bundle of edges starts from Θ in G. For each terminal node Θ, $E_G(\Theta) = \emptyset$. For each node Θ of the graph G, we define the set $Tree(G, \Theta)$ of decision trees in the following way. If Θ is a terminal node of G (in this case $U(\Theta) \leq \alpha$), then $Tree(G, \Theta) = \{tree(mcd(\Theta))\}$. Let Θ be a nonterminal node of G (in this case $U(\Theta) > \alpha$), $f_i \in E_G(\Theta)$, and $E(\Theta, f_i) = \{a_1, \ldots, a_t\}$. We denote $Tree(G, \Theta, f_i) = \{tree(f_i, a_1, \ldots, a_t, \Gamma_1, \ldots, \Gamma_t) : \Gamma_j \in Tree(G, \Theta(f_i, a_j)), j = 1, \ldots, t\}$. Then

$$Tree(G, \Theta) = \bigcup_{f_i \in E_G(T)} Tree(G, \Theta, f_i) .$$

Proposition 4.4 *Let U be an uncertainty measure, $\alpha \in \mathbb{R}_+$, and T be a decision table. Then, for any node Θ of the graph $\Delta_{U,\alpha}(T)$, the following equality holds:*

$$Tree(\Delta_{U,\alpha}(T), \Theta) = DT_{U,\alpha}(\Theta) .$$

Proof We prove this statement by induction on nodes of $\Delta_{U,\alpha}(T)$. Let Θ be a terminal node of $\Delta_{U,\alpha}(T)$. Then $Tree(\Delta_{U,\alpha}(T), \Theta) = \{tree(mcd(\Theta))\} = DT_{U,\alpha}(\Theta)$. Let now Θ be a nonterminal node of $\Delta_{U,\alpha}(T)$, and let us assume that

$$Tree(\Delta_{U,\alpha}(T), \Theta(f_i, a_j)) = DT_{U,\alpha}(\Theta(f_i, a_j))$$

for any $f_i \in E(\Theta)$ and $a_j \in E(\Theta, f_i)$. Then, for any $f_i \in E(\Theta)$, we have

$$Tree(\Delta_{U,\alpha}(T), \Theta, f_i) = DT_{U,\alpha}(\Theta, f_i) \ .$$

Using Proposition 4.3, we obtain $Tree(\Delta_{U,\alpha}(T), \Theta) = DT_{U,\alpha}(\Theta)$. □

4.3 Cardinality of the Set $Tree(G, \Theta)$

Let U be an uncertainty measure, $\alpha \in \mathbb{R}_+$, T be a decision table with n attributes $f_1, \ldots, f_n$, and G be a proper subgraph of the graph $\Delta_{U,\alpha}(T)$. We describe now an algorithm which counts, for each node Θ of the graph G, the cardinality $C(\Theta)$ of the set $Tree(G, \Theta)$, and returns the number $C(T) = |Tree(G, T)|$.

Algorithm $\mathcal{A}_4$ (counting the number of decision trees).

Input: A proper subgraph G of the graph $\Delta_{U,\alpha}(T)$ for some decision table T, uncertainty measure U, and number $\alpha \in \mathbb{R}_+$.
Output: The number $|Tree(G, T)|$.

1. If all nodes of the graph G are processed then return the number $C(T)$ and finish the work of the algorithm. Otherwise, choose a node Θ of the graph G which is not processed yet and which is either a terminal node of G or a nonterminal node of G such that, for each $f_i \in E_G(T)$ and $a_j \in E(\Theta, f_i)$, the node $\Theta(f_i, a_j)$ is processed.
2. If Θ is a terminal node then set $C(\Theta) = 1$, mark the node Θ as processed, and proceed to step 1.
3. If Θ is a nonterminal node then set

$$C(\Theta) = \sum_{f_i \in E_G(\Theta)} \prod_{a_j \in E(\Theta, f_i)} C(\Theta(f_i, a_j)) \ ,$$

mark the node Θ as processed, and proceed to step 1.

Proposition 4.5 *Let U be an uncertainty measure, $\alpha \in \mathbb{R}_+$, T be a decision table with n attributes $f_1, \ldots, f_n$, and G be a proper subgraph of the graph $\Delta_{U,\alpha}(T)$. Then the algorithm $\mathcal{A}_4$ returns the number $|Tree(G, T)|$ and makes at most*

$$nL(G)range(T)$$

operations of addition and multiplication.

Proof We prove by induction on the nodes of G that $C(\Theta) = |Tree(G, \Theta)|$ for each node Θ of G. Let Θ be a terminal node of G. Then $Tree(G, \Theta) = \{tree(mcd(\Theta))\}$ and $|Tree(G, \Theta)| = 1$. Therefore the considered statement holds for Θ. Let now Θ

be a nonterminal node of G such that the considered statement holds for its children. By definition, $Tree(G, \Theta) = \bigcup_{f_i \in E_G(\Theta)} Tree(G, \Theta, f_i)$, and, for $f_i \in E_G(\Theta)$,

$$Tree(G, \Theta, f_i) = \{tree(f_i, a_1, \ldots, a_t, \Gamma_1, \ldots, \Gamma_t) \\ : \Gamma_j \in Tree(G, \Theta(f_i, a_j)), j = 1, \ldots, t\}$$

where $\{a_1, \ldots, a_t\} = E(\Theta, f_i)$. One can show that, for any $f_i \in E_G(\Theta)$,

$$|Tree(G, \Theta, f_i)| = \prod_{a_j \in E(\Theta, f_i)} \left|Tree(G, \Theta(f_i, a_j))\right| ,$$

and $|Tree(G, \Theta)| = \sum_{f_i \in E_G(\Theta)} |Tree(G, \Theta, f_i)|$. By the induction hypothesis,

$$C(\Theta(f_i, a_j)) = \left|Tree(G, \Theta(f_i, a_j))\right|$$

for any $f_i \in E_G(\Theta)$ and $a_j \in E(\Theta, f_i)$. Therefore $C(\Theta) = |Tree(G, \Theta)|$. Hence, the considered statement holds. From here it follows that $C(T) = |Tree(G, T)|$, and the Algorithm $\mathcal{A}_4$ returns the cardinality of the set $Tree(G, T)$.

It is easy to see that the considered algorithm makes at most $nL(G)range(T)$ operations of addition and multiplication where $L(G)$ is the number of nodes in the graph G and $range(T) = \max\{|E(T, f_i)| : i = 1, \ldots, n\}$. □

Proposition 4.6 *Let $\mathcal{U}$ be a restricted information system. Then the algorithm $\mathcal{A}_4$ has polynomial time complexity for decision tables from $\mathcal{T}(\mathcal{U})$ depending on the number of conditional attributes in these tables.*

Proof All operations made by the algorithm $\mathcal{A}_4$ are basic numerical operations. From Proposition 4.5 it follows that the number of these operations is bounded from above by a polynomial depending on the size of input table T and on the number of separable subtables of T.

According to Proposition 3.5, the algorithm $\mathcal{A}_4$ has polynomial time complexity for decision tables from $\mathcal{T}(\mathcal{U})$ depending on the number of conditional attributes in these tables. □

4.4 $U^{\max}$-Decision Trees for T

Let U be an uncertainty measure, T be a decision table, and Γ be a decision tree for T. We denote by $V_t(\Gamma)$ the set of terminal nodes of Γ, and by $V_n(\Gamma)$ we denote the set of nonterminal nodes of Γ. We denote by $U^{\max}(T, \Gamma)$ the number $\max\{U(T(v)) : v \in V_t(\Gamma)\}$ which will be interpreted as a kind of uncertainty of Γ.

Let $V_n(\Gamma) \neq \emptyset$. For a nonterminal node v of Γ, we denote by Γ^v a decision tree for T which is obtained from Γ by removal all nodes and edges of the subtree $\Gamma(v)$ with the exception of v. Instead of an attribute we attach to v the number $mcd(T(v))$.

The operation of transformation of Γ into Γ^v will be called *pruning of the subtree* $\Gamma(v)$. Let $v_1, \ldots, v_m$ be nonterminal nodes of Γ such that, for any $i, j \in \{1, \ldots, m\}$, $i \neq j$, the node v_i does not belong to the subtree $\Gamma(v_j)$. We denote by $\Gamma^{v_1 \ldots v_m}$ the tree obtained from Γ by sequential pruning of subtrees $\Gamma(v_1), \ldots, \Gamma(v_m)$.

A decision tree Γ for T is called a $U^{\max}$*-decision tree for* T if either $V_n(\Gamma) = \emptyset$ or $U^{\max}(T, \Gamma^v) > U^{\max}(T, \Gamma)$ for any node $v \in V_n(\Gamma)$.

We denote by $DT_U^{\max}(T)$ the set of $U^{\max}$-decision trees for T. These trees can be considered as irredundant decision trees for T relative to the uncertainty $U^{\max}$ of decision trees. According to the definition, $DT_U^{\max}(T) \subseteq DT(T)$.

Proposition 4.7 *Let U be an uncertainty measure and T be a decision table. Then* $DT_U^{\max}(T) = \bigcup_{\alpha \in \mathbb{R}_+} DT_{U,\alpha}(T)$.

Proof From Propositions 4.1 and 4.3 it follows that $tree(mcd(T))$ is the only decision tree with one node in $DT_U^{\max}(T)$ and the only decision tree with one node in $\bigcup_{\alpha \in \mathbb{R}_+} DT_{U,\alpha}(T)$. We now consider decision trees Γ from $DT(T)$ that contain more than one node.

Let $\Gamma \in DT_U^{\max}(T)$ and $\alpha = U^{\max}(T, \Gamma)$. Let $v \in V_n(\Gamma)$. Since $U^{\max}(T, \Gamma^v) > \alpha$, we have $U(T(v)) > \alpha$. So, for each terminal node v of Γ, $U(T(v)) \leq \alpha$ and, for each nonterminal node v of Γ, $U(T(v)) > \alpha$. Taking into account that $\Gamma \in DT(T)$, we obtain $\Gamma \in DT_{U,\alpha}(T)$.

Let $\Gamma \in DT_{U,\alpha}(T)$ for some $\alpha \in \mathbb{R}_+$. Then $U^{\max}(T, \Gamma) \leq \alpha$ and, for each node $v \in V_n(\Gamma)$, the inequality $U(T(v)) > \alpha$ holds. Therefore, for each node $v \in V_n(\Gamma)$, we have $U^{\max}(T, \Gamma^v) > U^{\max}(T, \Gamma)$, and $\Gamma \in DT_U^{\max}(T)$. □

Proposition 4.8 *Let U be an uncertainty measure, T be a decision table, and $\Gamma \in DT(T) \setminus DT_U^{\max}(T)$. Then there exist nodes $v_1, \ldots, v_m \in V_n(\Gamma)$ such that, for any $i, j \in \{1, \ldots, m\}$, $i \neq j$, the node v_i does not belong to the subtree $\Gamma(v_j)$, $\Gamma^{v_1 \ldots v_m} \in DT_U^{\max}(T)$ and $U^{\max}(T, \Gamma^{v_1 \ldots v_m}) \leq U^{\max}(T, \Gamma)$.*

Proof Let $\alpha = U^{\max}(T, \Gamma)$ and $v_1, \ldots, v_m$ be all nonterminal nodes v in Γ such that $U(T(v)) \leq \alpha$ and there is no a nonterminal node v' such that $v \neq v'$, $U(T(v')) \leq \alpha$ and v belongs to $\Gamma(v')$. One can show that, for any $i, j \in \{1, \ldots, m\}$, $i \neq j$, the node v_i does not belong to the subtree $\Gamma(v_j)$, $\Gamma^{v_1 \ldots v_m} \in DT_U^{\max}(T)$ and $U^{\max}(T, \Gamma^{v_1 \ldots v_m}) \leq \alpha$. □

4.5 U^{sum}-Decision Trees

Let U be an uncertainty measure, T be a decision table, and Γ be a decision tree for T. We denote by $U^{\text{sum}}(T, \Gamma)$ the number $\sum_{v \in V_t(\Gamma)} U(T(v))$ which will be interpreted as a kind of uncertainty of Γ.

A decision tree Γ for T is called a U^{sum} *-decision tree for* T if either $V_n(\Gamma) = \emptyset$ or $U^{\text{sum}}(T, \Gamma^v) > U^{\text{sum}}(T, \Gamma)$ for any node $v \in V_n(\Gamma)$. We denote by $DT_U^{\text{sum}}(T)$ the set of U^{sum}-decision trees for T.

These trees can be considered as irredundant decision trees for T relative to the uncertainty U^{sum} of decision trees. According to the definition, $DT_U^{\text{sum}}(T) \subseteq DT(T)$.

Proposition 4.9 *Let U be a strictly sum-increasing uncertainty measure. Then $DT_U^{\text{sum}}(T) = DT(T)$ for any decision table T.*

Proof We know that $DT_U^{\text{sum}}(T) \subseteq DT(T)$. Let us show that $DT(T) \subseteq DT_U^{\text{sum}}(T)$. Let $\Gamma \in DT(T)$. If $V_n(\Gamma) = \emptyset$ then $\Gamma \in DT_U^{\text{sum}}(T)$. Let now $V_n(\Gamma) \neq \emptyset$, $v \in V_n(\Gamma)$, and t edges start from v and enter nodes $v_1, \ldots, v_t$. Taking into account that U is a strictly sum-increasing uncertainty measure and $\Gamma \in DT(T)$, we obtain $U(T(v)) > U(T(v_1)) + \ldots + U(T(v_t))$. Using this fact one can show that, for any $v \in V_n(\Gamma)$, $U(T(v)) > \sum_{u \in V_t(\Gamma(v))} U(T(u))$. From here it follows that $U^{\text{sum}}(T, \Gamma^v) > U^{\text{sum}}(T, \Gamma)$ for any node $v \in V_n(\Gamma)$. Therefore $\Gamma \in DT_U^{\text{sum}}(T)$. □

Note that from Proposition 3.2 and Remark 3.2 it follows that R and $R + me$ are strictly sum-increasing uncertainty measures.

Proposition 4.10 *Let U be an uncertainty measure, T be a decision table, and $\Gamma \in DT(T) \setminus DT_U^{\text{sum}}(T)$. Then there exist nodes $v_1, \ldots, v_m \in V_n(\Gamma)$ such that, for any $i, j \in \{1, \ldots, m\}$, $i \neq j$, the node v_i does not belong to the subtree $\Gamma(v_j)$, $\Gamma^{v_1 \ldots v_m} \in DT_U^{\text{sum}}(T)$ and $U^{\text{sum}}(T, \Gamma^{v_1 \ldots v_m}) \leq U^{\text{sum}}(T, \Gamma)$.*

Proof Let $v_1, \ldots, v_m$ be all nonterminal nodes v in Γ such that

$$U(T(v)) \leq \sum_{u \in V_t(\Gamma(v))} U(T(u))$$

and there is no a nonterminal node v' such that $U(T(v')) \leq \sum_{u \in V_t(\Gamma(v'))} U(T(u))$, $v \neq v'$, and v belongs to $\Gamma(v')$. It is clear that, for any $i, j \in \{1, \ldots, m\}$, $i \neq j$, the node v_i does not belong to the subtree $\Gamma(v_j)$ and, for any nonterminal node v of the tree $\Gamma' = \Gamma^{v_1 \ldots v_m}$, $\sum_{u \in V_t(\Gamma'(v))} U(T(u)) \leq \sum_{u \in V_t(\Gamma(v))} U(T(u))$. Using this fact, one can show that $\Gamma^{v_1 \ldots v_m} \in DT_U^{\text{sum}}(T)$ and $U^{\text{sum}}(T, \Gamma^{v_1 \ldots v_m}) \leq U^{\text{sum}}(T, \Gamma)$. □

4.6 Cost Functions for Decision Trees

Let n be a natural number. We consider a partial order $\leq$ on the set $\mathbb{R}^n$: $(x_1, \ldots, x_n) \leq (y_1, \ldots, y_n)$ if $x_1 \leq y_1, \ldots, x_n \leq y_n$. A function $g : \mathbb{R}_+^n \to \mathbb{R}_+$ is called *increasing* if $g(x) \leq g(y)$ for any $x, y \in \mathbb{R}_+^n$ such that $x \leq y$. A function $g : \mathbb{R}_+^n \to \mathbb{R}_+$ is called *strictly increasing* if $g(x) < g(y)$ for any $x, y \in \mathbb{R}_+^n$ such that $x \leq y$ and $x \neq y$. If g is strictly increasing then, evidently, g is increasing. For example, $\max(x_1, x_2)$ is increasing and $x_1 + x_2$ is strictly increasing.

Let f be a function from $\mathbb{R}_+^2$ to $\mathbb{R}_+$. We can extend f to a function with arbitrary number of variables in the following way: $f(x_1) = x_1$ and, if $n > 2$ then $f(x_1, \ldots, x_n) = f(f(x_1, \ldots, x_{n-1}), x_n)$.

Proposition 4.11 *Let f be an increasing function from $\mathbb{R}_+^2$ to $\mathbb{R}_+$. Then, for any natural n, the function $f(x_1, \ldots, x_n)$ is increasing.*

Proof We prove the considered statement by induction on n. If $n = 1$ then, evidently, the function $f(x_1) = x_1$ is increasing. We know that the function $f(x_1, x_2)$ is increasing. Let us assume that, for some $n \geq 2$, the function $f(x_1, \ldots, x_n)$ is increasing. We now show that the function $f(x_1, \ldots, x_{n+1})$ is increasing. Let $x = (x_1, \ldots, x_{n+1}), y = (y_1, \ldots, y_{n+1}) \in \mathbb{R}_+^{n+1}$ and $x \leq y$. By the inductive hypothesis,

$$f(x_1, \ldots, x_n) \leq f(y_1, \ldots, y_n) .$$

Since $x_{n+1} \leq y_{n+1}$, we obtain $f(f(x_1, \ldots, x_n), x_{n+1}) \leq f(f(y_1, \ldots, y_n), y_{n+1})$. Therefore the function $f(x_1, \ldots, x_{n+1})$ is increasing. □

Proposition 4.12 *Let f be a strictly increasing function from $\mathbb{R}_+^2$ to $\mathbb{R}_+$. Then, for any natural n, the function $f(x_1, \ldots, x_n)$ is strictly increasing.*

Proof We prove the considered statement by induction on n. If $n = 1$ then the function $f(x_1) = x_1$ is strictly increasing. We know that the function $f(x_1, x_2)$ is strictly increasing. Let us assume that, for some $n \geq 2$, the function $f(x_1, \ldots, x_n)$ is strictly increasing. We now show that the function $f(x_1, \ldots, x_{n+1})$ is strictly increasing. Let

$$x = (x_1, \ldots, x_{n+1}), y = (y_1, \ldots, y_{n+1}) \in \mathbb{R}_+^{n+1} ,$$

$x \leq y$ and $x \neq y$. It is clear that $x' = (x_1, \ldots, x_n) \leq y' = (y_1, \ldots, y_n)$. If $x' \neq y'$ then, by induction hypothesis, $f(x_1, \ldots, x_n) < f(y_1, \ldots, y_n)$ and, since $x_{n+1} \leq y_{n+1}$, $f(f(x_1, \ldots, x_n), x_{n+1}) < f(f(y_1, \ldots, y_n), y_{n+1})$. Let now $x' = y'$. Then $x_{n+1} < y_{n+1}$ and $f(f(x_1, \ldots, x_n), x_{n+1}) < f(f(y_1, \ldots, y_n), y_{n+1})$. Thus the function $f(x_1, \ldots, x_{n+1})$ is strictly increasing. □

A cost function for decision trees is a function $\psi(T, \Gamma)$ which is defined on pairs decision table T and a decision tree Γ for T, and has values from $\mathbb{R}_+$. The function ψ is given by three functions $\psi^0 : \mathscr{T} \to \mathbb{R}_+$, $F : \mathbb{R}_+^2 \to \mathbb{R}_+$ and $w : \mathscr{T} \to \mathbb{R}_+$.

The value of $\psi(T, \Gamma)$ is defined by induction:

- If $\Gamma = tree(mcd(T))$ then $\psi(T, \Gamma) = \psi^0(T)$.
- If $\Gamma = tree(f_i, a_1, \ldots, a_t, \Gamma_1, \ldots, \Gamma_t)$ then

$$\psi(T, \Gamma) = F(\psi(T(f_i, a_1), \Gamma_1), \ldots, \psi(T(f_i, a_t), \Gamma_t)) + w(T) .$$

The cost function ψ is called *increasing* if F is an increasing function. The cost function ψ is called *strictly increasing* if F is a strictly increasing function. The cost function ψ is called *integral* if, for any $T \in \mathscr{T}$ and any $x, y \in \omega = \{0, 1, 2, \ldots\}$, $\psi^0(T) \in \omega$, $F(x, y) \in \omega$, and $w(T) \in \omega$.

We now consider examples of cost functions for decision trees:

- *Depth* $h(T, \Gamma) = h(\Gamma)$ of a decision tree Γ for a decision table T is the maximum length of a path in Γ from the root to a terminal node. For this cost function, $\psi^0(T) = 0$, $F(x, y) = \max(x, y)$, and $w(T) = 1$. This is an increasing integral cost function.
- *Total path length* $tpl(T, \Gamma)$ of a decision tree Γ for a decision table T is equal to $\sum_{r \in Row(T)} l_\Gamma(r)$ where $Row(T)$ is the set of rows of T, and $l_\Gamma(r)$ is the length of a path in Γ from the root to a terminal node v such that the row r belongs to $T_\Gamma(v)$. For this cost function, $\psi^0(T) = 0$, $F(x, y) = x + y$, and $w(T) = N(T)$. This is a strictly increasing integral cost function. Let T be a nonempty decision table. The value $tpl(T, \Gamma)/N(T)$ is the *average depth* $h_{avg}(T, \Gamma)$ of a decision tree Γ for a decision table T.
- *Number of nodes* $L(T, \Gamma) = L(\Gamma)$ of a decision tree Γ for a decision table T. For this cost function, $\psi^0(T) = 1$, $F(x, y) = x + y$, and $w(T) = 1$. This is a strictly increasing integral cost function.
- *Number of nonterminal nodes* $L_n(T, \Gamma) = L_n(\Gamma)$ of a decision tree Γ for a decision table T. For this cost function, $\psi^0(T) = 0$, $F(x, y) = x + y$, and $w(T) = 1$. This is a strictly increasing integral cost function.
- *Number of terminal nodes* $L_t(T, \Gamma) = L_t(\Gamma)$ of a decision tree Γ for a decision table T. For this cost function, $\psi^0(T) = 1$, $F(x, y) = x + y$, and $w(T) = 0$. This is a strictly increasing integral cost function.

For each of the considered cost functions, corresponding functions $\psi^0(T)$ and $w(T)$ have polynomial time complexity depending on the size of decision tables.

Note that the functions $U^{\text{sum}}(T, \Gamma)$ and $U^{\max}(T, \Gamma)$ where U is an uncertainty measure can be represented in the form of cost functions for decision trees:

- For $U^{\max}(T, \Gamma)$, $\psi^0(T) = U(T)$, $F(x, y) = \max(x, y)$, and $w(T) = 0$.
- For $U^{\text{sum}}(T, \Gamma)$, $\psi^0(T) = U(T)$, $F(x, y) = x + y$, and $w(T) = 0$.

We will say that a cost function ψ is *bounded* if, for any decision table T and any decision tree Γ for T, $\psi(T, tree(mcd(T))) \leq \psi(T, \Gamma)$.

Lemma 4.1 *The cost functions h, tpl, L, L_t, and L_n are bounded cost functions for decision trees.*

Proof If T is a degenerate decision table then there is only one decision tree for T which is equal to $tree(mcd(T))$. Let T be a nondegenerate decision table, $\Gamma_0 = tree(mcd(T))$ and Γ be a decision tree for T such that $\Gamma \neq \Gamma_0$. Since Γ is a decision tree for T different from Γ_0, the root of Γ is a nonterminal node, and there are at least two edges starting from the root. From here it follows that $h(\Gamma) \geq 1$, $tpl(T, \Gamma) \geq 2$, $L(\Gamma) \geq 3$, $L_n(\Gamma) \geq 1$, and $L_t(\Gamma) \geq 2$. Since $h(\Gamma_0) = 0$, $tpl(T, \Gamma_0) = 0$, $L(\Gamma_0) = 1$, $L_n(\Gamma_0) = 0$, and $L_t(\Gamma_0) = 1$, we have h, tpl, L, L_t, and L_n are bounded cost functions. □

Let ψ be an integral cost function and T be a decision table. We denote

$$ub(\psi, T) = \max\{\psi(\Theta, \Gamma) : \Theta \in SEP(T), \Gamma \in DT(\Theta)\} .$$

It is clear that, for any separable subtable Θ of T and for any decision tree Γ for Θ, $\psi(\Theta, \Gamma) \in \{0, \ldots, ub(\psi, T)\}$.

Lemma 4.2 *Let T be a decision table with n conditional attributes. Then $ub(h, T) \leq n$, $ub(tpl, T) \leq nN(T)$, $ub(L, T) \leq 2N(T)$, $ub(L_n, T) \leq N(T)$, and $ub(L_t, T) \leq N(T)$.*

Proof Let Θ be a separable subtable of the table T and Γ be a decision tree for Θ. From the definition of a decision tree for a decision table it follows that, for any node v of Γ, the subtable $\Theta_\Gamma(v)$ is nonempty, for any nonterminal node v of Γ, at least two edges start in v, and, in any path from the root to a terminal node of Γ, nonterminal nodes are labeled with pairwise different attributes. Therefore $h(\Theta, \Gamma) \leq n$, $L_t(\Theta, \Gamma) \leq N(\Theta) \leq N(T)$, $tpl(\Theta, \Gamma) \leq nN(\Theta) \leq nN(T)$, $L_n(\Theta, \Gamma) \leq N(\Theta) \leq N(T)$, and $L(\Theta, \Gamma) \leq 2N(\Theta) \leq 2N(T)$. □

Proposition 4.13 *Let U be an uncertainty measure, ψ be bounded and increasing cost function for decision trees, T be a decision table, and $\Gamma \in DT(T) \setminus DT_U^{\max}(T)$. Then there is a decision tree $\Gamma' \in DT_U^{\max}(T)$ such that $U^{\max}(T, \Gamma') \leq U^{\max}(T, \Gamma)$ and $\psi(T, \Gamma') \leq \psi(T, \Gamma)$.*

Proof By Proposition 4.8, there are nodes $v_1, \ldots, v_m \in V_n(\Gamma)$ such that the decision tree $\Gamma' = \Gamma^{v_1 \ldots v_m}$ belongs to $DT_U^{\max}(T)$ and $U^{\max}(T, \Gamma') \leq U^{\max}(T, \Gamma)$. Since ψ is bounded and increasing, $\psi(T, \Gamma') \leq \psi(T, \Gamma)$. □

Proposition 4.14 *Let U be an uncertainty measure, ψ be bounded and increasing cost function for decision trees, T be a decision table, and $\Gamma \in DT(T) \setminus DT_U^{\text{sum}}(T)$. Then there is a decision tree $\Gamma' \in DT_U^{\text{sum}}(T)$ such that $U^{\text{sum}}(T, \Gamma') \leq U^{\text{sum}}(T, \Gamma)$ and $\psi(T, \Gamma') \leq \psi(T, \Gamma)$.*

Proof By Proposition 4.10, there are nodes $v_1, \ldots, v_m \in V_n(\Gamma)$ such that the decision tree $\Gamma' = \Gamma^{v_1 \ldots v_m}$ belongs to $DT_U^{\text{sum}}(T)$ and $U^{\text{sum}}(T, \Gamma') \leq U^{\text{sum}}(T, \Gamma)$. Since ψ is bounded and increasing, $\psi(T, \Gamma') \leq \psi(T, \Gamma)$. □

References

1. Alkhalid, A., Chikalov, I., Moshkov, M.: Comparison of greedy algorithms for α-decision tree construction. In: Yao, J., Ramanna, S., Wang, G., Suraj, Z. (eds.) Rough Sets and Knowledge Technology – 6th International Conference, RSKT 2011, Banff, Canada, October 9–12, 2011. Lecture Notes in Computer Science, vol. 6954, pp. 178–186. Springer, Berlin (2011)
2. Alkhalid, A., Chikalov, I., Moshkov, M.: Comparison of greedy algorithms for decision tree construction. In: Filipe J., Fred A.L.N. (eds.) International Conference on Knowledge Discovery and Information Retrieval, KDIR 2011, Paris, France, October 26–29, 2011, pp. 438–443. SciTePress (2011)

3. Alkhalid, A., Chikalov, I., Moshkov, M.: Decision tree construction using greedy algorithms and dynamic programming – comparative study. In: Szczuka M., Czaja L., Skowron A., Kacprzak M. (eds.) 20th International Workshop on Concurrency, Specification and Programming, CS&P 2011, Pultusk, Poland, September 28–30, 2011, pp. 1–9. Białystok University of Technology (2011)
4. Alkhalid, A., Amin, T., Chikalov, I., Hussain, S., Moshkov, M., Zielosko, B.: Optimization and analysis of decision trees and rules: dynamic programming approach. Int. J. Gen. Syst. **42**(6), 614–634 (2013)
5. Boryczka, U., Kozak, J.: New algorithms for generation decision trees – ant-miner and its modifications. In: Abraham A., Hassanien A.E, de Leon Ferreira de Carvalho A.C.P., Snášel V. (eds.) Foundations of Computational Intelligence – Volume 6: Data Mining, pp. 229–262. Springer, Berlin (2009)
6. Breiman, L., Friedman, J.H., Olshen, R.A., Stone, C.J.: Classification and Regression Trees. Wadsworth and Brooks, Monterey, CA (1984)
7. Breitbart, Y., Reiter, A.: A branch-and-bound algorithm to obtain an optimal evaluation tree for monotonic boolean functions. Acta Inf. **4**, 311–319 (1975)
8. Chai, B., Zhuang, X., Zhao, Y., Sklansky, J.: Binary linear decision tree with genetic algorithm. In: 13th International Conference on Pattern Recognition, ICPR 1996, Vienna, Austria, August 25–30, 1996, vol. 4, pp. 530–534. IEEE Computer Society (1996)
9. Chikalov, I., Hussain, S., Moshkov, M.: Average depth and number of misclassifications for decision trees. In: Popova-Zeugmann L. (ed.) 21st International Workshop on Concurrency, Specification and Programming, CS&P 2012, Berlin, Germany, September 26–28, 2012. CEUR Workshop Proceedings, vol. 928, pp. 160–169. CEUR-WS.org (2012)
10. Chikalov, I., Hussain, S., Moshkov, M.: On cost and uncertainty of decision trees. In: Yao, J., Yang, Y., Slowinski, R., Greco, S., Li, H., Mitra, S., Polkowski, L. (eds.) 8th International Conference Rough Sets and Current Trends in Computing, RSCTC 2012, Chengdu, China, August 17–20, 2012. Lecture Notes in Computer Science, vol. 7413, pp. 190–197. Springer, Berlin (2012)
11. Chikalov, I., Hussain, S., Moshkov, M.: Relationships between average depth and number of nodes for decision trees. In: Sun, F., Li, T., Li, H. (eds.) Knowledge Engineering and Management, 7th International Conference on Intelligent Systems and Knowledge Engineering, ISKE 2012, Beijing, China, December 15–17, 2012. Advances in Intelligent Systems and Computing, vol. 214, pp. 519–529. Springer, Berlin (2014)
12. Chikalov, I., Hussain, S., Moshkov, M.: Relationships between number of nodes and number of misclassifications for decision trees. In: Yao, J., Yang, Y., Slowinski, R., Greco, S., Li, H., Mitra, S., Polkowski, L. (eds.) 8th International Conference Rough Sets and Current Trends in Computing, RSCTC 2012, Chengdu, China, August 17–20, 2012. Lecture Notes in Computer Science, vol. 7413, pp. 212–218. Springer, Berlin (2012)
13. Chikalov, I., Hussain, S., Moshkov, M.: Totally optimal decision trees for Boolean functions. Discret. Appl. Math. **215**, 1–13 (2016)
14. Fayyad, U.M., Irani, K.B.: The attribute selection problem in decision tree generation. In: Swartout W.R. (ed.) 10th National Conference on Artificial Intelligence. San Jose, CA, July 12–16, 1992, pp. 104–110. AAAI Press/The MIT Press (1992)
15. Garey, M.R.: Optimal binary identification procedures. SIAM J. Appl. Math. **23**, 173–186 (1972)
16. Hastie, T., Tibshirani, R., Friedman, J.H.: The Elements of Statistical Learning: Data Mining, Inference, and Prediction. Springer, New York (2001)
17. Heath, D.G., Kasif, S., Salzberg, S.: Induction of oblique decision trees. In: Bajcsy R. (ed.) 13th International Joint Conference on Artificial Intelligence, IJCAI 1993, Chambéry, France, August 28–September 3, 1993, pp. 1002–1007. Morgan Kaufmann (1993)
18. Hussain, S.: Relationships among various parameters for decision tree optimization. In: Faucher, C., Jain, L.C. (eds.) Innovations in Intelligent Machines-4 – Recent Advances in Knowledge Engineering. Studies in Computational Intelligence, vol. 514, pp. 393–410. Springer, Berlin (2014)

19. Hyafil, L., Rivest, R.L.: Constructing optimal binary decision trees is NP-complete. Inf. Process. Lett. **5**(1), 15–17 (1976)
20. Kononenko, I.: On biases in estimating multi-valued attributes. In: 14th International Joint Conference on Artificial Intelligence, IJCAI 95, Montréal Québec, Canada, August 20–25, 1995, pp. 1034–1040. Morgan Kaufmann (1995)
21. Martelli, A., Montanari, U.: Optimizing decision trees through heuristically guided search. Commun. ACM **21**(12), 1025–1039 (1978)
22. Martin, J.K.: An exact probability metric for decision tree splitting and stopping. Mach. Learn. **28**(2–3), 257–291 (1997)
23. Mingers, J.: Expert systems - rule induction with statistical data. J. Oper. Res. Soc. **38**, 39–47 (1987)
24. Moret, B.M.E., Thomason, M.G., Gonzalez, R.C.: The activity of a variable and its relation to decision trees. ACM Trans. Program. Lang. Syst. **2**(4), 580–595 (1980)
25. Moshkov, M.: Time complexity of decision trees. In: Peters, J.F., Skowron, A. (eds.) Trans. Rough Sets III. Lecture Notes in Computer Science, vol. 3400, pp. 244–459. Springer, Berlin (2005)
26. Quinlan, J.R.: C4.5: Programs for Machine Learning. Morgan Kaufmann, San Francisco (1993)
27. Quinlan, J.R.: Induction of decision trees. Mach. Learn. **1**(1), 81–106 (1986)
28. Riddle, P., Segal, R., Etzioni, O.: Representation design and brute-force induction in a Boeing manufacturing domain. Appl. Artif. Intel. **8**, 125–147 (1994)
29. Rokach, L., Maimon, O.: Data Mining with Decision Trees: Theory and Applications. World Scientific Publishing, River Edge (2008)
30. Schumacher, H., Sevcik, K.C.: The synthetic approach to decision table conversion. Commun. ACM **19**(6), 343–351 (1976)

Chapter 5
Multi-stage Optimization of Decision Trees with Some Applications

In this chapter, we consider multi-stage optimization technique for decision trees and two of its applications: study of totally optimal (simultaneously optimal relative to a number of cost functions) decision trees for Boolean functions, and improvement of bounds on the depth of decision trees for diagnosis of circuits.

In Sect. 5.1, we discuss how to optimize (U, α)-decision trees represented by a proper subgraph of the graph $\Delta_{U,\alpha}(T)$ relative to a cost function for decision trees. We also explain possibilities of multi-stage optimization of decision trees for different cost functions, and consider the notion of a totally optimal decision tree relative to a number of cost functions.

In Sect. 5.2, we study totally optimal decision trees for Boolean functions, and in Sect. 5.3, we investigate the depth of decision trees for diagnosis of constant faults in iteration-free circuits over monotone basis.

5.1 Optimization of Decision Trees

Let ψ be an increasing cost function for decision trees given by the triple of functions ψ^0, F and w, U be an uncertainty measure, $\alpha \in \mathbb{R}_+$, T be a decision table with n conditional attributes $f_1, \ldots, f_n$, and G be a proper subgraph of the graph $\Delta_{U,\alpha}(T)$.

In Sect. 4.2, for each nonterminal node Θ of the graph G, we denoted by $E_G(\Theta)$ the set of attributes f_i from $E(\Theta)$ such that f_i-bundle of edges starts from Θ in G. For each node Θ of the graph G, we defined a set $Tree(G, \Theta)$ of (U, α)-decision trees for Θ in the following way. If Θ is a terminal node of G, then $Tree(G, \Theta) = \{tree(mcd(\Theta))\}$. Let Θ be a nonterminal node of G, $f_i \in E_G(\Theta)$, and $E(\Theta, f_i) = \{a_1, \ldots, a_t\}$. Then

H. AbouEisha et al., *Extensions of Dynamic Programming for Combinatorial Optimization and Data Mining*, Intelligent Systems Reference Library 146,
https://doi.org/10.1007/978-3-319-91839-6_5

$Tree(G, \Theta, f_i) = \{tree(f_i, a_1, \ldots, a_t, \Gamma_1, \ldots, \Gamma_t) : \Gamma_j \in Tree(G, \Theta(f_i, a_j)),$ $j = 1, \ldots, t\}$ and

$$Tree(G, \Theta) = \bigcup_{f_i \in E_G(T)} Tree(G, \Theta, f_i) .$$

Let Θ be a node of G, $\Gamma \in Tree(G, \Theta)$, and v be a node of Γ. In Sect. 4.1, we defined a subtree $\Gamma(v)$ of Γ for which v is the root, and a subtable $\Theta_\Gamma(v)$ of Θ. If v is the root of Γ then $\Theta_\Gamma(v) = \Theta$. Let v be not the root of Γ and $v_1, e_1, \ldots, v_m, e_m, v_{m+1} = v$ be the directed path from the root of Γ to v in which nodes $v_1, \ldots, v_m$ are labeled with attributes $f_{i_1}, \ldots, f_{i_m}$ and edges $e_1, \ldots, e_m$ are labeled with numbers $a_1, \ldots, a_m$, respectively. Then $\Theta_\Gamma(v) = \Theta(f_{i_1}, a_1) \ldots (f_{i_m}, a_m)$. One can show that the decision tree $\Gamma(v)$ belongs to the set $Tree(G, \Theta_\Gamma(v))$.

A decision tree Γ from $Tree(G, \Theta)$ is called an *optimal decision tree for Θ relative to ψ and G* if $\psi(\Theta, \Gamma) = \min\{\psi(\Theta, \Gamma') : \Gamma' \in Tree(G, \Theta)\}$.

A decision tree Γ from $Tree(G, \Theta)$ is called a *strictly optimal decision tree for Θ relative to ψ and G* if, for any node v of Γ, the decision tree $\Gamma(v)$ is an optimal decision tree for $\Theta_\Gamma(v)$ relative to ψ and G.

We denote by $Tree_\psi^{opt}(G, \Theta)$ the set of optimal decision trees for Θ relative to ψ and G. We denote by $Tree_\psi^{s-opt}(G, \Theta)$ the set of strictly optimal decision trees for Θ relative to ψ and G.

Let $\Gamma \in Tree_\psi^{opt}(G, \Theta)$ and $\Gamma = tree(f_i, a_1, \ldots, a_t, \Gamma_1, \ldots, \Gamma_t)$. Then

$$\Gamma \in Tree_\psi^{s-opt}(G, \Theta)$$

if and only if $\Gamma_j \in Tree_\psi^{s-opt}(G, \Theta(f_i, a_j))$ for $j = 1, \ldots, t$.

Proposition 5.1 *Let ψ be a strictly increasing cost function for decision trees, U be an uncertainty measure, $\alpha \in \mathbb{R}_+$, T be a decision table, and G be a proper subgraph of the graph $\Delta_{U,\alpha}(T)$. Then, for any node Θ of the graph G, $Tree_\psi^{opt}(G, \Theta) = Tree_\psi^{s-opt}(G, \Theta)$.*

Proof It is clear that $Tree_\psi^{s-opt}(G, \Theta) \subseteq Tree_\psi^{opt}(G, \Theta)$. Let $\Gamma \in Tree_\psi^{opt}(G, \Theta)$ and let us assume that $\Gamma \notin Tree_\psi^{s-opt}(G, \Theta)$. Then there is a node v of Γ such that $\Gamma(v) \notin Tree_\psi^{opt}(G, \Theta_\Gamma(v))$. Let $\Gamma_0 \in Tree_\psi^{opt}(G, \Theta_\Gamma(v))$ and Γ' be the decision tree obtained from Γ by replacing $\Gamma(v)$ with Γ_0. One can show that $\Gamma' \in Tree(G, \Theta)$. Since ψ is strictly increasing and $\psi(\Theta_\Gamma(v), \Gamma_0) < \psi(\Theta_\Gamma(v), \Gamma(v))$, we have $\psi(\Theta, \Gamma') < \psi(\Theta, \Gamma)$. Therefore $\Gamma \notin Tree_\psi^{opt}(G, \Theta)$ which is impossible. Thus $Tree_\psi^{opt}(G, \Theta) \subseteq Tree_\psi^{s-opt}(G, \Theta)$. □

We describe now an algorithm $\mathcal{A}_5$ (a *procedure of optimization relative to the cost function ψ*). The algorithm $\mathcal{A}_5$ attaches to each node Θ of G the number $c(\Theta) = \min\{\psi(\Theta, \Gamma) : \Gamma \in Tree(G, \Theta)\}$ and, probably, remove some f_i-bundles of edges starting from nonterminal nodes of G. As a result, we obtain a proper subgraph G^ψ of the graph G. It is clear that G^ψ is also a proper subgraph of the graph $\Delta_\alpha(T)$.

Algorithm $\mathcal{A}_5$ (procedure of decision tree optimization)

Input: A proper subgraph G of the graph $\Delta_{U,\alpha}(T)$ for some decision table T, uncertainty measure U, a number $\alpha \in \mathbb{R}_+$, and an increasing cost function ψ for decision trees given by the triple of functions ψ^0, F and w.
Output: The proper subgraph G^ψ of the graph G.

1. If all nodes of the graph G are processed then return the obtained graph as G^ψ and finish the work of the algorithm. Otherwise, choose a node Θ of the graph G which is not processed yet and which is either a terminal node of G or a nonterminal node of G for which all children are processed.
2. If Θ is a terminal node then set $c(\Theta) = \psi^0(\Theta)$, mark node Θ as processed and proceed to step 1.
3. If Θ is a nonterminal node then, for each $f_i \in E_G(\Theta)$, compute the value $c(\Theta, f_i) = F(c(\Theta(f_i, a_1)), \ldots, c(\Theta(f_i, a_t))) + w(\Theta)$ where

$$\{a_1, \ldots, a_t\} = E(\Theta, f_i) ,$$

and set $c(\Theta) = \min\{c(\Theta, f_i) : f_i \in E_G(\Theta)\}$. Remove all f_i-bundles of edges starting from Θ for which $c(\Theta) < c(\Theta, f_i)$. Mark the node Θ as processed and proceed to step 1.

Proposition 5.2 *Let G be a proper subgraph of the graph $\Delta_{U,\alpha}(T)$ for some decision table T with n conditional attributes $f_1, \ldots, f_n$, uncertainty measure U, and a number $\alpha \in \mathbb{R}_+$, and ψ be an increasing cost function for decision trees given by the triple of functions ψ^0, F and w. Then, to construct the graph G^ψ, the algorithm $\mathcal{A}_5$ makes*

$$O(nL(G)range(T))$$

elementary operations (computations of F, w, ψ^0, comparisons, and additions).

Proof In each terminal node of the graph G, the algorithm $\mathcal{A}_5$ computes the value of ψ^0. In each nonterminal node of G, the algorithm $\mathcal{A}_5$ computes the value of F (as function with two variables) at most $range(T)n$ times, where $range(T) = \max\{|E(T, f_i)| : i = 1, \ldots, n\}$, and the value of w at most n times, makes at most n additions and at most $2n$ comparisons. Therefore the algorithm $\mathcal{A}_5$ makes

$$O(nL(G)range(T))$$

elementary operations. □

Proposition 5.3 *Let $\psi \in \{h, tpl, L, L_n, L_t\}$ and $\mathcal{U}$ be a restricted information system. Then the algorithm $\mathcal{A}_5$ has polynomial time complexity for decision tables from $\mathcal{T}(\mathcal{U})$ depending on the number of conditional attributes in these tables.*

Proof Since $\psi \in \{h, tpl, L, L_n, L_t\}$, ψ^0 is a constant, F is either $\max(x, y)$ or $x + y$, and w is either a constant or $N(T)$. Therefore the elementary operations used by the algorithm $\mathscr{A}_5$ are either basic numerical operations or computations of numerical parameters of decision tables which have polynomial time complexity depending on the size of decision tables. From Proposition 5.2 it follows that the number of elementary operations is bounded from above by a polynomial depending on the size of input table T and on the number of separable subtables of T.

According to Proposition 3.5, the algorithm $\mathscr{A}_5$ has polynomial time complexity for decision tables from $\mathscr{T}(\mathscr{U})$ depending on the number of conditional attributes in these tables. □

For any node Θ of the graph G and for any $f_i \in E_G(\Theta)$, we denote $\psi_G(\Theta) = \min\{\psi(\Theta, \Gamma) : \Gamma \in Tree(G, \Theta)\}$ and

$$\psi_G(\Theta, f_i) = \min\{\psi(\Theta, \Gamma) : \Gamma \in Tree(G, \Theta, f_i)\} .$$

Lemma 5.1 *Let G be a proper subgraph of the graph $\Delta_{U,\alpha}(T)$ for some decision table T with n conditional attributes, uncertainty measure U, and number $\alpha \in \mathbb{R}_+$, and ψ be an increasing cost function for decision trees given by the triple of functions ψ^0, F and w. Then, for any node Θ of the graph G and for any attribute $f_i \in E_G(\Theta)$, the algorithm $\mathscr{A}_5$ computes values $c(\Theta) = \psi_G(\Theta)$ and $c(\Theta, f_i) = \psi_G(\Theta, f_i)$.*

Proof We prove the considered statement by induction on the nodes of the graph G . Let Θ be a terminal node of G. Then $Tree(G, \Theta) = \{tree(mcd(\Theta))\}$ and $\psi_G(\Theta) = \psi^0(\Theta)$. Therefore $c(\Theta) = \psi_G(\Theta)$. Since $E_G(\Theta) = \emptyset$, the considered statement holds for Θ.

Let now Θ be a nonterminal node of G such that the considered statement holds for each node $\Theta(f_i, a_j)$ with $f_i \in E_G(\Theta)$ and $a_j \in E(\Theta, f_i)$. By definition, $Tree(G, \Theta) = \bigcup_{f_i \in E_G(\Theta)} Tree(G, \Theta, f_i)$ and, for each $f_i \in E_G(\Theta)$, $Tree(G, \Theta, f_i) = \{tree(f_i, a_1, \ldots, a_t, \Gamma_1, \ldots, \Gamma_t) : \Gamma_j \in Tree(G, \Theta(f_i, a_j)), j = 1, \ldots, t\}$ where

$$\{a_1, \ldots, a_t\} = E(\Theta, f_i) .$$

Since ψ is an increasing cost function,

$$\psi_G(\Theta, f_i) = F(\psi_G(\Theta(f_i, a_1)), \ldots, \psi_G(\Theta(f_i, a_t))) + w(\Theta)$$

where $\{a_1, \ldots, a_t\} = E(\Theta, f_i)$. It is clear that $\psi_G(\Theta) = \min\{\psi_G(\Theta, f_i) : f_i \in E_G(\Theta)\}$. By the induction hypothesis, $\psi_G(\Theta(f_i, a_j)) = c(\Theta(f_i, a_j))$ for each $f_i \in E_G(\Theta)$ and $a_j \in E(\Theta, f_i)$. Therefore $c(\Theta, f_i) = \psi_G(\Theta, f_i)$ for each $f_i \in E_G(\Theta)$, and $c(\Theta) = \psi_G(\Theta)$. □

Theorem 5.1 *Let ψ be an increasing cost function for decision trees, U be an uncertainty measure, $\alpha \in \mathbb{R}_+$, T be a decision table, and G be a proper subgraph of the graph $\Delta_{U,\alpha}(T)$. Then, for any node Θ of the graph G^{ψ}, the following equality holds: $Tree(G^{\psi}, \Theta) = Tree_{\psi}^{s-opt}(G, \Theta)$.*

Proof We prove the considered statement by induction on nodes of G^ψ. We use Lemma 5.1 which shows that, for any node Θ of the graph G and for any $f_i \in E_G(\Theta)$, $c(\Theta) = \psi_G(\Theta)$ and $c(\Theta, f_i) = \psi_G(\Theta, f_i)$.

Let Θ be a terminal node of G^ψ. Then $Tree(G^\psi, \Theta) = \{tree(mcd(\Theta))\}$. It is clear that $Tree(G^\psi, \Theta) = Tree_\psi^{s-opt}(G, \Theta)$. Therefore the considered statement holds for Θ.

Let Θ be a nonterminal node of G^ψ such that the considered statement holds for each node $\Theta(f_i, a_j)$ with $f_i \in E_G(\Theta)$ and $a_j \in E(\Theta, f_i)$. By definition,

$$Tree(G^\psi, \Theta) = \bigcup_{f_i \in E_{G^\psi}(\Theta)} Tree(G^\psi, \Theta, f_i)$$

and, for each $f_i \in E_{G^\psi}(\Theta)$, $Tree(G^\psi, \Theta, f_i) = \{tree(f_i, a_1, \ldots, a_t, \Gamma_1, \ldots, \Gamma_t) : \Gamma_j \in Tree(G^\psi, \Theta(f_i, a_j)), j = 1, \ldots, t\}$, where $\{a_1, \ldots, a_t\} = E(\Theta, f_i)$.

We know that $E_{G^\psi}(\Theta) = \{f_i : f_i \in E_G(\Theta), \psi_G(\Theta, f_i) = \psi_G(\Theta)\}$. Let $f_i \in E_{G^\psi}(\Theta)$ and $\Gamma \in Tree(G^\psi, \Theta, f_i)$. Then $\Gamma = tree(f_i, a_1, \ldots, a_t, \Gamma_1, \ldots, \Gamma_t)$, where

$$\{a_1, \ldots, a_t\} = E(\Theta, f_i)$$

and $\Gamma_j \in Tree(G^\psi, \Theta(f_i, a_j))$ for $j = 1, \ldots, t$. According to the induction hypothesis,

$$Tree(G^\psi, \Theta(f_i, a_j)) = Tree_\psi^{s-opt}(G, \Theta(f_i, a_j))$$

and $\Gamma_j \in Tree_\psi^{s-opt}(G^\psi, \Theta(f_i, a_j))$ for $j = 1, \ldots, t$. In particular,

$$\psi(\Theta(f_i, a_j), \Gamma_j) = \psi_G(\Theta(f_i, a_j))$$

for $j = 1, \ldots, t$. Since $\psi_G(\Theta, f_i) = \psi_G(\Theta)$ we have

$$F(\psi_G(\Theta(f_i, a_1)), \ldots, \psi_G(\Theta(f_i, a_t))) + w(\Theta) = \psi_G(\Theta)$$

and $\psi(\Theta, \Gamma) = \psi_G(\Theta)$. Therefore $\Gamma \in Tree_\psi^{opt}(G, \Theta)$, $\Gamma \in Tree_\psi^{s-opt}(G, \Theta)$ and

$$Tree(G^\psi, \Theta) \subseteq Tree_\psi^{s-opt}(G, \Theta) .$$

Let $\Gamma \in Tree_\psi^{s-opt}(G, \Theta)$. Since Θ is a nonterminal node, Γ can be represented in the form $\Gamma = tree(f_i, a_1, \ldots, a_t, \Gamma_1, \ldots, \Gamma_t)$, where $f_i \in E_G(\Theta)$, $\{a_1, \ldots, a_t\} = E(\Theta, f_i)$, and

$$\Gamma_j \in Tree_\psi^{s-opt}(G, \Theta(f_i, a_j))$$

for $j = 1, \ldots, t$. Since $\Gamma \in Tree_\psi^{s-opt}(G, \Theta)$, $\psi_G(\Theta, f_i) = \psi_G(\Theta)$ and $f_i \in E_{G^\psi}(T)$. According to the induction hypothesis,

$$Tree(G^{\psi}, \Theta(f_i, a_j)) = Tree_{\psi}^{s-opt}(G, \Theta(f_i, a_j))$$

for $j = 1, \ldots, t$. Therefore $\Gamma \in Tree(G^{\psi}, \Theta, f_i) \subseteq Tree(G^{\psi}, \Theta)$. As a result, we have $Tree_{\psi}^{s-opt}(G, \Theta) \subseteq Tree(G^{\psi}, \Theta)$. □

Corollary 5.1 *Let ψ be a strictly increasing cost function, U be an uncertainty measure, $\alpha \in \mathbb{R}_+$, T be a decision table, and G be a proper subgraph of the graph $\Delta_{U,\alpha}(T)$. Then, for any node Θ of the graph G^{ψ}, $Tree(G^{\psi}, \Theta) = Tree_{\psi}^{opt}(G, \Theta)$.*

This corollary follows immediately from Proposition 5.1 and Theorem 5.1.

We can make multi-stage optimization of (U, α)-decision trees for T relative to a sequence of strictly increasing cost functions $\psi_1, \psi_2, \ldots$. We begin from the graph $G = \Delta_{U,\alpha}(T)$ and apply to it the procedure of optimization relative to the cost function ψ_1 (the algorithm $\mathcal{A}_5$). As a result, we obtain a proper subgraph G^{ψ_1} of the graph G.

By Proposition 4.4, the set $Tree(G, T)$ is equal to the set $DT_{U,\alpha}(T)$ of all (U, α)-decision trees for T. Using Corollary 5.1, we obtain that the set $Tree(G^{\psi_1}, T)$ coincides with the set $Tree_{\psi_1}^{opt}(G, T)$ of all decision trees from $Tree(G, T)$ which have minimum cost relative to ψ_1 among all trees from the set $Tree(G, T)$. Next we apply to G^{ψ_1} the procedure of optimization relative to the cost function ψ_2. As a result, we obtain a proper subgraph G^{ψ_1,ψ_2} of the graph G^{ψ_1} (and of the graph $G = \Delta_{\alpha}(T)$). By Corollary 5.1, the set $Tree(G^{\psi_1,\psi_2}, T)$ coincides with the set $Tree_{\psi_2}^{opt}(G^{\psi_1}, T)$ of all decision trees from $Tree(G^{\psi_1}, T)$ which have minimum cost relative to ψ_2 among all trees from $Tree(G^{\psi_1}, T)$, etc.

If one of the cost functions ψ_i is increasing and not strictly increasing then the set $Tree(G^{\psi_1,\ldots,\psi_i}, T)$ coincides with the set $Tree_{\psi_i}^{s-opt}(G^{\psi_1,\ldots,\psi_{i-1}}, T)$ which is a subset of the set of all decision trees from $Tree(G^{\psi_1,\ldots,\psi_{i-1}}, T)$ that have minimum cost relative to ψ_i among all trees from $Tree(G^{\psi_1,\ldots,\psi_{i-1}}, T)$.

For a cost function ψ, we denote $\psi^{U,\alpha}(T) = \min\{\psi(T, \Gamma) : \Gamma \in DT_{U,\alpha}(T)\}$, i.e., $\psi^{U,\alpha}(T)$ is the minimum cost of a (U, α)-decision tree for T relative to the cost function ψ. Let $\psi_1, \ldots, \psi_m$ be cost functions and $m \geq 2$. A (U, α)-decision tree Γ for T is called a *totally optimal (U, α)-decision tree for T relative to the cost functions* $\psi_1, \ldots, \psi_m$ if $\psi_1(T, \Gamma) = \psi_1^{U,\alpha}(T), \ldots, \psi_m(T, \Gamma) = \psi_m^{U,\alpha}(T)$, i.e., Γ is optimal relative to $\psi_1, \ldots, \psi_m$ simultaneously.

Assume that $\psi_1, \ldots, \psi_{m-1}$ are strictly increasing cost functions and ψ_m is increasing or strictly increasing. We now describe how to recognize the existence of a (U, α)-decision tree for T which is a totally optimal (U, α)-decision tree for T relative to the cost functions $\psi_1, \ldots, \psi_m$.

First, we construct the graph $G = \Delta_{U,\alpha}(T)$ using the algorithm $\mathcal{A}_3$. For $i = 1, \ldots, m$, we apply to G the procedure of optimization relative to ψ_i (the Algorithm $\mathcal{A}_5$). As a result, we obtain for $i = 1, \ldots, m$, the graph G^{ψ_i} and the number $\psi_i^{U,\alpha}(T)$ attached to the node T of G^{ψ_i}. Next, we apply to G sequentially the procedures of optimization relative to the cost functions $\psi_1, \ldots, \psi_m$. As a result, we obtain graphs $G^{\psi_1}, G^{\psi_1,\psi_2}, \ldots, G^{\psi_1,\ldots,\psi_m}$ and numbers $\varphi_1, \varphi_2, \ldots, \varphi_m$ attached to the node T of these graphs. It is clear that $\varphi_1 = \psi_1^{U,\alpha}(T)$. For $i = 2, \ldots, m$,

$\varphi_i = \min\{\psi_i(T, \Gamma) : \Gamma \in Tree(G^{\psi_1,\ldots,\psi_{i-1}}, T)\}$. One can show that a totally optimal (U, α)-decision tree for T relative to the cost functions $\psi_1, \ldots, \psi_m$ exists if and only if $\varphi_i = \psi_i^{U,\alpha}(T)$ for $i = 1, \ldots, m$.

5.2 Totally Optimal Decision Trees for Boolean Functions

In this section, we study totally optimal decision trees (which are optimal relative to a number of cost functions, simultaneously) for computing Boolean functions. We consider depth and total path length (average depth) of decision trees as time complexity in the worst- and in the average-case, respectively, and the number of nodes in decision trees as space complexity. We perform experiments on decision tables representing Boolean functions, we formulate hypotheses based on the experimental results, and prove them. We also present results regarding complexity of optimization of decision trees for total and partial Boolean and pseudo-Boolean functions relative to depth, number of nodes, and average depth.

Note that there are many papers which discuss the complexity of Boolean functions in various contexts and for different parameters (see, for example, the surveys by Korshunov [13] and Buhrman and de Wolf [2]). This section is based on the paper [4].

5.2.1 *On Optimization of Decision Trees for Boolean Functions*

Here we discuss time complexity of optimization of decision trees for total and partial Boolean functions relative to depth, total path length (average depth), and number of nodes. We assume that Boolean functions are represented in the form of decision tables. We also mention some results for pseudo-Boolean functions $f : \{0, 1\}^n \to \mathbb{R}$.

A *partial Boolean function* $f(x_1, \ldots, x_n)$ is a partial function of the kind $f : \{0, 1\}^n \to \{0, 1\}$. We work with a *decision table representation of the function f* (*table* for short) which is a rectangular table T_f with n columns filled with numbers from the set $\{0, 1\}$. Columns of the table are labeled with variables $x_1, \ldots, x_n$ which are considered as attributes. Rows of the table are pairwise different, and the set of rows coincides with the set of n-tuples from $\{0, 1\}^n$ on which the value of f is defined. Each row is labeled with the value of function f on this row.

We use relative misclassification error *rme* as uncertainty measure and study both exact decision trees computing f ((rme, 0)-decision trees for T_f) and approximate decision trees computing f ((rme, α)-decision trees for T_f where $\alpha > 0$).

We consider total Boolean functions as a special case of partial Boolean functions, and denote by $\mathscr{F}_2$ the set of table representations for all partial Boolean functions.

5.2.1.1 Total Boolean Functions

Let $0 \leq \alpha < 1$, and $T \in \mathscr{F}_2$. We use the Algorithm $\mathscr{A}_3$ for the construction of the directed acyclic graph $\Delta_{rme,\alpha}(T)$.

The uncertainty measure *rme* has polynomial time complexity depending on the size of the input decision table T. Using Proposition 3.3 we obtain that the time complexity of the algorithm $\mathscr{A}_3$ is bounded from above by a polynomial on the size of the input table $T \in \mathscr{F}_2$ and the number $|SEP(T)|$ of different separable subtables of T. For a decision table T corresponding to a total Boolean function with n variables, $N(T) = 2^n$ and $|SEP(T)| \leq 3^n$. From here, the next statement follows.

Proposition 5.4 *The time complexity of the algorithm $\mathscr{A}_3$ for decision tables corresponding to total Boolean functions is bounded from above by a polynomial on the size of the input table.*

Let T be a table from $\mathscr{F}_2$, $0 \leq \alpha < 1$, and G be a proper subgraph of the graph $\Delta_{rme,\alpha}(T)$. In our study, we use the algorithm $\mathscr{A}_4$ which counts the cardinality of the set $Tree(G, T)$. Using Proposition 4.5 we can prove the following statement.

Proposition 5.5 *The time complexity of the algorithm $\mathscr{A}_4$ for decision tables corresponding to total Boolean functions is bounded from above by a polynomial on the size of the input table.*

Let ψ be an increasing cost function for decision trees, T be a table from $\mathscr{F}_2$, and G be a proper subgraph of the graph $\Delta_{rme,\alpha}(T)$. We use the algorithm $\mathscr{A}_5$ to optimize decision trees from $Tree(G, T)$ relative to the cost function ψ. Using Proposition 5.2 and properties of cost functions h, tpl, and L we obtain the following statement.

Proposition 5.6 *For any cost function $\psi \in \{h, tpl, L\}$, for decision tables corresponding to total Boolean functions, the time complexity of the algorithm $\mathscr{A}_5$ is bounded from above by a polynomial on the size of the input table.*

From Propositions 5.4 and 5.6 it follows that there exist polynomial time algorithms for optimization of decision trees relative to depth, total path length, and number of nodes for total Boolean functions represented by decision tables (note that this result for the number of nodes was obtained earlier in [8]). The considered results can easily be extended to total pseudo-Boolean functions as well. However, the situation with partial Boolean and partial pseudo-Boolean functions is different.

5.2.1.2 Partial Boolean Functions

For decision tables which are representations of partial Boolean functions, the number of separable subtables and the number of nodes in the graph $\Delta_{rme,\alpha}(T)$ can grow exponentially with the number of rows in table T. Let us consider a table T with n columns labeled with variables $x_1, \ldots, x_n$ and $n + 1$ rows

$$(0, \ldots, 0), (1, 0, \ldots, 0), \ldots, (0, \ldots, 0, 1)$$

labeled with values $0, 1, \ldots, 1$, respectively. It is not difficult to show that, for any subset $\{i_1, \ldots, i_m\}$ of the set $\{1, \ldots, n\}$ which is different from $\{1, \ldots, n\}$, the separable subtable $T(x_{i_1}, 0) \ldots (x_{i_m}, 0)$ of the table T is nondegenerate. The considered subtables are pairwise different. Therefore $|SEP(T)| \geq 2^n - 1$. One can show that any nondegenerate separable subtable of the table T is a node of the graph $\Delta_{rme,0}(T)$. Therefore the number of nodes in $\Delta_{rme,0}(T)$ is at least $2^n - 1$. From here it follows that, for partial Boolean functions, algorithms $\mathscr{A}_3$, $\mathscr{A}_4$, and $\mathscr{A}_5$ can have exponential time complexity.

We now reduce the *set cover problem* (Richard Karp in [12] shown that this problem is NP-hard) to the problem of minimization of depth (number of nodes) for decision trees computing partial Boolean functions.

Let $A = \{a_1, \ldots, a_N\}$ be a set with N elements, and $F = \{S_1, \ldots, S_p\}$ be a family of subsets of A such that $A = \bigcup_{i=1}^{p} S_i$. A subfamily $\{S_{i_1}, \ldots, S_{i_t}\}$ of the family F is called a *cover* if $\bigcup_{j=1}^{t} S_{i_j} = A$. The problem of searching for cover with minimum cardinality t is the *set cover problem*.

We correspond to the considered set cover problem a partial Boolean function $f_{A,F}$ with p variables $x_1, \ldots, x_p$ represented by a decision table $T(A, F)$. This table contains p columns labeled with variables $x_1, \ldots, x_p$ corresponding to the sets $S_1, \ldots, S_p$, respectively, and $N + 1$ rows. The first N rows correspond to elements $a_1, \ldots, a_N$, respectively. The table contains the value 0 at the intersection of jth row and ith column, for $j = 1, \ldots, N$ and $i = 1, \ldots, p$, if and only if $a_j \in S_i$. The last $(N + 1)$st row is filled with 1's. The value of $f_{A,F}$ corresponding to the last row is equal to 0. All other rows are labeled with the value 1 of the function $f_{A,F}$.

One can show that each decision tree computing $f_{A,F}$ is as in Fig. 5.1 where $\{S_{i_1}, \ldots, S_{i_m}\}$ is a cover for the considered set cover problem. From here it follows that there is a polynomial time reduction of the set cover problem to the problem of minimization of decision tree depth for partial Boolean functions, and there exists a polynomial time reduction of the set cover problem to the problem of minimization of number of nodes in decision trees computing partial Boolean functions. So, we have the following statements.

Proposition 5.7 *The problem of minimization of depth for decision trees computing partial Boolean functions given by decision tables is NP-hard.*

Proposition 5.8 *The problem of minimization of number of nodes for decision trees computing partial Boolean functions given by decision tables is NP-hard.*

To the best of our knowledge, the mentioned results for depth and number of nodes are folklore results.

Similar result for minimization of average depth of decision trees computing partial pseudo-Boolean functions was obtained by Hyafil and Rivest in [10].

Proposition 5.9 [10] *The problem of minimization of average depth of decision trees computing partial pseudo-Boolean functions given by decision tables is NP-hard.*

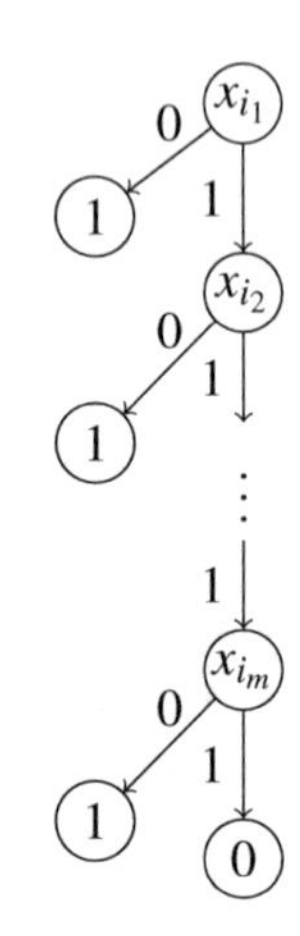

Fig. 5.1 Decision tree computing the function $f_{A,F}$

5.2.2 Experiments with Three Total Boolean Functions

We consider three known total Boolean functions, conjunction, linear, and majority (with slight modification):

$$\mathrm{con}_n(x_1, \dots, x_n) = \bigwedge_{i=1}^{n} x_i \ ,$$

$$\mathrm{lin}_n(x_1, \dots, x_n) = \sum_{i=1}^{n} x_i \mod 2 \ ,$$

$$\mathrm{maj}_n(x_1, \dots, x_n) = \begin{cases} 0, \text{ if } \sum_{i=1}^{n} x_i < n/2 \ , \\ 1, \text{ otherwise} \ . \end{cases}$$

We found that, for each $n = 2, \dots, 16$ and for each of the functions con_n, lin_n, and maj_n, there exists a totally optimal $(rme, 0)$-decision tree relative to h, h_{avg} and L. Figures 5.2, 5.3, and 5.4 show totally optimal $(rme, 0)$-decision trees relative to h, h_{avg}, and L for the functions con_4, lin_4, and maj_4, respectively.

Table 5.1 contains minimum values of average depth and number of nodes for exact decision trees ($(rme, 0)$-decision trees) computing Boolean functions con_n, lin_n, and maj_n, $n = 2, \dots, 16$. The minimum depth of $(rme, 0)$-decision trees computing functions con_n, lin_n, and maj_n is equal to n for $n = 2, \dots, 16$.

We now consider some results for $n = 10$. The number of different $(rme, 0)$-decision trees for the function con_{10} is equal to 3,628,800. Each such tree is a totally optimal $(rme, 0)$-decision tree for con_{10} relative to h, h_{avg}, and L. The same situation is with the functions lin_{10} and maj_{10}. The number of different $(rme, 0)$-decision trees for the functions lin_{10} and maj_{10} is 5.84×10^{353} and 2.90×10^{251}, respectively.

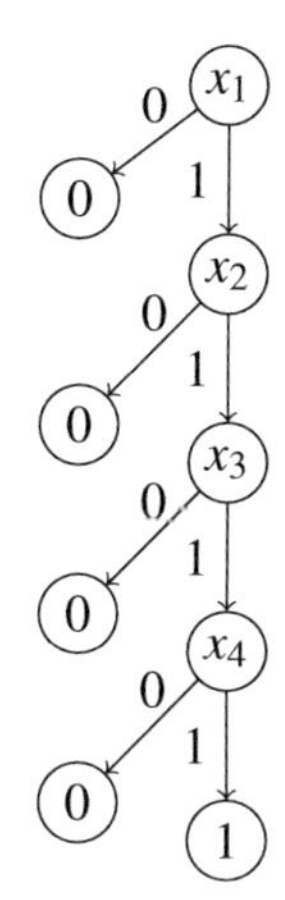

Fig. 5.2 Totally optimal decision tree for con_4

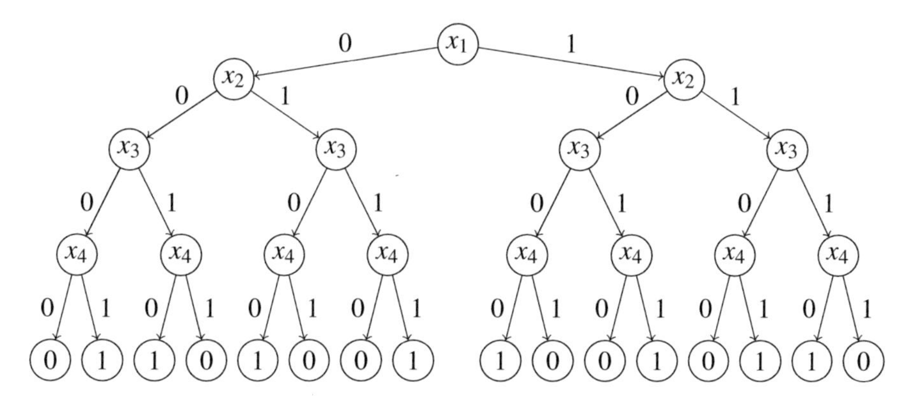

Fig. 5.3 Totally optimal decision tree for lin_4

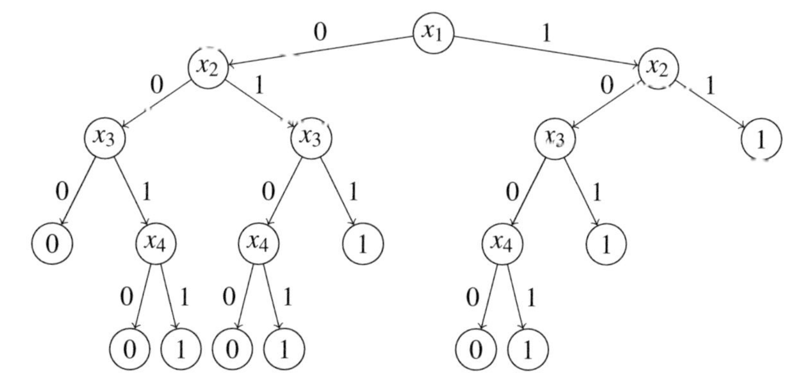

Fig. 5.4 Totally optimal decision tree for maj_4

Table 5.1 Minimum average depth and minimum number of nodes for $(rme, 0)$-decision trees computing the functions con_n, lin_n, and maj_n, $n = 2, \ldots, 16$

n	con_n		lin_n		maj_n	
	h_{avg}	L	h_{avg}	L	h_{avg}	L
2	1.5000	5	2	7	1.5000	5
3	1.7500	7	3	15	2.5000	11
4	1.8750	9	4	31	3.1250	19
5	1.9375	11	5	63	4.1250	39
6	1.9687	13	6	127	4.8125	69
7	1.9844	15	7	255	5.8125	139
8	1.9922	17	8	511	6.5391	251
9	1.9961	19	9	1,023	7.5391	503
10	1.9980	21	10	2,047	8.2930	923
11	1.9990	23	11	4,095	9.2930	1,847
12	1.9995	25	12	8,191	10.0674	3,431
13	1.9998	27	13	16,383	11.0674	6,863
14	1.9999	29	14	32,767	11.8579	12,869
15	1.9999	31	15	65,535	12.8579	25,739
16	1.9999	33	16	131,071	13.6615	48,619

Table 5.2 Minimum values of h, h_{avg}, and L for (rme, α)-decision trees ($\alpha \in \{0, 0.1, \ldots, 0.6\}$) computing functions con_{10}, lin_{10}, and maj_{10}

α	con_{10}			lin_{10}			maj_{10}		
	h	h_{avg}	L	h	h_{avg}	L	h	h_{avg}	L
0	10	1.9980	21	10	10	2,047	10	8.2930	923
0.1	0	0	1	10	10	2,047	10	7.1602	717
0.2	0	0	1	10	10	2,047	10	4.9766	277
0.3	0	0	1	10	10	2,047	8	2.8125	61
0.4	0	0	1	10	10	2,047	0	0	1
0.5	0	0	1	0	0	1	0	0	1
0.6	0	0	1	0	0	1	0	0	1

Table 5.2 contains minimum values of depth, average depth, and number of nodes for (rme, α)-decision trees ($\alpha \in \{0, 0.1, \ldots, 0.6\}$) computing functions con_{10}, lin_{10}, and maj_{10}. In particular, there is only one $(rme, 0.3)$-decision tree for the function con_{10} and this tree is totally optimal relative to h, h_{avg}, and L. The number of totally optimal $(rme, 0.3)$-decision trees for the function lin_{10} remain the same as 5.84×10^{353}. However, the number of totally optimal $(rme, 0.3)$-decision trees for the function maj_{10} reduces to 9.83×10^{20}.

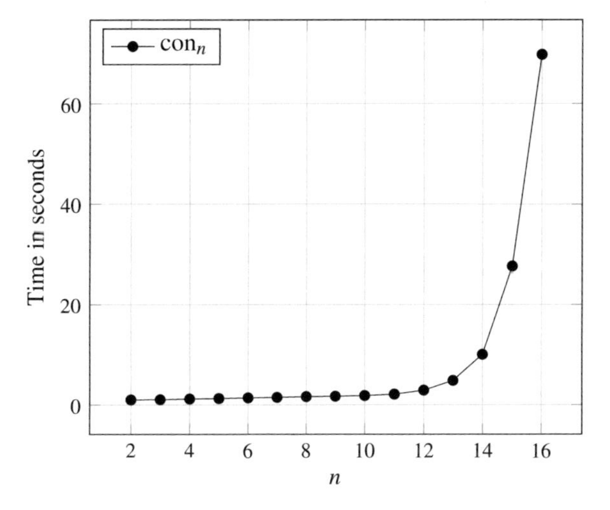

Fig. 5.5 Computation time for the function con_n for $n = 2, 3, \ldots, 16$

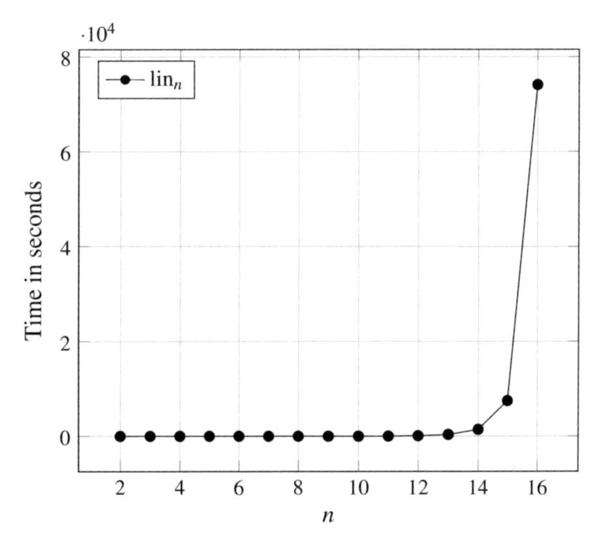

Fig. 5.6 Computation time for the function lin_n for $n = 2, 3, \ldots, 16$

All experimental results presented in this section were performed on a fairly powerful machine with four Intel Xeon E7-4870 (2.3GHz) processors and 1.5 Terabyte of onboard system memory. Figures 5.5, 5.6, and 5.7 show the time required to compute minimum values for cost functions h, h_{avg}, and L for functions con_n, lin_n, and maj_n, respectively, where $n = 2, \ldots, 16$. We can see that the function con_n is the easiest of the three functions to optimize decision trees while the function lin_n takes huge amount of time. The total time for all these experiments was almost 26 h.

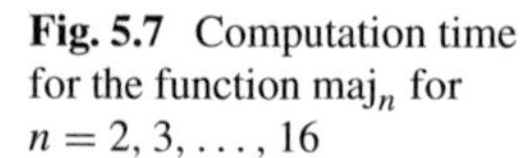
Fig. 5.7 Computation time for the function maj_n for $n = 2, 3, \ldots, 16$

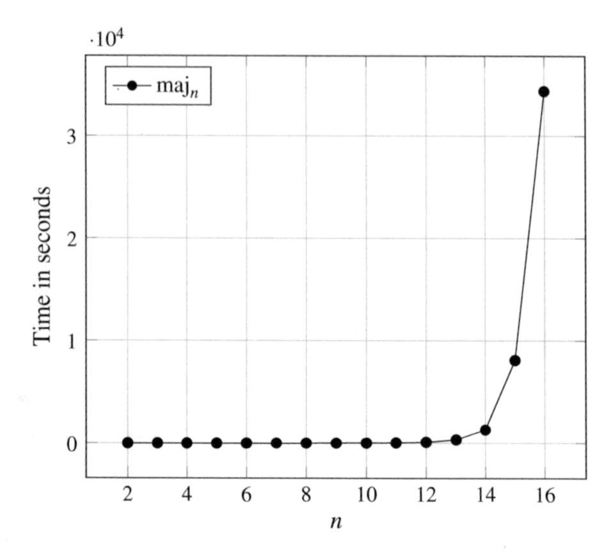

5.2.3 Experiments with Total Boolean Functions

We studied the existence of totally optimal decision trees relative to different combinations of the depth, average depth, and number of nodes for monotone and arbitrary Boolean functions with $n = 0, 1, 2, \ldots$ variables.

The obtained results can be found in Tables 5.3 and 5.4. In particular, for each monotone Boolean function with at most 6 variables, there exists a totally optimal decision tree relative to the depth and number of nodes (relative to h and L) computing this function. However, for each n, $n \geq 7$, there exists a monotone Boolean function with n variables which does not have a totally optimal decision tree relative to h and L. We also give an example of such function, f_1 in this case.

List of functions, f_1, f_2, g_1, g_2, and g_3 mentioned in Tables 5.3 and 5.4:

$$
\begin{aligned}
f_1 &= x_1x_2x_5x_7 \vee x_1x_2x_6x_7 \vee x_1x_3x_6x_7 \vee x_1x_4x_6x_7 \vee x_2x_3x_6x_7 \\
&\quad \vee x_2x_5x_6x_7 \vee x_1x_4x_5 \vee x_2x_4x_5 \vee x_3x_4x_5 \\
f_2 &= x_1x_2x_4 \vee x_1x_4x_5 \vee x_5x_6 \vee x_3x_4 \vee x_3x_6 \\
g_1 &= x_1\bar{x}_2\bar{x}_3\bar{x}_4 \vee \bar{x}_1\bar{x}_2x_3 \vee \bar{x}_1x_3x_5 \vee \bar{x}_1x_4 \vee x_2x_4 \vee x_3x_4x_5 \\
g_2 &= \bar{x}_1x_2\bar{x}_4 \vee \bar{x}_1x_3x_4 \vee \bar{x}_2\bar{x}_3 \\
g_3 &= x_1\bar{x}_2\bar{x}_3x_5 \vee x_1x_3\bar{x}_4\bar{x}_5 \vee x_1x_4x_5 \vee \bar{x}_1x_2x_3x_5 \vee \bar{x}_1\bar{x}_2x_3\bar{x}_5 \\
&\quad \vee \bar{x}_1\bar{x}_2\bar{x}_3x_4 \vee \bar{x}_1x_4\bar{x}_5 \vee x_2\bar{x}_3x_4\bar{x}_5
\end{aligned}
$$

Table 5.5 shows results for experiments on total Boolean functions. For $n = 4, \ldots, 10$, we randomly generated 1,000 total Boolean functions. For different combinations of cost functions, we counted the number of Boolean functions for which

Table 5.3 The existence (example f_i) or nonexistence (–) of a monotone Boolean function with n variables which does not have totally optimal decision trees relative to different combinations of cost functions h, h_{avg}, and L

n	h, L	h, h_{avg}	h_{avg}, L	h, h_{avg}, L
0	–	–	–	–
1	–	–	–	–
2		–	–	–
3	–	–	–	–
4	–	–	–	–
5	–	–	–	–
6	–	f_2	f_2	f_2
7	f_1	f_2	f_2	f_2
>7	f_1	f_2	f_2	f_2

Table 5.4 The existence (example g_i) or nonexistence (–) of a Boolean function with n variables which does not have totally optimal decision trees relative to different combinations of cost functions h, h_{avg}, and L

n	h, L	h, h_{avg}	h_{avg}, L	h, h_{avg}, L
0	–	–	–	–
1	–	–	–	–
2	–	–	–	–
3	–	–	–	–
4	–	g_2	–	g_2
5	g_1	g_2	g_3	g_2
6	g_1	g_2	g_3	g_2
7	g_1	g_2	g_3	g_2
>7	g_1	g_2	g_3	g_2

Table 5.5 Number of total Boolean functions with totally optimal trees

n	h, h_{avg}	h, L	h_{avg}, L	L, h_{avg}, h
4	992	1,000	1,000	992
5	985	997	994	982
6	997	999	927	926
7	1,000	1,000	753	753
8	1,000	1,000	426	426
9	1,000	1,000	100	100
10	1,000	1,000	14	14

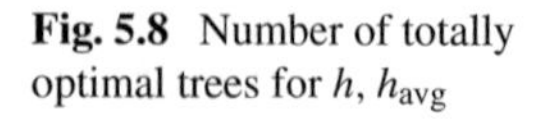
Fig. 5.8 Number of totally optimal trees for h, h_{avg}

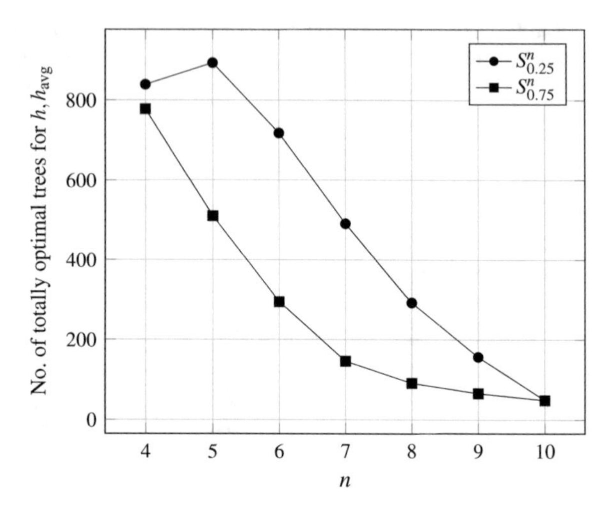

there exist totally optimal decision trees. We can see that the number of total Boolean functions with totally optimal trees, relative to some cost functions, decreases whenever the cost functions contain two strictly increasing functions. Based on these results, we can formulate the following two hypotheses. That is, for almost all total Boolean functions there exist (i) totally optimal decision trees relative to the depth and number of nodes, and (ii) totally optimal decision trees relative to the depth and average depth (see following Sect. 5.2.5 for proofs for these hypotheses).

5.2.4 *Experiments with Partial Boolean Functions*

For each $n = 4, \ldots, 10$, we randomly generated two sets $S^n_{0.25}, S^n_{0.75}$ of partial Boolean functions. Both sets $S^n_t, t \in \{0.25, 0.75\}, n = 4, \ldots, 10$, contain 1,000 table representations of partial Boolean functions with $t \times 2^n$ rows. The obtained results (see Figs. 5.8, 5.9, and 5.10) show that the number of partial Boolean functions with totally optimal trees is decreasing quickly, with the growth of the number of variables.

5.2.5 *Proofs of Hypotheses*

In our experiments with randomly generated total Boolean functions, we noted that the minimum depth of decision trees was always equal to the number of variables in the considered Boolean functions. It helped us to prove the two formulated hypotheses.

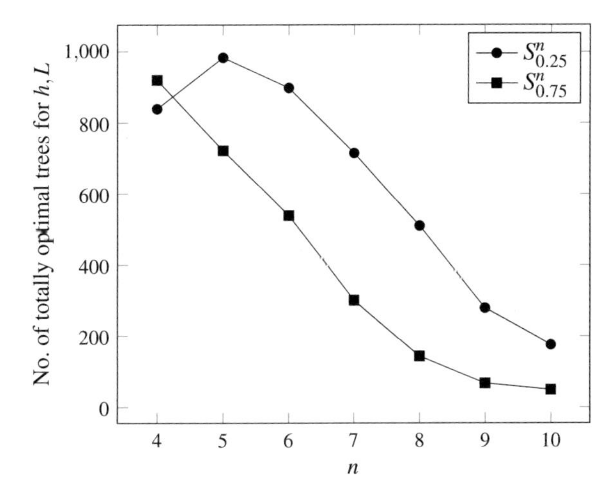

Fig. 5.9 Number of totally optimal trees for h, L

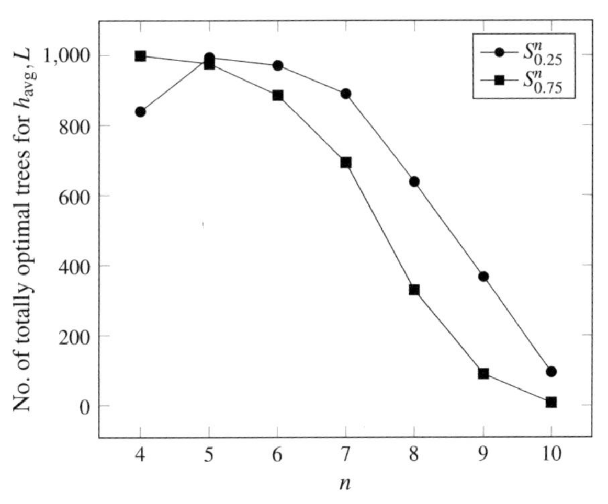

Fig. 5.10 Number of totally optimal trees for h_{avg}, L

Theorem 5.2 *As $n \to \infty$, almost all Boolean functions with n variables have totally optimal decision trees relative to depth and number of nodes.*

Proof A Boolean function with n variables is called *exhaustive* if the minimum depth of the decision tree computing this function is equal to n. Rivest and Vuillemin (see Corollary 3.4 in [18]) proved that as $n \to \infty$, almost all Boolean functions with n variables are exhaustive.

Let f be an exhaustive Boolean function with n variables and Γ_f be a decision tree which computes f and has the minimum number of nodes. From the definition of decision tree computing f it follows that the depth of Γ_f is at most n. Since f is an exhaustive Boolean function, the depth of Γ_f is exactly n. Therefore, Γ_f is a totally optimal tree for f relative to depth and number of nodes. Taking into account that f is an arbitrary exhaustive function we have as $n \to \infty$, almost all Boolean functions with n variables have totally optimal decision trees relative to depth and number of nodes.

We can prove the following theorem in a similar way.

Theorem 5.3 *As $n \to \infty$, almost all Boolean functions with n variables have totally optimal decision trees relative to depth and average depth.*

We now clarify the words about almost all Boolean functions with n variables. Rivest and Vuillemin proved (see proof of Corollary 3.4 in [18]) the following: the probability $P(n)$ that a Boolean function with n variables is not exhaustive is at most $\frac{1}{2^{2k}} \cdot \binom{2k}{k}$ where $k = 2^{n-1}$. It is known (see [14]) that $\binom{2k}{k} < \frac{2^{2k}}{\sqrt{2k}}$ for $k \geq 2$. Therefore $P(n) < \frac{1}{\sqrt{2^n}}$ for $n \geq 2$.

Based on this bound and the proof of Theorem 5.2, we obtain the following statement: the probability that a Boolean function with $n \geq 2$ variables has no totally optimal decision trees relative to depth and number of nodes is less than $\frac{1}{\sqrt{2^n}}$. We can prove the following statement in similar way: the probability that a Boolean function with $n \geq 2$ variables has no totally optimal decision trees relative to depth and average depth is less than $\frac{1}{\sqrt{2^n}}$.

We shown that there exist Boolean functions which have no totally optimal decision trees relative to different combinations of cost functions. From results mentioned in Table 5.4 it follows, in particular, that

- For $n \leq 4$, each Boolean function with n variables has a totally optimal decision tree relative to depth and number of nodes. Boolean function g_1 with five variables has no totally optimal decision trees relative to depth and number of nodes.
- For $n \leq 3$, each Boolean function with n variables has a totally optimal decision tree relative to depth and average depth. Boolean function g_2 with four variables has no totally optimal decision trees relative to depth and average depth.

5.3 Diagnosis of Constant Faults in Iteration-Free Circuits over Monotone Basis

Diagnosis of faults is a well-established area of research in computer science with a long enough history: the early results were published in the fifties of the last century [3, 6, 15, 19]. This section is related to the study of decision trees for diagnosis of constant faults on inputs of gates in iteration-free (tree-like) combinatorial circuits over arbitrary finite basis B of Boolean functions.

A Boolean function is called *linear* if it can be represented in the form $\delta_1 x_1 + \cdots + \delta_n x_n + \delta_{n+1} \pmod 2$ where $\delta_1, \ldots, \delta_{n+1} \in \{0, 1\}$. A Boolean function is called *unate* if it can be represented in the form $f(x_1^{\sigma_1}, \ldots, x_n^{\sigma_n})$ where $f(x_1, \ldots, x_n)$ is a monotone Boolean function, $\sigma_1, \ldots, \sigma_n \in \{0, 1\}$, and $x^1 = x$ and $x^0 = \neg x$.

From the results obtained in [16, 17], which generalize essentially the results presented in [7, 11], it follows that if B contains only linear or only unate Boolean

Table 5.6 Number $M(n)$ of monotone Boolean functions with n, $1 \leq n \leq 5$, variables

n	1	2	3	4	5
$M(n)$	3	6	20	168	7,581

functions then the minimum depth of decision trees in the worst case grows at most linearly depending on the number of gates in the circuits, and it grows exponentially if B contains a non-linear function and a function which is not unate.

In this section, we consider only monotone Boolean functions. Similar results can be obtained for unate Boolean functions.

The results obtained in [9, 16, 17] imply that, for each iteration-free combinatorial circuit S over a basis B containing only monotone Boolean functions with at most n variables, there exists a decision tree for diagnosis of constant faults on inputs of gates with depth at most $\varphi(n)L(S)$, where $\varphi(n) = \binom{n}{\lfloor n/2 \rfloor} + \binom{n}{\lfloor n/2 \rfloor + 1}$ and $L(S)$ is the number of gates in S. We decrease the coefficient from $\varphi(n)$ to $n + 1$ for $n = 1, 2, 3, 4$ and to $n + 2$ for $n = 5$ (note that $\varphi(n) = n + 1$ for $n = 1, 2$). The obtained bounds are sharp.

The results considered here were published in [1].

5.3.1 Basic Notions

A *monotone Boolean function* is a Boolean function $f(x_1, \ldots, x_n)$ satisfying the following condition: for any $(a_1, \ldots, a_n), (b_1, \ldots, b_n) \in \{0, 1\}^n$ if $a_1 \leq b_1, \ldots, a_n \leq b_n$ then $f(a_1, \ldots, a_n) \leq f(b_1, \ldots, b_n)$. We will consider only monotone Boolean functions with at least one variable. The number $M(n)$ of monotone Boolean functions with n, $1 \leq n \leq 5$, variables can be found in Table 5.6 (see [5]).

A finite set of monotone Boolean functions B will be called a *monotone basis*. An *iteration-free circuit over the basis B* is a tree-like combinatorial circuit in which inputs are labeled with pairwise different variables and gates are labeled with functions from B. Figures 5.11 and 5.12 show examples of iteration-free circuits.

Let S be a circuit over B with m circuit inputs and k inputs of gates. We consider 0 and 1 *constant faults* on the inputs of the gates of S, i.e., we assume that some of the inputs of gates are faulty and have fixed constant values (0 or 1), regardless of the values supplied to the inputs. There are 3^k possible k-tuples of faults for a circuit with k inputs of gates (since for each input we have a normal state and two different faults). The *problem of diagnosis* of S is to recognize the function implemented by the circuit S which has one of the 3^k possible k-tuples of faults.

We solve this problem using decision trees. At each step, a decision tree gives a tuple from $\{0, 1\}^m$ on the inputs of S and observes the output of S. We denote by $h(S)$ the minimum depth of a decision tree for diagnosis of S.

Figure 5.13 shows a decision tree of depth 3 for diagnosis of the one-gate circuit $S_\vee$ depicted in Fig. 5.12. First, the decision tree asks about the output of faulty circuit $S_\vee$ for the input tuple (0, 1). Let the output be equal to 0. Then the decision tree asks

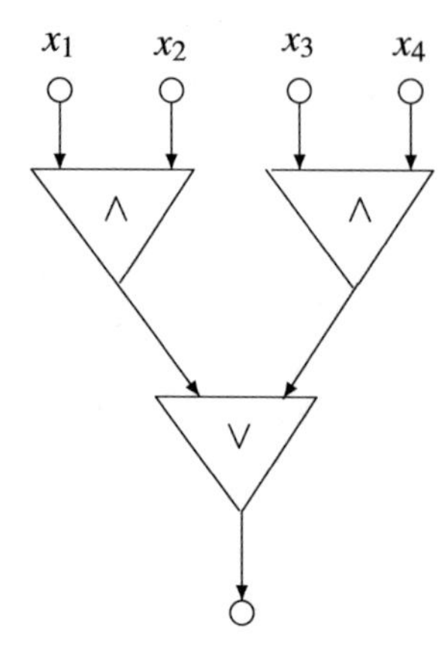

Fig. 5.11 Iteration-free circuit with 4 inputs, 3 gates and 6 inputs of gates over the basis $\{x \wedge y, x \vee y\}$

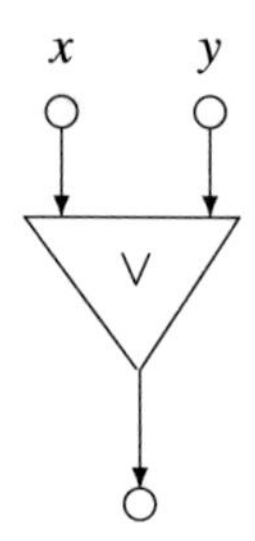

Fig. 5.12 Iteration-free circuit $S_\vee$ with one gate

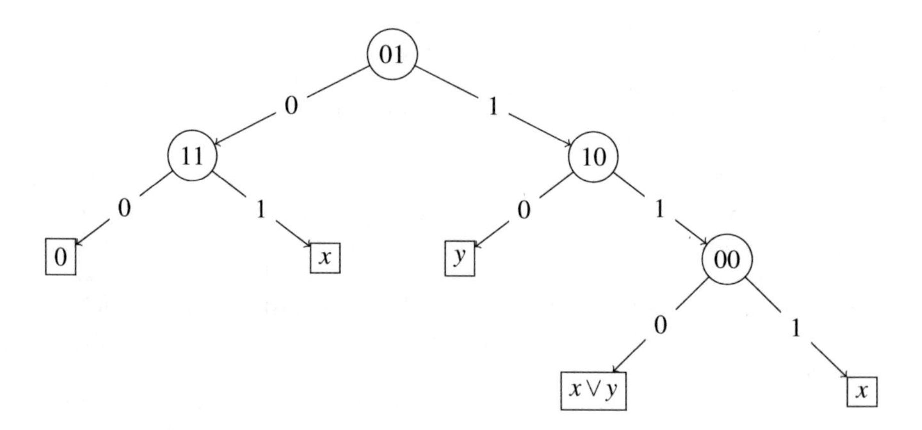

Fig. 5.13 Decision tree for the circuit $S_\vee$ diagnosis

Fig. 5.14 Decision table $T_\vee$ for the function $x \vee y$

00	01	10	11	
0	1	1	1	$x \vee y$
0	0	1	1	x
0	1	0	1	y
1	1	1	1	1
0	0	0	0	0

about the output of the circuit $S_\vee$ for the input tuple $(1, 1)$. If the output is equal 0 then the faulty circuit $S_\vee$ implements the function 0. Note that in Figs. 5.13 and 5.14 we write ab instead of (a, b) for $a, b \in \{0, 1\}$.

5.3.2 *Main Results*

For a given Boolean function f, we denote by $h(f)$ the minimum depth of a decision tree for diagnosis of iteration-free circuit S_f with one gate labeled with the function f. For example, $h(x \vee y) = 3$ since we cannot recognize five functions using decision tree of the depth two, which contains only four terminal nodes (decision tree with depth three for the diagnosis of $S_\vee$ is depicted in Fig. 5.13).

For a monotone basis B, we denote $h_B = \max\{h(f) : f \in B\}$. In [16, 17] it is shown that

$$h(S) \leq h_B L(S)$$

for an arbitrary iteration-free circuit S over B where $L(S)$ is the number of gates in S.

We illustrate the work of a decision tree which gives us this bound on the example of the circuit S depicted in Fig. 5.11. We begin with the diagnosis of the $\vee$ gate. According to the description of the decision tree depicted in Fig. 5.13, we should first give $(0, 1)$ on the inputs of the $\vee$ gate. To this end, we give $(0, 0, 1, 1)$ on the inputs of the circuit S. If, for example, the value on the output of the left $\wedge$ gate is not 0 but 1, the decision tree will interpret it as the constant fault 1 on the left input of the $\vee$ gate.

Let us assume that the decision tree has recognized that the $\vee$ gate implements the function $x \vee y$. Now we need to recognize the behavior of the two $\wedge$ gates. We begin from the left gate and "block" the right one: we take $(0, 0)$ on the inputs x_3 and x_4 of the circuit S. In this case, the output of S will repeat the output of the left $\wedge$ gate.

One can show that, for the circuit $S_\wedge$ with one gate labeled with $\wedge$, there exists a decision tree with depth three for the diagnosis of constant faults on the gate inputs. Therefore the maximum number of steps of the described diagnosis procedure will be at most $3 \times 3 = 9$.

Table 5.7 Values of $H(n)$ and $\varphi(n)$ for $n = 1, \ldots, 5$

n	1	2	3	4	5
$H(n)$	2	3	4	5	7
$\varphi(n)$	2	3	6	10	20

Note that we do not need to memorize the decision tree corresponding to the described procedure. We only need the description of the circuit over basis B under consideration and the descriptions of decision trees for diagnosis of one-gate circuits implementing functions from B.

To our knowledge, the best known upper bound on h_B is

$$\varphi(n) = \binom{n}{\lfloor n/2 \rfloor} + \binom{n}{\lfloor n/2 \rfloor + 1},$$

where n is the maximum number of variables in functions from monotone basis B. This bound follows from the results of Hansel [9] for recognition of monotone Boolean functions using only membership queries. We decrease this upper bound for $n = 3, 4, 5$.

We have found, for each monotone Boolean function f with n, $1 \le n \le 5$, variables, the value $h(f)$. We construct all subfunctions g of the function f obtained by substituting some of the variables with constants 0 and 1. After that, we construct a decision table T_f for the function f. This table contains 2^n columns labeled with n-tuples from $\{0, 1\}^n$, considered as attributes. For each subfunction g of the function f, this table contains a row which is the 2^n-tuple of values of g on n-tuples from $\{0, 1\}^n$, and this row is labeled with the function g. Decision table $T_\vee$ for the function $x \vee y$ is depicted in Fig. 5.14.

We apply to the table T_f the algorithm $\mathscr{A}_3$ and construct the graph $\Delta(T_f) = \Delta_{U,0}(T_f)$ for some uncertainty measure U. We apply to $\Delta(T_f)$ and depth h the algorithm $\mathscr{A}_5$ which returns the graph $\Delta(T_f)^h$ and the number $h(f)$ attached to the node T of this graph.

We denote by $H(n)$ the maximum value of $h(f)$ among all monotone Boolean functions f with n variables. For $n = 1, \ldots, 5$, the value $H(n)$ can be found in Table 5.7. We added for comparison the values of $\varphi(n)$ for $n = 1, \ldots, 5$. Note that $H(n) \le \varphi(n)$ for any natural n.

From here it follows that if B contains only monotone Boolean functions with at most n, $1 \le n \le 5$, variables then

$$h_B \le \begin{cases} n+1, \ 1 \le n \le 4\,, \\ n+2, \ n = 5\,, \end{cases}$$

and for any iteration-free circuit S over B,

$$h(S) \le \begin{cases} (n+1)L(S), \ 1 \le n \le 4\,, \\ (n+2)L(S), \ n = 5\,. \end{cases}$$

It is clear that the obtained bounds are sharp.

References

1. Alrawaf, S., Chikalov, I., Hussain, S., Moshkov, M.: Diagnosis of constant faults in iteration-free circuits over monotone basis. Discret. Appl. Math. **166**, 287–291 (2014)
2. Buhrman, H., de Wolf, R.: Complexity measures and decision tree complexity: A survey. Theor. Comput. Sci. **288**(1), 21–43 (2002)
3. Chegis, I.A., Yablonskii, S.V.: Logical methods of control of work of electric schemes. Trudy Mat. Inst. Steklov (in Russian) **51**, 270–360 (1958)
4. Chikalov, I., Hussain, S., Moshkov, M.: Totally optimal decision trees for Boolean functions. Discret. Appl. Math. **215**, 1–13 (2016)
5. Church, R.: Numerical analysis of certain free distributive structures. Duke Math. J. **6**, 732–734 (1940)
6. Eldred, R.D.: Test routines based on symbolic logical statements. J. ACM **6**(1), 33–37 (1959)
7. Goldman, R.S., Chipulis, V.P.: Diagnosis of iteration-free combinatorial circuits. Diskret. Analiz (in Russian) **14**, 3–15 (1969)
8. Guijarro, D., Lavín, V., Raghavan, V.: Exact learning when irrelevant variables abound. Inf. Process. Lett. **70**(5), 233–239 (1999)
9. Hansel, G.: Sur le nombre des fonctions booléennes monotones de n variables. C. R. Acad. Sci. Paris Sér. A-B (in French) **262**, A1088–A1090 (1966)
10. Hyafil, L., Rivest, R.L.: Constructing optimal binary decision trees is NP-complete. Inf. Process. Lett. **5**(1), 15–17 (1976)
11. Karavai, M.F.: Diagnosis of tree-form circuits having an arbitrary basis. Autom. Remote Control **34**(1), 154–161 (1973)
12. Karp, R.M.: Reducibility among combinatorial problems. In: Miller, R.E., Thatcher, J.W. (eds.) Symposium on the Complexity of Computer Computations. The IBM Research Symposia Series, pp. 85–103. Plenum Press, New York (1972)
13. Korshunov, A.D.: Computational complexity of Boolean functions. Russ. Math. Surv. **67**(1), 93–165 (2012)
14. Koshy, T.: Catalan Numbers with Applications. Oxford University Press, New York (2008)
15. Moore, E.F.: Gedanken-experiments on sequential machines. Automata Studies, Ann. Math. Stud. **34**, 129–153 (1956)
16. Moshkov, M.: Diagnosis of constant faults in circuits. Math. Probl. Cybern. (in Russian) **9**, 79–100 (2000)
17. Moshkov, M.: Time complexity of decision trees. In: Peters, J.F., Skowron, A. (eds.) Trans. Rough Sets III. Lecture Notes in Computer Science, vol. 3400, pp. 244–459. Springer, Berlin (2005)
18. Rivest, R.L., Vuillemin, J.: On recognizing graph properties from adjacency matrices. Theor. Comput. Sci. **3**(3), 371–384 (1976)
19. Yablonskii, S.V., Chegis, I.A.: On tests for electric circuits. Uspekhi Mat. Nauk (in Russian) **10**, 182–184 (1955)

Chapter 6
More Applications of Multi-stage Optimization of Decision Trees

There are many ways of studying the complexity of a given problem. One way is to consider an algorithm that solves this problem correctly and analyze its time complexity. This provides an upper bound on the complexity of the studied problem.

Proving lower bounds is a different method for tackling complexity analysis. The aim of such technique is to show that all algorithms that solve the problem correctly have a minimum complexity threshold. In order to be able to have such statements, we need to specify a concrete model of computation that represents all algorithms for a given problem. There are many available models of computation such as Turing machines, circuits, decision tests, decision trees, etc.

This chapter starts with a description of some results for the eight element sorting problem under the decision tree model of computation in Sect. 6.1. Those results are obtained using the techniques of multi-stage optimization of decision trees discussed previously. In addition, hypotheses about the complexity of a modified majority problem are formulated with the help of computer experiments and proved analytically in Sect. 6.2. We study two classes of algorithms: adaptive algorithms modeled as decision trees and non-adaptive algorithms modeled as decision tests.

One common step of our work with the sorting and majority problems is to represent the problem in the form of a decision table. Attributes of the table denote possible steps of algorithms solving the problem. For example, such steps may be comparing two elements when studying comparison-based algorithms for sorting. Rows represent possible inputs for the problem and they are labeled with the corresponding output.

Finally, Sect. 6.3 concludes the chapter with an algorithm to optimize decision tests for a given decision table. The last step of this algorithm may reduce the considered problem to the problem of decision tree optimization.

H. AbouEisha et al., *Extensions of Dynamic Programming for Combinatorial Optimization and Data Mining*, Intelligent Systems Reference Library 146,
https://doi.org/10.1007/978-3-319-91839-6_6

6.1 Sorting of Eight Elements

We prove that the minimum average depth of a decision tree for sorting 8 pairwise different elements is equal to 620, 160/8!. We show also that each decision tree for sorting 8 elements, which has minimum average depth (the number of such trees is approximately equal to $8.548 \times 10^{326,365}$), has also minimum depth. Both problems were considered by Knuth in [9]. This section is based on the work [3].

6.1.1 Introduction

The problem of sorting n pairwise different elements from a linearly ordered set is a fundamental problem in the theory of algorithms [9]. As a model of computation, we use binary decision trees [6, 12] where each step is a comparison of two elements. The minimum number of nodes in any such tree is equal to $2(n!) - 1$. For $n = 2, \ldots, 11$, the minimum depth of a decision tree for sorting n elements is equal to the well known lower bound $\lceil \log_2(n!) \rceil$ [17]. For $n = 2, 3, 4, 5, 6, 9, 10$, the minimum average depth of a decision tree for sorting n elements is equal to the known lower bound $\varphi(n)/n!$, where $\varphi(n) = (\lceil \log_2 n! \rceil + 1) \cdot n! - 2^{\lceil \log_2 n! \rceil}$ is the minimum external path length in an extended binary tree with $n!$ terminal nodes [9]. Césary [5] proved that, for $n = 7$ and $n = 8$, there are no decision trees for sorting n elements whose average depth is equal to $\varphi(n)/n!$. Kollár [10] found that the minimum average depth of a decision tree for sorting 7 elements is equal to 62, 416/7!. We find that the minimum average depth of a decision tree for sorting 8 elements is equal to 620, 160/8!.

Another open problem considered by Knuth [9] is the existence of decision trees for sorting n elements which have simultaneously minimum average depth and minimum depth. As it was mentioned by Knuth in [9] , if a decision tree for sorting n elements has average depth equal to $\varphi(n)/n!$ then this tree has depth equal to $\lceil \log_2 n! \rceil$. Therefore, for $n = 2, 3, 4, 5, 6, 9, 10$, each decision tree for sorting n elements, which has minimum average depth, has also minimum depth. We extended this result to the cases $n = 7$ (Kollár in [10] did not consider this question) and $n = 8$. For $n = 2, \ldots, 8$, we compute the number of decision trees for sorting n elements which have minimum average depth. In particular, for $n = 8$, the number of such trees is approximately equal to $8.548 \times 10^{326,365}$. We recalculate known values of the minimum depth for $n = 2, \ldots, 8$ and minimum average depth for $n = 2, \ldots, 7$ to make sure that the new results are valid.

6.1.2 Main Results

Let $x_1, \ldots, x_n$ be a sequence of pairwise different elements from a linearly ordered set. We should find a permutation $(p_1, \ldots, p_n)$ from the set $P_n = \{\pi_1, \ldots, \pi_{n!}\}$ of all permutations of the set $\{1, \ldots, n\}$ such that $x_{p_1} < \ldots < x_{p_n}$. To solve this problem, we use decision trees with attributes from the set $F_n = \{f_{i,j} : 1 \leq i < j \leq n\}$ where

$$f_{i,j}(x_1, \ldots, x_n) = \begin{cases} 1, \; x_i < x_j \; , \\ 0, \; x_i > x_j \; . \end{cases}$$

To study such decision trees, we consider the decision table T_n with $\frac{n(n-1)}{2}$ columns labeled with attributes from the set F_n and with $n!$ rows labeled with permutations $\pi_1, \ldots, \pi_{n!}$. The table T_n is filled by numbers from the set $\{0, 1\}$. The intersection of the column $f_{i,j}$ and the row labeled with π_k has value 1 if and only if i precedes j in π_k.

It is convenient for us to consider some definitions not for T_n but for an arbitrary separable subtable Θ of the table T_n (note that T_n is considered as a separable subtable of T_n).

We study only exact decision trees for the table Θ, i.e., $(U, 0)$ -decision trees for Θ where U is an uncertainty measure. This definition does not depend on the choice of U. We denote by $DT_{U,0}(\Theta)$ the set of $(U, 0)$-decision trees for Θ.

As cost functions for decision trees we use depth h, and average depth h_{avg}. The average depth can be computed based on the total path length *tpl*: for any $\Gamma \in DT_{U,0}(\Theta)$, $h_{\text{avg}}(\Theta, \Gamma) = tpl(\Theta, \Gamma)/N(\Theta)$. We denote by $h(\Theta)$ the minimum depth, by $tpl(\Theta)$ the minimum total path length, and by $h_{\text{avg}}(\Theta)$ the minimum average depth of decision trees from $DT_{U,0}(\Theta)$.

For $n = 2, \ldots, 8$, the values $h(T_n)$ and $h_{\text{avg}}(T_n)$ can be found in Tables 6.1 and 6.2. For the considered values of n, the parameter $h(T_n)$ is equal to its lower bound $\lceil \log_2(n!) \rceil$, and the parameter $h_{\text{avg}}(T_n)$ is equal to its lower bound $\varphi(n)/n!$ for $n = 2, \ldots, 6$.

Let $\Gamma \in DT_{U,0}(\Theta)$ and $\psi \in \{h, h_{\text{avg}}\}$. We will say that Γ is *optimal relative to* ψ if $\psi(\Theta, \Gamma) = \psi(\Theta)$. We denote by $Opt_\psi(\Theta)$ the set of decision trees from $DT_{U,0}(\Theta)$ which are optimal relative to ψ.

Based on results of computer experiments we obtain that $Opt_{h_{avg}}(T_n) \subseteq Opt_h(T_n)$ for $n = 2, \ldots, 8$. For $n = 2, \ldots, 8$, we count also the cardinality of the set $Opt_{h_{\text{avg}}}(T_n)$.

Table 6.1 Results for sorting $n = 2, 3, 4, 5$ elements

	n			
	2	3	4	5
$h(T_n)$	1	3	5	7
$h_{\text{avg}}(T_n)$	2/2!	16/3!	112/4!	832/5!
$\varphi(n)/n!$	2/2!	16/3!	112/4!	832/5!
$\lvert Opt_{h_{\text{avg}}}(T_n)\rvert$	1	12	27, 744	2, 418, 647, 040
$\lvert SEP(T_n)\rvert$	3	19	219	4231

Table 6.2 Results for sorting $n = 6, 7, 8$ elements

	n		
	6	7	8
$h(T_n)$	10	13	16
$h_{\text{avg}}(T_n)$	$6,896/6!$	$62,416/7!$	$620,160/8!$
$\varphi(n)/n!$	$6,896/6!$	$62,368/7!$	$619,904/8!$
$\left\|Opt_{h_{\text{avg}}}(T_n)\right\|$	1.968×10^{263}	$4.341 \times 10^{6,681}$	$8.548 \times 10^{326,365}$
$\|SEP(T_n)\|$	$130,023$	$6,129,859$	$431,723,379$

We denote by $SEP(T_n)$ the set of all separable subtables of T_n. The cardinality of the set $SEP(T_n)$ for $n = 1, \ldots, 8$ can be found in Tables 6.1 and 6.2.

6.1.3 Computer Experiments

First, we construct the decision table T_n for $n = 2, \ldots, 8$. After that, we construct the graph $\Delta_n = \Delta_{U,0}(T_n)$ using the algorithm $\mathscr{A}_3$ and find the number of its nodes which is equal to $|SEP(T_n)|$. Then we apply to the graph Δ_n the procedure of decision tree optimization (the algorithm $\mathscr{A}_5$) relative to *tpl* and h separately. We obtain proper subgraphs Δ_n^{tpl} and Δ_n^h of the graph Δ_n in which nodes T_n are labeled with numbers $tpl(T_n)$ and $h(T_n)$, respectively (see Proposition 4.4 and Lemma 5.1). As a result, we obtain the value $h_{\text{avg}}(T_n) = tpl(T_n)/n!$. Using the algorithm $\mathscr{A}_4$ we count the cardinality of the set of decision trees $Tree(\Delta_n^{tpl}, T_n)$ corresponding to the node T_n in the graph Δ_n^{tpl}. By Corollary 5.1, this number is equal to $\left|Opt_{h_{\text{avg}}}(T_n)\right|$.

After that, we apply to the graph Δ_n^{tpl} the procedure of decision tree optimization (the algorithm $\mathscr{A}_5$) relative to h. As a result, we obtain proper subgraph $\left(\Delta_n^{tpl}\right)^h$ of the graph Δ_n^{tpl}. We find that the graphs $\left(\Delta_n^{tpl}\right)^h$ and Δ_n^{tpl} are equal if we do not consider numbers attached to the nodes of these graphs, and the number attached to the node T_n of the graph $\left(\Delta_n^{tpl}\right)^h$ is equal to $h(T_n)$. Using Theorem 5.1, we obtain that all decision trees for sorting n elements with minimum average depth have the same depth equal to $h(T_n)$. Therefore $Opt_{h_{\text{avg}}}(T_n) \subseteq Opt_h(T_n)$.

For $n = 2, \ldots, 7$, all computations were done on a Mac Pro desktop with two 2.40 GHz 6-core Intel Xeon processors and 64 GB of RAM. For $n = 8$, we used Amazon cr1.8xlarge instance with two 2.60 GHz 8-core Intel Xeon processors and 244 GB of RAM (see http://aws.amazon.com/ec2/instance-types/ for details).

For $n = 2, \ldots, 7$, all computations were done directly as it was described above. For $n = 8$, we used a notion of equivalent separable subtables of T_8 and studied only nonequivalent separable subtables defined by one condition of the kind $f_{i,j} = a$,

$a \in \{0, 1\}$ (the number of such subtables is equal to one) and defined by two such conditions (the number of such subtables is equal to three).

We say that two separable subtables Θ_1 and Θ_2 of the table T_n are equivalent if $N(\Theta_1) = N(\Theta_2)$ and there exists a bijection (a one-to-one correspondence) $\varphi : DT_{U,0}(\Theta_1) \rightarrow DT_{U,0}(\Theta_2)$ which preserves depth and total path length of decision trees.

First, we show that, for any $i, j \in \{1, \ldots, n\}$, $i < j$, and $a \in \{0, 1\}$, the subtable $T_n(f_{i,j}, a)$ is equivalent to the subtable $T_n(f_{1,2}, 1)$. Next, we show that, for any $i, j, k, l \in \{1, \ldots, n\}$, $i < j$, $k < l$, $\{i, j\} \neq \{k, l\}$, and $a, b \in \{0, 1\}$, the subtable $T_n(f_{i,j}, a)(f_{k,l}, b)$ is equivalent to one of the three subtables: $S_1^n = T_n(f_{1,2}, 1)(f_{3,4}, 1)$, $S_2^n = T_n(f_{1,2}, 1)(f_{2,3}, 1)$, and $S_3^n = T_n(f_{1,2}, 1)(f_{1,3}, 1)$.

We studied subtables S_1^8, S_2^8, and S_3^8 in a way similar to the described above for T_n, and combined the obtained results. The numbers of separable subtables for S_1^8, S_2^8, and S_3^8 are equal to 29,668,833, 12,650,470, and 50,444,492, respectively. The time of computation is 1848 s for S_1^8, 326 s for S_2^8, and 3,516 s for S_3^8.

The study of separable subtables of T_8 described by at most two conditions allows us to understand the structure of the first two levels of decision trees for T_8 having minimum average depth. The root (the only node in the first level) can be labeled with an arbitrary attribute $f_{i,j}$ such that $i, j \in \{1, \ldots, n\}$, $i < j$. Any node in the second level can be labeled with an arbitrary attribute $f_{k,l}$ such that $k, l \in \{1, \ldots, n\}$, $k < l$ and $\{k, l\} \cap \{i, j\} = \emptyset$.

6.2 Modified Majority Problem

In this section, we study the complexity of adaptive and non-adaptive algorithms for a modified majority problem.

6.2.1 Introduction

Given an n-tuple of elements colored by one of two colors, we should find an element of the majority color if it exists or report that there is no majority color. To this end we can use questions about whether two elements have the same color or not. This problem is known as the *majority problem.* Saks and Werman [18] showed that the minimum number of questions for adaptive algorithms solving the majority problem is equal to $n - B(n)$ where $B(n)$ is the number of 1's in the binary representation of n. Aigner [4] proved that the minimum number of questions for non-adaptive algorithms solving the majority problem is equal to $n - q(n)$ where $q(n) = 2$ if n is odd and $q(n) = 1$ if n is even. Adaptive algorithms are algorithms that can decide their work based on each answer of a given question. On the other hand, non-adaptive (oblivious) algorithms depend only on the collective answers of a set of predetermined questions asked.

We consider a *modified majority problem*: given an n-tuple of elements colored by one of two colors, we should find an element of the majority color with minimum index if it exists or report that there is no majority color. We studied this problem experimentally and formulated a hypothesis based on the results of experiments. We proved that the minimum number of questions for solving the modified majority problem is equal to $n-1$ both for adaptive and non-adaptive algorithms.

6.2.2 Main Notions and Results

Let $\bar{a} = (a_1, \ldots, a_n) \in \{0, 1\}^n$ be an input of the *modified majority problem* (a_i represents the color of ith element), and $S(\bar{a}) = a_1 + \ldots + a_n$. If $S(\bar{a}) = n/2$ then the solution of the considered problem $mmp(\bar{a})$ is -1. If $S(\bar{a}) < n/2$ then $mmp(\bar{a})$ is the minimum index $i \in \{1, \ldots, n\}$ such that $a_i = 0$. If $S(\bar{a}) > n/2$ then $mmp(\bar{a})$ is the minimum index $i \in \{1, \ldots, n\}$ such that $a_i = 1$.

To solve this problem we use decision trees (adaptive algorithms) and tests (non-adaptive algorithms) with attributes (questions) $g_{i,j}$ where $i, j \in \{1, \ldots, n\}$, $i \neq j$ and

$$g_{i,j}(a_1, \ldots, a_n) = \begin{cases} 0, \ a_i \neq a_j \ , \\ 1, \ a_i = a_j \ . \end{cases}$$

For any $a \in \{0, 1\}$, we define the value $\neg a$ in the following way: $\neg a = 0$ if $a = 1$, and $\neg a = 1$ if $a = 0$. It is clear that $g_{i,j}(a_1, \ldots, a_n) = g_{i,j}(\neg a_1, \ldots, \neg a_n)$ for any $(a_1, \ldots, a_n) \in \{0, 1\}^n$ and $i, j \in \{1, \ldots, n\}$, $i \neq j$. So as the set of inputs for the modified majority problem we will consider not the set $\{0, 1\}^n$ but the set $In(n) = \{1\} \times \{0, 1\}^{n-1}$.

To study decision trees and tests, we consider decision table Q_n with $n^2 - n$ columns corresponding to the attributes $g_{i,j}$, $i, j \in \{1, \ldots, n\}$, $i \neq j$ and 2^{n-1} rows corresponding to n-tuples from $In(n)$. The row corresponding to a tuple $\bar{a} \in In(n)$ is filled by values of attributes $g_{i,j}$ on $\bar{a}$ and is labeled with the decision $mmp(\bar{a})$ – the solution of the modified majority problem for the input $\bar{a}$.

We study $(U, 0)$-decision trees for Q_n where U is an uncertainty measure (this definition does not depend on the choice of U). We denote $DT_{U,0}(Q_n)$ the set of $(U, 0)$-decision trees for Q_n and $h(Q_n)$ the minimum depth of a decision tree from the set $DT_{U,0}(Q_n)$.

A *test for* Q_n is a set B of attributes (columns) on which any two rows of Q_n with different decisions have different values for at least one attribute from B. An equivalent definition is the following: a set B of attributes is a test for Q_n if, for any row of Q_n, the values of attributes from B on the considered row are enough to determine the decision attached to this row. We denote by $R(Q_n)$ the minimum cardinality of a test for Q_n. It is easy to show that $h(Q_n) \leq R(Q_n)$ (see Corollary 2.24 from [13]): we can construct a $(U, 0)$-decision trees for Q_n which sequentially compute values of attributes from a test for Q_n with minimum cardinality. The depth of this decision tree is equal to $R(Q_n)$.

For $n = 2, \ldots, 10$, we constructed the decision table Q_n and the directed acyclic graph $\Delta_{U,0}(Q_n)$ by the algorithm $\mathscr{A}_3$. Using the algorithm $\mathscr{A}_5$, we computed the value $h(Q_n)$ which was equal to $n - 1$. The obtained experimental results allowed us to formulate the following hypothesis: $h(Q_n) = R(Q_n) = n - 1$ for any $n \geq 2$. We prove now that this hypothesis is true.

Theorem 6.1 *For any $n \geq 2$, the following inequality holds:*

$$R(Q_n) \leq n - 1 \ .$$

Proof We prove now that the set $\{g_{1,2}, \ldots, g_{1,n}\}$ is a test for the table Q_n. Let $\bar{a} = (1, a_2, \ldots, a_n) \in In(n)$. Then the tuple $(g_{1,2}(\bar{a}), \ldots, g_{1,n}(\bar{a}))$ is equal to $(a_2, \ldots, a_n)$. Therefore all rows of Q_n are pairwise different on the columns $g_{1,2}, \ldots, g_{1,n}$, and $\{g_{1,2}, \ldots, g_{1,n}\}$ is a test for Q_n. □

Theorem 6.2 *For any $n \geq 2$, the following inequality holds:*

$$h(Q_n) \geq n - 1 \ .$$

Proof Let $\Gamma \in DT_{U,0}(Q_n)$, and $p(n) = 1$ if n is odd and 0 otherwise. It is easy to see that there exists exactly one tuple $\bar{\gamma} = (1, \gamma_2, \ldots, \gamma_n) \in In(n)$ for which $mmp(\bar{\gamma}) = \frac{n+p(n)}{2}$. In the tuple $\bar{\gamma}$, the first $\frac{n+p(n)}{2} - 1$ digits are equal to 1 and all other digits are equal to 0. Therefore Q_n contains exactly one row labeled with the decision $\frac{n+p(n)}{2}$, and Γ contains a terminal node labeled with the decision $\frac{n+p(n)}{2}$.

We study the path π from the root of Γ to the terminal node labeled with $\frac{n+p(n)}{2}$. Since $n \geq 2$, the root of Γ is not a terminal node. Let π contain m nonterminal nodes $v_1, \ldots, v_m$ labeled with attributes $g_{i_1,j_1}, \ldots, g_{i_m,j_m}$, and m edges starting in nodes $v_1, \ldots, v_m$ and labeled with numbers $b_1, \ldots, b_m$, respectively. Since $\Gamma \in DT_{U,0}(Q_n)$, the subtable $Q_n^{'} = Q_n(g_{i_1,j_1}, b_1) \ldots (g_{i_m,j_m}, b_m)$ is not empty and contains only one row corresponding to the tuple $\bar{\gamma}$.

We consider undirected graph G containing nodes $1, \ldots, n$ and edges

$$\{i_1, j_1\}, \ldots, \{i_m, j_m\} \ .$$

Let us assume that the graph G is not connected. Let C be the set of nodes of the connected component of G containing the node 1, and $D = \{1, \ldots, n\} \setminus C$. Without loss of generality, we can assume that $C = \{1, \ldots, k\}$ and $D = \{k + 1, \ldots, n\}$ for some $k \in \{1, \ldots, n - 1\}$. It is clear that the tuple $\bar{\gamma} = (1, \gamma_2, \ldots, \gamma_n)$ is a solution of the system of equations

$$\{g_{i_1,j_1}(\bar{x}) = b_1, \ldots, g_{i_m,j_m}(\bar{x}) = b_m\} \ .$$

Since there are no edges connecting sets of nodes C and D in the graph G, the tuple $(1, \gamma_2, \ldots, \gamma_k, \neg\gamma_{k+1}, \ldots, \neg\gamma_n)$ is also a solution of the considered system of equations. Therefore $Q_n^{'}$ contains at least two rows which is impossible. Thus the graph G is connected and contains at least $n - 1$ edges. As a result, the length of π is at least $n - 1$ and the depth Γ is at least $n - 1$. □

The next statement follows immediately from Theorems 6.1 and 6.2, and from the inequality $h(Q_n) \leq R(Q_n)$.

Corollary 6.1 *For any $n \geq 2$, the following equalities hold*

$$h(Q_n) = R(Q_n) = n - 1 .$$

6.3 Optimization of Reducts

This section discusses the problem of finding optimal reducts (reducts with minimum cardinality) for decision tables. We present an algorithm to find an optimal reduct and show the results of experiments on datasets from the UCI Machine Learning Repository [11]. This section is based on the work [1] which simplifies essentially the algorithm for reduct optimization proposed in [2].

6.3.1 *Introduction*

Reducts are related to knowledge discovery, feature selection, and data mining. They are key objects in the rough set theory [15, 16, 19]. Finding optimal reducts allows us to identify the complexity of non-adaptive algorithms for different problems (see, in particular, Sect. 6.2).

It is well known that the problem of finding an optimal reduct is NP-hard as the set cover problem can be reduced to it (see, for example, [13]). Many approximation algorithms for reduct optimization has been presented in literature [8, 20]. However, based on results of Feige for the set cover problem [7], it is possible to show that, under some natural assumptions about the class NP, the approximation ratio of the best approximate polynomial algorithm for reduct optimization is near to the natural logarithm on the number of pairs of rows with different decisions in the decision table [14]. Therefore, the improvement of exact algorithms for reduct optimization continues to be an important issue.

6.3.2 *Algorithm for Reduct Optimization*

Let T be a decision table with n columns labeled with conditional attributes $f_1, \ldots, f_n$. We denote by $P(T)$ the set of unordered pairs of rows of T with different decisions. A conditional attribute f_i of T *separates* a pair of rows from $P(T)$ if these rows have different values at the intersection with the column f_i.

A *test* (*superreduct*) *for* T is a subset of columns (conditional attributes) of T such that any pair of rows from $P(T)$ is separated by at least one attribute from the considered subset. A *reduct for* T is a test for T for which each proper subset is not a test for T. We denote by $R(T)$ the minimum cardinality of a reduct for T. Reducts for T with cardinality $R(T)$ will be called *optimal*. It is clear that $R(T)$ is the minimum cardinality of a test for T.

Let the values of conditional attributes be nonnegative integers. For any conditional attributes f_i and f_j, we will write $f_i \preceq f_j$ if, for any row of T, the number at the intersection of this row and the column f_i is at most the number at the intersection of this row and the column f_j.

We describe an algorithm $\mathscr{A}_{\text{reduct}}$ for the construction of an optimal reduct that uses the algorithms $\mathscr{A}_3$ and $\mathscr{A}_5$. The algorithm $\mathscr{A}_5$ (the procedure of decision tree optimization) allows us not only to find the minimum cost of a decision tree for a given decision table but also to derive optimal decision trees from the constructed DAG. Let U be an uncertainty measure.

Algorithm $\mathscr{A}_{\text{reduct}}$ (construction of optimal reduct).

Input: A decision table T with conditional attributes $f_1, \ldots, f_n$.
Output: An optimal reduct for the decision table T.

1. Transform the decision table T into a decision table $T^{(1)}$ which has n columns labeled with conditional attributes $f_1, \ldots, f_n$, and $|P(T)| + 1$ rows. The first $|P(T)|$ rows $r_1, \ldots, r_{|P(T)|}$ are filled by 0's and 1's, and correspond to unordered pairs of rows of T with different decisions. The row of $T^{(1)}$ corresponding to a pair of rows r', r'' from $P(T)$ contains 1 at the intersection with the column f_i, $i = 1, \ldots, n$, if and only if r' and r'' have different values in the column f_i, i.e., f_i separates this pair of rows. The last row $r_{|P(T)|+1}$ in $T^{(1)}$ is filled by 0's. This row is labeled with the decision 1. All other rows are labeled with the decision 0.
2. For each pair of columns f_i and f_j of $T^{(1)}$ such that $i \neq j$ and $f_i \preceq f_j$, remove the column f_i.
3. Set $T^{(2)}$ the decision table obtained after the step 2.
4. Set $A = \emptyset$.
5. For each row r_i of $T^{(2)}$, $1 \preceq i \preceq |P(T)|$, that is separated from the last row by a unique attribute f_j, add the attribute f_j to the set A and remove all rows that are separated by this attribute from the row $r_{|P(T)|+1}$.
6. Set $T^{(3)}$ the decision table obtained after the step 5.
7. Construct the directed acyclic graph $\Delta_{U,0}(T^{(3)})$ by the algorithm $\mathscr{A}_3$.
8. Construct a $(U, 0)$-decision tree Γ with minimum depth for the decision table $T^{(3)}$ using the algorithm $\mathscr{A}_5$ and set B to the set of attributes in the path from the root of Γ to the terminal node of Γ labeled with 1.
9. Return $A \cup B$.

Proposition 6.1 *For any decision table* T, *the algorithm* $\mathscr{A}_{\text{reduct}}$ *returns an optimal reduct for* T.

Proof It is easy to show that the decision tables T and $T^{(1)}$ have the same set of tests and, therefore, the same set of reducts.

It is clear that each test for $T^{(2)}$ is a test for $T^{(1)}$. If $f_i \leq f_j$ in $T^{(1)}$ then the attribute f_j separates all pairs of rows from $P(T^{(1)})$ separated by f_i. As a result, we can replace f_i by f_j in any test for $T^{(1)}$ containing f_i and we still have a test for $T^{(1)}$. Using this fact one can show that the minimum cardinality of a test for $T^{(2)}$ is equal to the minimum cardinality of a test for $T^{(1)}$. Therefore each optimal reduct for $T^{(2)}$ is also an optimal reduct for $T^{(1)}$ and, hence, for T.

If a pair of rows $(r_i, r_{|P(T)|+1})$, $i \in \{1, \ldots, |P(T)|\}$, in the table $T^{(2)}$ is separated by only one attribute then this attribute must belong to any test for $T^{(2)}$. Using this fact, one can prove that the union of each optimal reduct for $T^{(3)}$ with the set A is an optimal reduct for the table $T^{(2)}$.

It is not difficult to show that the set B of attributes attached to the path in Γ from the root to the terminal node labeled with 1 is an optimal reduct for the table $T^{(3)}$. Therefore, the set $A \cup B$ is an optimal reduct for $T^{(2)}$ and, hence, for T. □

The algorithm $\mathscr{A}_{\text{reduct}}$ consists of the two phases: the first one is connected with the construction of the set A (steps 1–6), and the second one is connected with the construction of the set B (steps 7–9). The first phase has polynomial time complexity depending on the size of input decision table. The second phase has exponential time complexity in the worst case depending on the size of input decision table. More precisely, the second phase complexity depends polynomially on the number of separable subtables of $T^{(3)}$ and its size.

6.3.3 Experimental Results

In this section, we present experimental results for 23 datasets (decision tables) from UCI ML Repository [11]. Some decision tables contain conditional attributes that take unique value for each row. Such attributes were removed. In some tables there are equal rows with, possibly, different decisions. In this case, each group of identical rows is replaced with a single row from the group with the most common decision for this group. In some tables there were missing values. Each such value was replaced with the most common value of the corresponding attribute.

Table 6.3 contains information about decision tables and results of experiments: the column "Decision table" contains the name of the decision table T from [11], the columns "Rows" and "Attrs" contain the number of rows and conditional attributes in T, the columns "$|A|$" and "$|B|$" contain the cardinalities of the sets A and B constructed by the algorithm $\mathscr{A}_{\text{reduct}}$ for the table T. Note that $R(T) = |A| + |B|$.

All experiments were done using Mac Pro desktop with 16 GB of RAM memory and dual Intel(R) Xeon(R) processors of 2.67 GHz. The algorithm $\mathscr{A}_{\text{reduct}}$ was executed sequentially. The time of each phase (construction of A and construction of B) was measured on average of ten executions for each data set. The maximum average time for construction of the set A was 79 s for the table "Nursery". The maximum

Table 6.3 Characteristics of decision tables and results of experiments

Decision table	Rows	Attrs	$\|A\|$	$\|B\|$
Adult-stretch	16	4	2	0
Balance-scale	625	4	4	0
Breast-cancer	266	9	8	0
Cars	1728	6	6	0
Hayes-roth-data	69	4	4	0
House-votes-84	279	16	10	1
kr-vs-kp	3196	36	27	2
Lenses	24	4	4	0
Lymphography	148	18	0	6
Monks-1-test	432	6	3	0
Monks-1-train	124	6	3	0
Monks-2-test	432	6	6	0
Monks-2-train	169	6	6	0
Monks-3-test	432	6	3	0
Monks-3-train	122	6	4	0
Mushroom	8124	22	0	4
Nursery	12960	8	8	0
Shuttle-landing	15	6	5	0
Soybean-small	47	35	0	2
Spect-test	169	22	8	3
Teeth	23	8	6	0
Tic-tac-toe	958	9	0	8
Zoo-data	59	16	2	3

average time for construction of the set B was 330 s for the table "Mushroom". Note that for 15 out of the 23 considered decision tables (for which $|B| = 0$), an optimal reduct was constructed during the first phase of the algorithm that has polynomial time complexity depending on the size of input decision table.

References

1. AbouEisha, H.: Finding optimal exact reducts. In: Fred, A.L.N., Filipe, J. (eds.) International Conference on Knowledge Discovery and Information Retrieval, KDIR 2014, Rome, Italy, October 21–24, 2014, pp. 149–153. SciTePress (2014)
2. AbouEisha, H., Farhan, M.A., Chikalov, I., Moshkov, M.: An algorithm for reduct cardinality minimization. In: Wang, S., Zhu, X., He, T. (eds.) 2013 IEEE International Conference on Granular Computing, GrC 2013, Beijing, China, December 13–15, 2013, pp. 1–3. IEEE (2013)
3. AbouEisha, H., Chikalov, I., Moshkov, M.: Decision trees with minimum average depth for sorting eight elements. Discret. Appl. Math. **204**, 203–207 (2016)

4. Aigner, M.: Variants of the majority problem. Discret. Appl. Math. **137**(1), 3–25 (2004)
5. Césari, Y.: Questionnaire, codage et tris. Ph.D. thesis, Institut Blaise Pascal, Centre National de la Recherche (1968)
6. Chikalov, I.: Average Time Complexity of Decision Trees, Intelligent Systems Reference Library, vol. 21. Springer, Heidelberg (2011)
7. Feige, U.: A threshold of ln n for approximating set cover (preliminary version). In: Miller, G.L. (ed.) 28th Annual ACM Symposium on the Theory of Computing, Philadelphia, Pennsylvania, USA, May 22–24, 1996, pp. 314–318. ACM (1996)
8. Hoa, N.S., Son, N.H.: Some efficient algorithms for rough set methods. In: Information Processing and Management of Uncertainty on Knowledge Based Systems, IPMU 1996, Granada, Spain, Universidad de Granada, July 1–5, 1996, vol. III, pp. 1451–1456 (1996)
9. Knuth, D.E.: The Art of Computer Programming: Sorting and Searching, vol. 3, 2nd edn. Pearson Education, Boston (1998)
10. Kollár, L.: Optimal sorting of seven element sets. In: Gruska, J., Rovan, B., Wiedermann, J. (eds.) Mathematical Foundations of Computer Science 1986, Bratislava, Czechoslovakia, August 25–29, 1986, Lecture Notes in Computer Science, vol. 233, pp. 449–457. Springer (1986)
11. Lichman, M.: UCI Machine Learning Repository. University of California, Irvine, School of Information and Computer Sciences (2013). http://archive.ics.uci.edu/ml
12. Moshkov, M.: Time complexity of decision trees. In: Peters, J.F., Skowron, A. (eds.) Trans. Rough Sets III, Lecture Notes in Computer Science, vol. 3400, pp. 244–459. Springer, Berlin (2005)
13. Moshkov, M., Zielosko, B.: Combinatorial Machine Learning - A Rough Set Approach, Studies in Computational Intelligence, vol. 360. Springer, Heidelberg (2011)
14. Moshkov, M., Piliszczuk, M., Zielosko, B.: Partial Covers, Reducts and Decision Rules in Rough Sets - Theory and Applications, Studies in Computational Intelligence, vol. 145. Springer, Heidelberg (2008)
15. Pawlak, Z.: Rough Sets - Theoretical Aspect of Reasoning About Data. Kluwer Academic Publishers, Dordrecht (1991)
16. Pawlak, Z., Skowron, A.: Rudiments of rough sets. Inf. Sci. **177**(1), 3–27 (2007)
17. Peczarski, M.: New results in minimum-comparison sorting. Algorithmica **40**(2), 133–145 (2004)
18. Saks, M.E., Werman, M.: On computing majority by comparisons. Combinatorica **11**(4), 383–387 (1991)
19. Skowron, A., Rauszer, C.: The discernibility matrices and functions in information systems. In: Słowiński, R. (ed.) Intelligent Decision Support: Handbook of Applications and Advances of the Rough Sets Theory, pp. 331–362. Kluwer Academic Publishers, Dordrecht (1992)
20. Wroblewski, J.: Finding minimal reducts using genetic algorithms. In: 2nd Joint Annual Conference on Information Sciences, Wrightsville Beach, North Carolina, USA, 28 September–1 October, 1995, pp. 186–189 (1995)

Chapter 7
Bi-criteria Optimization Problem for Decision Trees: Cost Versus Cost

In this chapter, we consider an algorithm which constructs the sets of Pareto optimal points for bi-criteria optimization problems for decision trees relative to two cost functions (Sect. 7.1). We also show how the constructed sets of Pareto optimal points can be transformed into the graphs of functions which describe the relationships between the considered cost functions (Sect. 7.2).

We study different cost functions such as depth, average depth, and number of nodes (terminal and nonterminal). Algorithms for analysis of relationships between various pairs of these functions have been previously discussed and presented in papers such as [2, 7, 9] and implemented in the Dagger system [1]. In this chapter, we propose a universal approach to the study of such relationships based on the construction of the set of Pareto optimal points.

At the end of chapter (Sects. 7.3–7.5) we show three applications of bi-criteria optimization for two cost functions: comparison of different greedy algorithms for construction of decision trees (some initial results presented in [3, 5, 8]), analysis of trade-offs for decision trees for corner point detection [4] (used in computer vision), and study of derivation of decision rules from decision trees.

Note that the paper [6] contains some similar theoretical results and additional experimental results for the three applications discussed in this chapter.

7.1 Pareto Optimal Points: Cost Versus Cost

Let ψ and φ be increasing cost functions for decision trees given by triples of functions ψ^0, F, w and φ^0, H, u, respectively, U be an uncertainty measure, $\alpha \in \mathbb{R}_+$, T be a decision table with n conditional attributes $f_1, \ldots, f_n$, and G be a proper subgraph of the graph $\Delta_{U,\alpha}(T)$ (it is possible that $G = \Delta_{U,\alpha}(T)$). Interesting cases are when $G = \Delta_{U,\alpha}(T)$ or G is a result of application of the procedure of optimization

H. AbouEisha et al., *Extensions of Dynamic Programming for Combinatorial Optimization and Data Mining*, Intelligent Systems Reference Library 146,
https://doi.org/10.1007/978-3-319-91839-6_7

of decision trees (algorithm $\mathscr{A}_5$) relative to cost functions different from ψ and φ to the graph $\Delta_{U,\alpha}(T)$.

For each node Θ of the graph G, we denote $t_{\psi,\varphi}(G, \Theta) = \{(\psi(\Theta, \Gamma), \varphi(\Theta, \Gamma)) : \Gamma \in Tree(G, \Theta)\}$. Note that, by Proposition 4.4, if $G = \Delta_{U,\alpha}(T)$ then the set $Tree(G, \Theta)$ is equal to the set of (U, α)-decision trees for Θ. We denote by $Par(t_{\psi,\varphi}(G, \Theta))$ the set of Pareto optimal points for $t_{\psi,\varphi}(G, \Theta)$.

We now describe an algorithm $\mathscr{A}_6$ which constructs the set $Par(t_{\psi,\varphi}(G, T))$. In fact, this algorithm constructs, for each node Θ of the graph G, the set $B(\Theta) = Par(t_{\psi,\varphi}(G, \Theta))$.

Algorithm $\mathscr{A}_6$ (construction of POPs for decision trees, cost versus cost).

Input: Increasing cost functions for decision trees ψ and φ given by triples of functions ψ^0, F, w and φ^0, H, u, respectively, a decision table T with n conditional attributes $f_1, \ldots, f_n$, and a proper subgraph G of the graph $\Delta_{U,\alpha}(T)$ where U is an uncertainty measure and $\alpha \in \mathbb{R}_+$.

Output: The set $Par(t_{\psi,\varphi}(G, T))$ of Pareto optimal points for the set of pairs $t_{\psi,\varphi}(G, T) = \{(\psi(T, \Gamma), \varphi(T, \Gamma)) : \Gamma \in Tree(G, T)\}$.

1. If all nodes in G are processed, then return the set $B(T)$. Otherwise, choose a node Θ in the graph G which is not processed yet and which is either a terminal node of G or a nonterminal node of G such that, for any $f_i \in E_G(\Theta)$ and any $a_j \in E(\Theta, f_i)$, the node $\Theta(f_i, a_j)$ is already processed, i.e., the set $B(\Theta(f_i, a_j))$ is already constructed.
2. If Θ is a terminal node, then set $B(\Theta) = \{(\psi^0(\Theta), \varphi^0(\Theta))\}$. Mark the node Θ as processed and proceed to step 1.
3. If Θ is a nonterminal node then, for each $f_i \in E_G(\Theta)$, apply the algorithm $\mathscr{A}_2$ to the functions F, H and the sets $B(\Theta(f_i, a_1)), \ldots, B(\Theta(f_i, a_t))$, where
$$\{a_1, \ldots, a_t\} = E(\Theta, f_i) .$$
Set $C(\Theta, f_i)$ the output of the algorithm $\mathscr{A}_2$ and
$$\begin{aligned} B(\Theta, f_i) &= C(\Theta, f_i) \langle ++ \rangle \{(w(\Theta), u(\Theta))\} \\ &= \{(a + w(\Theta), b + u(\Theta)) : (a, b) \in C(\Theta, f_i)\} . \end{aligned}$$
4. Construct the multiset $A(\Theta) = \bigcup_{f_i \in E_G(\Theta)} B(\Theta, f_i)$ by simple transcription of elements from the sets $B(\Theta, f_i)$, $f_i \in E_G(\Theta)$. Apply to the obtained multiset $A(\Theta)$ the algorithm $\mathscr{A}_1$ which constructs the set $Par(A(\Theta))$. Set $B(\Theta) = Par(A(\Theta))$. Mark the node Θ as processed and proceed to step 1.

Proposition 7.1 *Let ψ and φ be increasing cost functions for decision trees given by triples of functions ψ^0, F, w and φ^0, H, u, respectively, U be an uncertainty measure, $\alpha \in \mathbb{R}_+$, T be a decision table with n conditional attributes $f_1, \ldots, f_n$, and G be a proper subgraph of the graph $\Delta_{U,\alpha}(T)$. Then, for each node Θ of the graph G, the algorithm $\mathscr{A}_6$ constructs the set $B(\Theta) = Par(t_{\psi,\varphi}(G, \Theta))$.*

Proof We prove the considered statement by induction on nodes of G. Let Θ be a terminal node of G. Then $Tree(G, \Theta) = \{tree(mcd(\Theta))\}$,

$$t_{\psi,\varphi}(G, \Theta) = Par(t_{\psi,\varphi}(G, \Theta)) = \{(\psi^0(\Theta), \varphi^0(\Theta))\} ,$$

and $B(\Theta) = Par(t_{\psi,\varphi}(G, \Theta))$.

Let Θ be a nonterminal node of G such that, for any $f_i \in E_G(\Theta)$ and any $a_j \in E(\Theta, f_i)$, the considered statement holds for the node $\Theta(f_i, a_j)$, i.e., $B(\Theta(f_i, a_j)) = Par(t_{\psi,\varphi}(G, \Theta(f_i, a_j)))$.

Let $f_i \in E_G(\Theta)$ and $E(\Theta, f_i) = \{a_1, \ldots, a_t\}$. We denote

$$\begin{aligned} P(f_i) = \{&(F(b_1, \ldots, b_t) + w(\Theta), H(c_1, \ldots, c_t) + u(\Theta)) \\ &: (b_j, c_j) \in t_{\psi,\varphi}(G, \Theta(f_i, a_j)), j = 1, \ldots, t\} , \end{aligned}$$

and, for $j = 1, \ldots, t$, we denote $P_j = t_{\psi,\varphi}(G, \Theta(f_i, a_j))$.

If we apply the algorithm $\mathscr{A}_2$ to the functions F, H and the sets

$$Par(P_1), \ldots, Par(P_t) ,$$

we obtain the set $Par(Q_t)$ where $Q_1 = P_1$, and, for $j = 2, \ldots, t$, $Q_j = Q_{j-1}\langle FH\rangle P_j$. It is not difficult to show that $P(f_i) = Q_t \langle{+}{+}\rangle \{(w(\Theta), u(\Theta))\} = \{(a + w(\Theta), b + u(\Theta)) : (a, b) \in Q_t\}$ and

$$Par(P(f_i)) = Par(Q_t) \langle{+}{+}\rangle \{(w(\Theta), u(\Theta))\} .$$

According to the induction hypothesis, $B(\Theta(f_i, a_j)) = Par(P_j)$ for $j = 1, \ldots, t$. Therefore $C(\Theta, f_i) = Par(Q_t)$ and $B(\Theta, f_i) = Par(P(f_i))$.

One can show that $t_{\psi,\varphi}(G, \Theta) = \bigcup_{f_i \in E_G(\Theta)} P(f_i)$. By Lemma 2.5,

$$\begin{aligned} Par(t_{\psi,\varphi}(G, \Theta)) &= Par\left(\bigcup_{f_i \in E_G(\Theta)} P(f_i)\right) \\ &\subseteq \bigcup_{f_i \in E_G(\Theta)} Par(P(f_i)) . \end{aligned}$$

Using Lemma 2.4 we obtain $Par(t_{\psi,\varphi}(G, \Theta)) = Par\left(\bigcup_{f_i \in E_G(\Theta)} Par(P(f_i))\right)$. Since $B(\Theta, f_i) = Par(P(f_i))$ for any $f_i \in E_G(\Theta)$, $Par(t_{\psi,\varphi}(G, \Theta)) = Par\ (A(\Theta)) = B(\Theta)$. □

We now analyze the number of elementary operations made by the algorithm $\mathscr{A}_6$ during the construction of the set $Par(t_{\psi,\varphi}(G, T))$ for integral cost functions, for the cases when $F(x, y) = \max(x, y)$ (h is an example of such cost function) and when $F(x, y) = x + y$ (tpl, L, L_n, and L_t are examples of such cost functions).

Let us recall that $range(T) = \max\{|E(T, f_i)| : i = 1, \ldots, n\}$,

$$ub(\psi, T) = \max\{\psi(\Theta, \Gamma) : \Theta \in SEP(T), \Gamma \in DT(\Theta)\},$$

and $\psi(\Theta, \Gamma) \in \{0, 1, \ldots, ub(\psi, T)\}$ for any separable subtable Θ of T and for any decision tree Γ for Θ. Upper bounds on the value $ub(\psi, T)$ for $\psi \in \{h, tpl, L, L_n, L_t\}$ are given in Lemma 4.2: $ub(h, T) \leq n$, $ub(tpl, T) \leq nN(T)$, $ub(L, T) \leq 2N(T)$, $ub(L_n, T) \leq N(T)$, and $ub(L_t, T) \leq N(T)$.

Proposition 7.2 *Let ψ and φ be increasing integral cost functions for decision trees given by triples of functions ψ^0, F, w and φ^0, H, u, respectively, $F \in \{\max(x, y), x + y\}$, U be an uncertainty measure, $\alpha \in \mathbb{R}_+$, T be a decision table with n conditional attributes $f_1, \ldots, f_n$, and G be a proper subgraph of the graph $\Delta_{U,\alpha}(T)$. Then, to construct the set $Par(t_{\psi,\varphi}(G, T))$, the algorithm $\mathcal{A}_6$ makes*

$$O(L(G)range(T)ub(\psi, T)^2 n \log(ub(\psi, T)n))$$

elementary operations (computations of F, H, w, u, ψ^0, φ^0, additions, and comparisons) if $F = \max(x, y)$, and

$$O(L(G)range(T)^2 ub(\psi, T)^2 n \log(range(T)ub(\psi, T)n))$$

elementary operations (computations of F, H, w, u, ψ^0, φ^0, additions, and comparisons) if $F = x + y$.

Proof It is clear that, for any node Θ of the graph G, $t_{\psi,\varphi}(G, \Theta)^{(1)} = \{a : (a, b) \in t_{\psi,\varphi}(G, \Theta)\} \subseteq \{0, \ldots, ub(\psi, T)\}$. From Proposition 7.1 it follows that $B(\Theta) = Par(t_{\psi,\varphi}(G, \Theta))$ for any node Θ of the graph G. Therefore, for any node Θ of the graph G, $B(\Theta)^{(1)} = \{a : (a, b) \in B(\Theta)\} \subseteq \{0, \ldots, ub(\psi, T)\}$.

Let $F(x, y) = \max(x, y)$. To process a terminal node Θ, the algorithm $\mathcal{A}_6$ makes two elementary operations (computations of ψ^0 and φ^0).

We now consider the processing of a nonterminal node Θ (see description of the algorithm $\mathcal{A}_6$). We know that $B(\Theta(f_i, a_j))^{(1)} \subseteq \{0, \ldots, ub(\psi, T)\}$ for $j = 1, \ldots, t$, and $t \leq range(T)$. From Proposition 2.3 it follows that $|C(\Theta, f_i)| \leq ub(\psi, T) + 1$, and to construct the set $C(\Theta, f_i)$, the algorithm $\mathcal{A}_2$ makes

$$O(range(T)ub(\psi, T)^2 \log ub(\psi, T))$$

elementary operations (computations of F, H and comparisons). To construct the set $B(\Theta, f_i)$ from the set $C(\Theta, f_i)$, the algorithm makes computations of w and u, and at most $2ub(\psi, T) + 2$ additions. From here it follows that, to construct the set $B(\Theta, f_i)$, the algorithm makes

$$O(range(T)ub(\psi, T)^2 \log ub(\psi, T))$$

elementary operations. It is clear that $|E_G(\Theta)| \leq n$. Therefore, to construct the set $B(\Theta, f_i)$ for each $f_i \in E_G(\Theta)$, the algorithm makes

$$O(range(T)ub(\psi, T)^2 n \log ub(\psi, T))$$

elementary operations.

Since $|C(\Theta, f_i)| \leq ub(\psi, T) + 1$, we have $|B(\Theta, f_i)| \leq ub(\psi, T) + 1$. Since

$$|E_G(\Theta)| \leq n ,$$

we have $|A(\Theta)| \leq n(ub(\psi, T) + 1)$ where $A(\Theta) = \bigcup_{f_i \in E(\Theta)} B(\Theta, f_i)$. From Proposition 2.1 it follows that, to construct the set $B(\Theta) = Par(A(\Theta))$, the algorithm $\mathscr{A}_1$ makes $O(ub(\psi, T)n \log(ub(\psi, T)n))$ comparisons. So, to process a nonterminal node Θ (to construct $B(\Theta) = Par(t_{\psi,\varphi}(G, \Theta))$ if

$$B(\Theta(f_i, a_j)) = Par(t_{\psi,\varphi}(G, \Theta(f_i, a_j)))$$

is known for all $f_i \in E_G(\Theta)$ and $a_j \in E(\Theta, f_i)$), the algorithm $\mathscr{A}_6$ makes

$$O(range(T)ub(\psi, T)^2 n \log(ub(\psi, T)n))$$

elementary operations.

To construct the set $Par(t_{\psi,\varphi}(G, T))$ for given decision table T with n conditional attributes and the graph G, it is enough to make

$$O(L(G)range(T)ub(\psi, T)^2 n \log(ub(\psi, T)n))$$

elementary operations (computations of F, H, w, u, ψ^0, φ^0, additions, and comparisons).

Let $F(x, y) = x + y$. To process a terminal node Θ, the algorithm $\mathscr{A}_6$ makes two elementary operations (computations of ψ^0 and φ^0).

We now consider the processing of a nonterminal node Θ (see description of the algorithm $\mathscr{A}_6$). We know that $B(\Theta(f_i, a_j))^{(1)} \subseteq \{0, \ldots, ub(\psi, T)\}$ for $j = 1, \ldots, t$, and $t \leq range(T)$. From Proposition 2.3 it follows that $|C(\Theta, f_i)| \leq t \times ub(\psi, T) + 1 \leq range(T)ub(\psi, T) + 1$, and to construct the set $C(\Theta, f_i)$, the algorithm $\mathscr{A}_2$ makes

$$O(range(T)^2 ub(\psi, T)^2 \log(range(T)ub(\psi, T)))$$

elementary operations (computations of F, H and comparisons). To construct the set $B(\Theta, f_i)$ from the set $C(\Theta, f_i)$, the algorithm makes computations of w and u, and at most $2range(T)ub(\psi, T) + 2$ additions. From here it follows that, to construct the set $B(\Theta, f_i)$ the algorithm makes

$$O(range(T)^2 ub(\psi, T)^2 \log(range(T)ub(\psi, T)))$$

elementary operations. It is clear that $|E_G(\Theta)| \leq n$. Therefore, to construct the set $B(\Theta, f_i)$ for each $f_i \in E_G(\Theta)$, the algorithm makes

$$O(range(T)^2 ub(\psi, T)^2 n \log(range(T) ub(\psi, T)))$$

elementary operations.

Since $|C(\Theta, f_i)| \leq range(T) ub(\psi, T) + 1$, we have

$$|B(\Theta, f_i)| \leq range(T) ub(\psi, T) + 1 .$$

Since $|E_G(\Theta)| \leq n$, we have $|A(\Theta)| \leq n(range(T) ub(\psi, T) + 1)$ where $A(\Theta) = \bigcup_{f_i \in E_G(\Theta)} B(\Theta, f_i)$. From Proposition 2.1 it follows that, to construct the set $B(\Theta) = Par(A(\Theta))$, the algorithm $\mathscr{A}_1$ makes

$$O(range(T) ub(\psi, T) n \log(range(T) ub(\psi, T) n))$$

comparisons. So, to process a nonterminal node Θ (to construct

$$B(\Theta) = Par(t_{\psi,\varphi}(G, \Theta))$$

if $B(\Theta(f_i, a_j)) = Par(t_{\psi,\varphi}(G, \Theta(f_i, a_j)))$ is known for all $f_i \in E_G(\Theta)$ and $a_j \in E(\Theta, f_i)$), the algorithm $\mathscr{A}_6$ makes

$$O(range(T)^2 ub(\psi, T)^2 n \log(range(T) ub(\psi, T) n))$$

elementary operations.

To construct the set $Par(t_{\psi,\varphi}(G, T))$ for given decision table T with n conditional attributes and the graph G, it is enough to make

$$O(L(G) range(T)^2 ub(\psi, T)^2 n \log(range(T) ub(\psi, T) n))$$

elementary operations (computations of F, H, w, u, ψ^0, φ^0, additions, and comparisons). □

Note that similar results can be obtained if $H \in \{\max(x, y), x + y\}$.

Proposition 7.3 *Let ψ and φ be cost functions for decision trees given by triples of functions ψ^0, F, w and φ^0, H, u, respectively, $\psi, \varphi \in \{h, tpl, L, L_n, L_t\}$, and $\mathscr{U}$ be a restricted information system. Then the algorithm $\mathscr{A}_6$ has polynomial time complexity for decision tables from $\mathscr{T}(\mathscr{U})$ depending on the number of conditional attributes in these tables.*

Proof Since $\psi, \varphi \in \{h, tpl, L, L_n, L_t\}$, ψ^0 and φ^0 are constants, each of the functions F, H is either $\max(x, y)$ or $x + y$, and each of the functions w, u is either a constant or $N(T)$. From Proposition 7.2 and Lemma 4.2 it follows that, for the algorithm $\mathscr{A}_6$, the number of elementary operations (computations of F, H, w, u, ψ^0, φ^0, additions, and comparisons) is bounded from above by a polynomial depending on the size of input table T and on the number of separable subtables of T. All operations with numbers are basic ones. The computations of numerical parameters of decision tables used by the algorithm $\mathscr{A}_6$ (constants and $N(T)$) have polynomial time complexity depending on the size of decision tables.

According to Proposition 3.5, the algorithm $\mathscr{A}_6$ has polynomial time complexity for decision tables from $\mathscr{T}(\mathscr{U})$ depending on the number of conditional attributes in these tables. □

7.2 Relationships Between Two Cost Functions

Let ψ and φ be increasing cost functions for decision trees, U be an uncertainty measure, $\alpha \in \mathbb{R}_+$, T be a decision table, and G be a proper subgraph of the graph $\Delta_{U,\alpha}(T)$ (it is possible that $G = \Delta_{U,\alpha}(T)$).

To study relationships between cost functions ψ and φ on the set of decision trees $Tree(G, T)$ we consider partial functions $\mathscr{T}_{G,T}^{\psi,\varphi} : \mathbb{R} \to \mathbb{R}$ and $\mathscr{T}_{G,T}^{\varphi,\psi} : \mathbb{R} \to \mathbb{R}$ defined as follows:

$$\mathscr{T}_{G,T}^{\psi,\varphi}(x) = \min\{\varphi(T, \Gamma) : \Gamma \in Tree(G, T), \psi(T, \Gamma) \le x\} ,$$

$$\mathscr{T}_{G,T}^{\varphi,\psi}(x) = \min\{\psi(T, \Gamma) : \Gamma \in Tree(G, T), \varphi(T, \Gamma) \le x\} .$$

Let $(a_1, b_1), \ldots, (a_k, b_k)$ be the normal representation of the set $Par(t_{\psi,\varphi}(G, T))$ where $a_1 < \ldots < a_k$ and $b_1 > \ldots > b_k$. By Lemma 2.8 and Remark 2.2, for any $x \in \mathbb{R}$,

$$\mathscr{T}_{G,T}^{\psi,\varphi}(x) = \begin{cases} \textit{undefined}, & x < a_1 \\ b_1, & a_1 \le x < a_2 \\ \ldots & \ldots \\ b_{k-1}, & a_{k-1} \le x < a_k \\ b_k, & a_k \le x \end{cases} ,$$

$$\mathscr{T}_{G,T}^{\varphi,\psi}(x) = \begin{cases} \textit{undefined}, & x < b_k \\ a_k, & b_k \le x < b_{k-1} \\ \ldots & \ldots \\ a_2, & b_2 \le x < b_1 \\ a_1, & b_1 \le x \end{cases} .$$

7.3 Comparison of Greedy Algorithms for Decision Tree Construction

In this section, we compare 20 greedy algorithms for decision tree construction as single-criterion and bi-criteria optimization algorithms.

We consider five uncertainty measures for decision tables: entropy ent, Gini index $gini$, function R (in this section, we will denote it rt), misclassification error me, and relative misclassification error rme (see Sect. 3.2). Let T be a decision table, $f_i \in E(T)$ and $E(T, f_i) = \{a_1, \ldots, a_t\}$. The attribute f_i divides the table T into subtables $T_1 = T(f_i, a_1), \ldots, T_t = T(f_i, a_t)$. For a fixed uncertainty measure U, we can define *impurity functions* $I(T, f_i)$ of four types which give "impurity"of this partition: $\sum_{j=1}^{t} U(T_j)$ (*sum* type denoted by *s*), $\max_{1 \le j \le t} U(T_j)$ (*max* type denoted by *m*), $\sum_{j=1}^{t} U(T_j)N(T_j)$ (*weighted sum* type denoted by *ws*), and $\max_{1 \le j \le t} U(T_j)N(T_j)$ (*weighted max* type denoted by *wm*).

As a result, we have 20 impurity functions defined by pairs *type of impurity function-uncertainty measure*, for example, *wm-gini*. For each impurity function I, we describe a greedy algorithm $\mathscr{A}_I$ which, for a given decision table T and real number α, $0 \le \alpha < 1$, constructs a (rme, α)-decision tree for the table T.

Algorithm $\mathscr{A}_I$ (greedy algorithm for decision tree construction).

Input: Decision table T with n conditional attributes $f_1, \ldots, f_n$, and real number $\alpha, 0 \le \alpha < 1$.
Output: An (rme, α)-decision tree $\Gamma_{I,\alpha}(T)$ for the table T.

1. Construct a tree G consisting of a single node labeled with T.
2. If no node of the tree G is labeled with a table then the algorithm ends and returns the tree G.
3. Choose a node v in G which is labeled with a subtable Θ of the table T.
4. If $rme(\Theta) \le \alpha$ then we label the node v by $mcd(\Theta)$ instead of Θ and proceed to the Step 2.
5. If $rme(\Theta) > \alpha$ then, for each $f_i \in E(\Theta)$, we compute the value $I(\Theta, f_i)$, remove the label Θ, and label the node v with the attribute $f_{i_0} \in E(\Theta)$ with minimum index i_0 such that $I(\Theta, f_{i_0}) = \min\{I(\Theta, f_i) : f_i \in E(\Theta)\}$. For each $\delta \in E(\Theta, f_{i_0})$, we add to the tree G a node $v(\delta)$ and an edge e_δ connecting v and $v(\delta)$. We label the node $v(\delta)$ with the subtable $\Theta(f_{i_0}, \delta)$, and label the edge e_δ with the number δ. We proceed to the Step 2.

We compare experimentally the obtained 20 greedy algorithms. Let

$$V = \{0, 0.1, \ldots, 0.9\} .$$

By S we denote the set of 23 decision tables from UCI ML Repository [10] (information about these decision tables can be found in Table 7.4). We study the following pairs of increasing cost functions (ψ, φ): (h, L), (h, h_{avg}), (h_{avg}, L), and (h, L_t) where $h_{\text{avg}}(T, \Gamma) = tpl(T, \Gamma)/N(T)$.

For each such pair (ψ, φ), each $T \in S$ and each $\alpha \in V$, we construct the set $Par_{\psi,\varphi,\alpha}(T) = Par(t_{\psi,\varphi}(\Delta_{rme,\alpha}(T), T))$ of Pareto optimal points for the problem of bi-criteria optimization of (rme, α)-decision trees for T relative to ψ and φ. Next we find the values $\psi_\alpha(T) = \min\{\psi(T, \Gamma) : \Gamma \in DT_{rme,\alpha}(T)\}$ and $\varphi_\alpha(T) = \min\{\varphi(T, \Gamma) : \Gamma \in DT_{rme,\alpha}(T)\}$ in the following way: $\psi_\alpha(T) = \min\{a : (a, b) \in Par_{\psi,\varphi,\alpha}(T)\}$ and $\varphi_\alpha(T) = \min\{b : (a, b) \in Par_{\psi,\varphi,\alpha}(T)\}$. For each $T \in S$, each $\alpha \in V$, and each of the considered 20 impurity functions I, we apply the algorithm A_I to the table T and the number α, and construct an (rme, α) decision tree $\Gamma_{I,\alpha}(T)$ for T.

We compute

$$\psi_{I,\alpha}(S) = \frac{1}{23}\sum_{T \in S}(\psi_\alpha(T, \Gamma_{I,\alpha}(T)) - \psi_\alpha(T))/\psi_\alpha(T) ,$$

$$\varphi_{I,\alpha}(S) = \frac{1}{23}\sum_{T \in S}(\varphi_\alpha(T, \Gamma_{I,\alpha}(T)) - \varphi_\alpha(T))/\varphi_\alpha(T) .$$

The values $\psi_{I,\alpha}(S)$ and $\varphi_{I,\alpha}(S)$ characterize the accuracy of the algorithm $\mathscr{A}_I$ as an algorithm for optimization of (rme, α)-decision trees for tables from the set S relative to ψ and φ, respectively. We also compute the values

$$\psi_I(S) = \frac{1}{10}\sum_{\alpha \in V}\psi_{I,\alpha}(S) ,$$

$$\varphi_I(S) = \frac{1}{10}\sum_{\alpha \in V}\varphi_{I,\alpha}(S) ,$$

which characterize the accuracy of the algorithm $\mathscr{A}_I$ as an algorithm for optimization of decision trees for tables from the set S relative to ψ and φ, respectively.

Let $A_\alpha = \max\{a : (a, b) \in Par_{\psi,\varphi,\alpha}(T)\}$, $B_\alpha = \max\{b : (a, b) \in Par_{\psi,\varphi,\alpha}(T)\}$, and $Par^*_{\psi,\varphi,\alpha}(T) = \{(a/A_\alpha, b/B_\alpha) : (a, b) \in Par_{\psi,\varphi,\alpha}(T)\}$. We denote by $d^{\psi,\varphi}_{I,\alpha}(T)$ the Euclidean distance between the point

$$(\psi(T, \Gamma_{I,\alpha}(T))/A_\alpha, \varphi(T, \Gamma_{I,\alpha}(T))/B_\alpha)$$

and the set $Par^*_{\psi,\varphi,\alpha}(T)$ (the distance between a point p and a finite set P of points is the minimum distance between the point p and a point from the set P). We compute the value

$$d^{\psi,\varphi}_{I,\alpha}(S) = \frac{1}{23}\sum_{T \in S} d^{\psi,\varphi}_{I,\alpha}(T) ,$$

which characterizes the accuracy of the algorithm $\mathscr{A}_I$ as an algorithm for bi-criteria optimization of (rme, α)-decision trees for tables from the set S relative to ψ and φ. We also compute the value

Table 7.1 Best impurity functions for single-criterion optimization of decision trees

Rank	Cost functions			
	h	h_{avg}	L	L_t
1-st	*s-rt*	*ws-gini*	*ws-gini*	*ws-gini*
2-nd	*ws-gini*	*ws-ent*	*ws-ent*	*ws-ent*
3-rd	*ws-ent*	*s-rt*	*s-me*	*s-me*

Table 7.2 Best impurity functions for bi-criteria optimization of decision trees

Rank	Pairs of cost functions			
	(h, L)	(h, h_{avg})	(h_{avg}, L)	(h, L_t)
1-st	*ws-gini*	*ws-rt*	*ws-gini*	*s-rt*
2-nd	*ws-ent*	*wm-rt*	*ws-ent*	*ws-rt*
3-rd	*s-rt*	*s-rt*	*s-rt*	*ws-me*

$$d_I^{\psi,\varphi}(S) = \frac{1}{10}\sum_{\alpha \in V} d_{I,\alpha}^{\psi,\varphi}(S) ,$$

which characterizes the accuracy of the algorithm $\mathscr{A}_I$ as an algorithm for bi-criteria optimization of decision trees for tables from the set S relative to ψ and φ.

For each cost function $\psi \in \{h, h_{\text{avg}}, L, L_t\}$, we order 20 greedy algorithms $\mathscr{A}_I$ according to the value $\psi_I(S)$ and choose three impurity functions corresponding to the three best algorithms (see Table 7.1). For each pair of cost functions $(\psi, \varphi) \in \{(h, L), (h, h_{\text{avg}}), (h_{\text{avg}}, L), (h, L_t)\}$, we order 20 greedy algorithms $\mathscr{A}_I$ according to the value $d_I^{\psi,\varphi}(S)$ and choose three impurity functions corresponding to the three best algorithms (see Table 7.2).

Results for pairs (h, L) and (h_{avg}, L) are predictable. For example, the best impurity function *ws-gini* for bi-criteria optimization relative to (h, L) is the first for single-criterion optimization relative to L. Similar situation is for all three best impurity functions for bi-criteria optimization relative to (h, L) and relative to (h_{avg}, L).

Results for pairs (h, h_{avg}) and (h, L_t) are more interesting. For example, the best impurity function *ws-rt* for bi-criteria optimization relative to (h, h_{avg}) is the 12th for single-criterion optimization relative to h and the 5th for single-criterion optimization relative to h_{avg}. The second impurity function *wm-rt* for bi-criteria optimization relative to (h, h_{avg}) is the 14th for single-criterion optimization relative to h and the 14th for single-criterion optimization relative to h_{avg}.

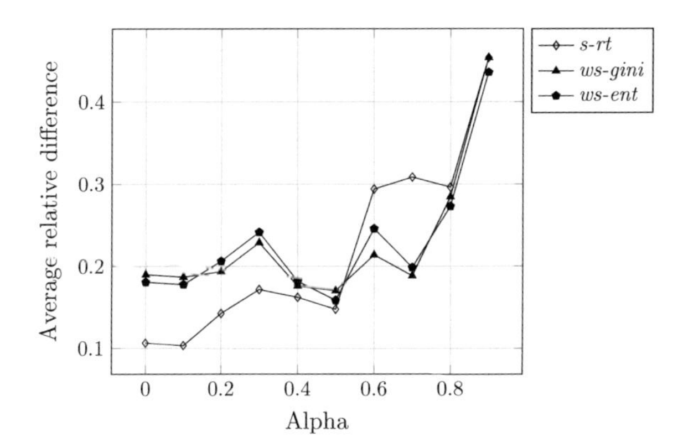

Fig. 7.1 Comparison of greedy algorithms for h

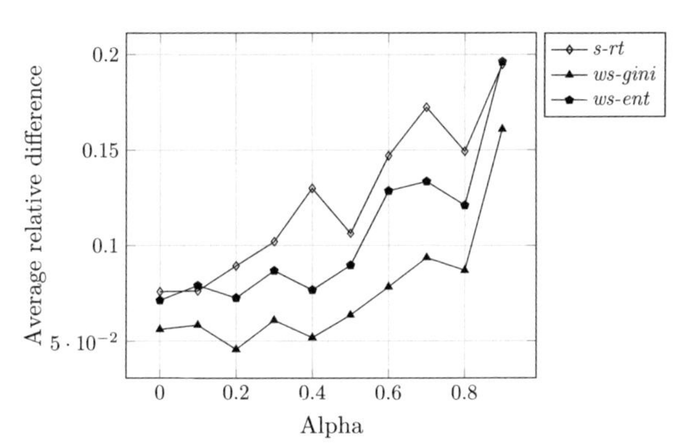

Fig. 7.2 Comparison of greedy algorithms for h_{avg}

The second impurity function *ws-rt* for bi-criteria optimization relative to (h, L_t) is the 12th for single-criterion optimization relative to h and the 16th for single-criterion optimization relative to L_t. The third impurity function *ws-me* for bi-criteria optimization relative to (h, L_t) is the 4th for single-criterion optimization relative to h and the 7th for single-criterion optimization relative to L_t.

Table 7.1 contains, for each cost function $\psi \in \{h, h_{\mathrm{avg}}, L, L_t\}$, the three best impurity functions I. For these impurity functions I, the graphs of the functions $\psi_{I,\alpha}(S)$ (values of these functions are called *average relative difference*) depending on α can be found on Figs. 7.1, 7.2, 7.3, and 7.4 for the cost functions h, h_{avg}, L, and L_t, respectively. The graphs of the functions $d_{I,\alpha}^{\psi,\varphi}(S)$ (values of these functions are called *average minimum distance*) depending on α for each $(\psi, \varphi) \in \{(h, L), (h, L_t), (h, h_{\mathrm{avg}}), (h_{\mathrm{avg}}, L)\}$ and the three best impurity functions I for (ψ, φ) (see Table 7.2) can be found on Figs. 7.5, 7.6, 7.7, and 7.8.

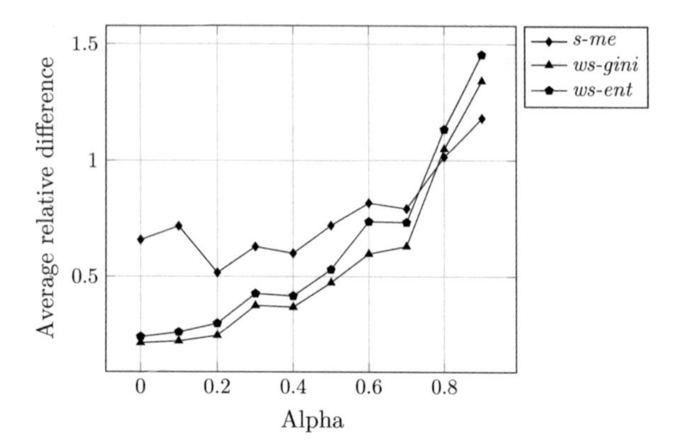

Fig. 7.3 Comparison of greedy algorithms for L

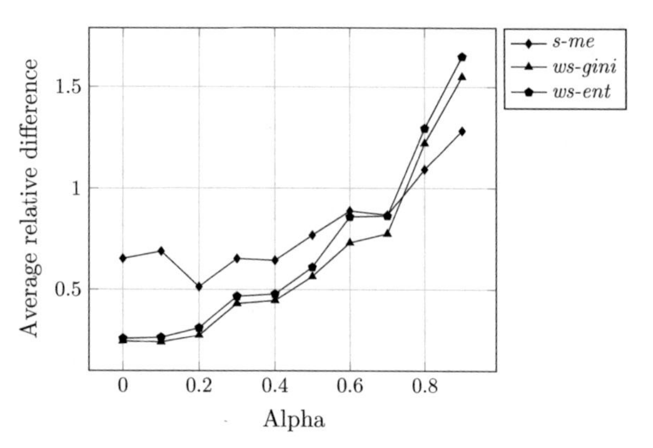

Fig. 7.4 Comparison of greedy algorithms for L_t

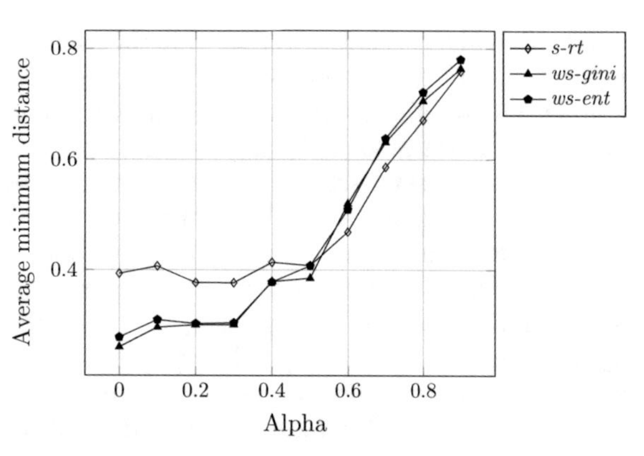

Fig. 7.5 Comparison of greedy algorithms for (h, L)

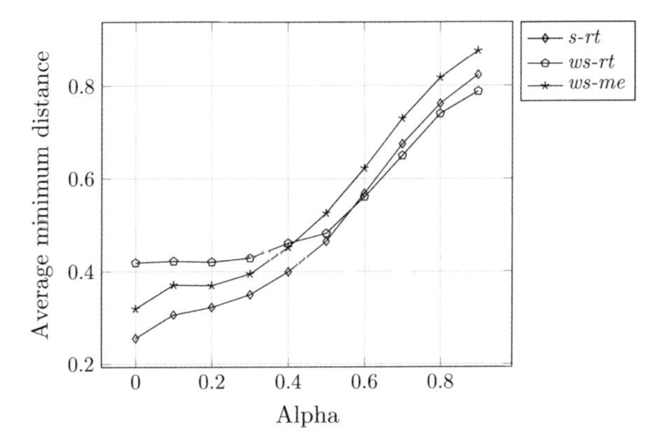

Fig. 7.6 Comparison of greedy algorithms for (h, L_t)

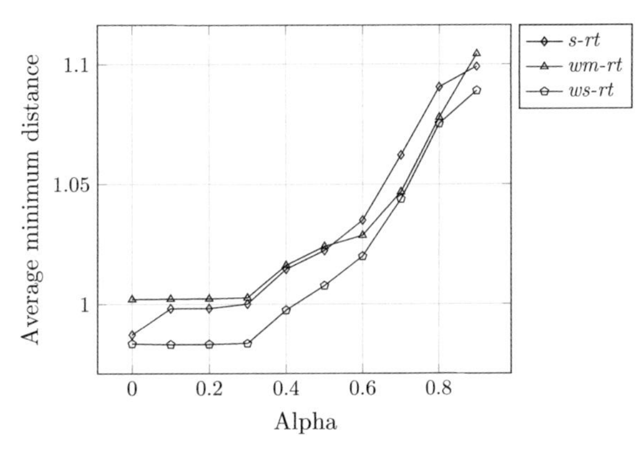

Fig. 7.7 Comparison of greedy algorithms for (h, h_{avg})

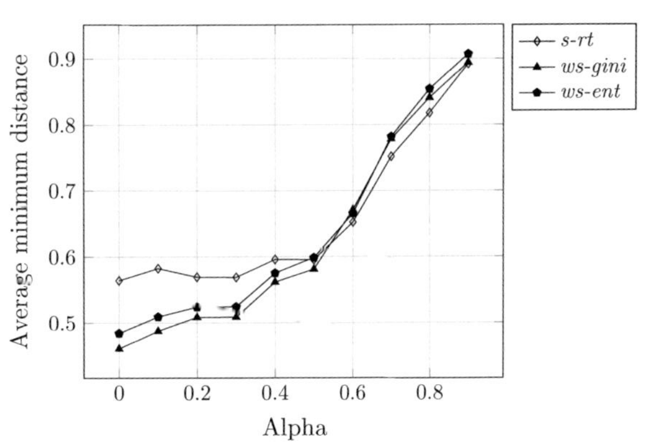

Fig. 7.8 Comparison of greedy algorithms for (h_{avg}, L)

Fig. 7.9 Bresenham circle of radius three surrounding pixel a

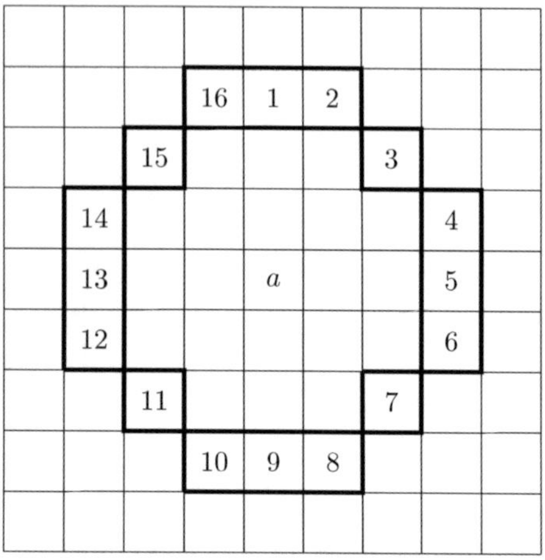

7.4 Decision Trees for Corner Point Detection

Object tracking (locating an object and determining its position) in a video stream is an important problem in computer vision. One of the approaches to solve this problem is based on the use of distinctive points located on the image. FAST algorithm devised by Rosten and Drummond [14, 15] is an example of such a solution. This algorithm iterates through all pixels and detects *corner points* by comparing the intensity of the current pixel and surrounding pixels.

In order to determine if an image pixel a is a corner point, a circle of 16 pixels (a Bresenham circle of radius three) surrounding a is examined (see Fig. 7.9). The intensity of each pixel of the circle is compared with the intensity of a. The pixel a is assumed to be a corner point if at least 12 contiguous pixels on the circle are all either brighter or darker than a by a given threshold $\gamma > 0$.

For an arbitrary pixel a and for $i = 1, \ldots, 16$, $\phi_i(a)$ denotes the intensity of the i-th pixel in the circle surrounding a (ordering as shown in Fig. 7.9) and $\phi(a)$ denotes the intensity of the pixel a. The pixel a can be represented as an object that is characterized by the attributes $f_1, \ldots, f_{16}$ where, for $i = 1, \ldots, 16$,

$$f_i(a) = \begin{cases} 0, \text{ if } \phi(a) - \phi_i(a) & > \gamma \,, \\ 1, \text{ if } |\phi_i(a) - \phi(a)| & \leq \gamma \,, \\ 2, \text{ if } \phi_i(a) - \phi(a) & > \gamma \,. \end{cases}$$

Then the problem of corner point detection for a given set of images can be represented as a decision table T with conditional attributes $f_1, \ldots, f_{16}$. The table T contains rows (tuples of attribute values) for all pixels from the considered images

Table 7.3 Corner point detection experiments: number of Pareto optimal points

Decision table	Rows	# POPs		
		(L, tpl)	(h, L)	(h, tpl)
BOX-20	81,900	280	1	2
BOX-50	10,972	37	3	2
BOX-70	3,225	8	3	4
JUNK-50	3,509	12	2	3
JUNK-70	980	1	2	2
MAZE-70	5,303	15	4	4
MAZE-100	1,343	1	2	3

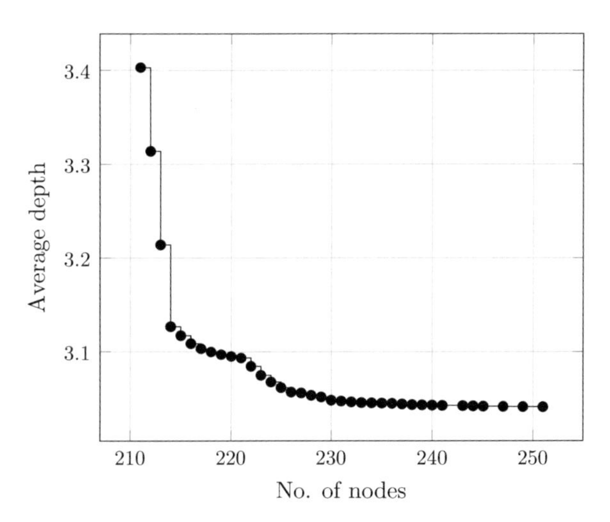

Fig. 7.10 Relationship for (L, h_{avg}) and BOX- 50

with the exception of outer boundaries. A row of T is labeled with the decision 1 if this tuple of attribute values describes a corner point, and the decision 0 otherwise.

Study of relationships among different cost functions for decision trees for the table T allows us to find a decision tree for corner point detection problem which is adapted to the considered set of images and has appropriate values of cost functions.

In our experiments, we use decision tables corresponding to the sets of images BOX, MAZE, and JUNK considered in [15] (see also [4]) and different thresholds γ. For each such table T, we construct the set of Pareto optimal points $Par(t_{\psi,\varphi}(\Delta_{rme,0}(T), T))$ for the following pairs of cost functions (ψ, φ): (L, tpl), (h, tpl), and (h, L). Some results of experiments including the number of rows in the considered decision tables and the number of Pareto optimal points (POPs) for the three pairs of cost functions can be found in Table 7.3. The name of each table consists of the name of a set of images and a threshold γ. In the case when the number of POPs is equal to one, there exists a totally optimal decision tree for the considered decision table relative to the considered pair of cost functions.

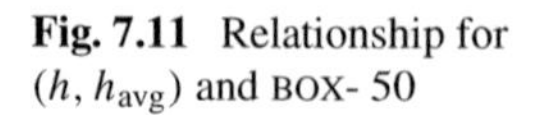

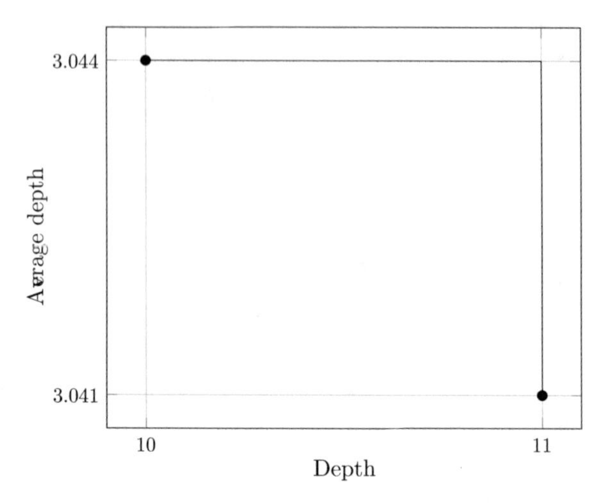

Fig. 7.11 Relationship for (h, h_{avg}) and BOX- 50

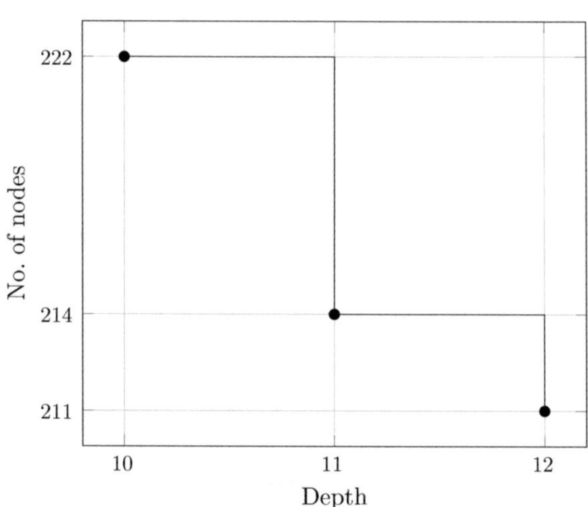

Fig. 7.12 Relationship for (h, L) and BOX- 50

Figures 7.10, 7.11, 7.12, and 7.13 show in detail relationships (see Sect. 7.2) for the decision table BOX- 50 and pairs of cost functions (L, h_{avg}), (h, h_{avg}), (h, L), and (h, L_t) $(h_{\text{avg}}(T) = tpl(T)/N(T))$, where the small filled circles are Pareto optimal points. The study of the last pair (h, L_t) allows us to understand the relationship between the maximum length and the number of decision rules derived from decision trees for BOX- 50 (see Sect. 7.5).

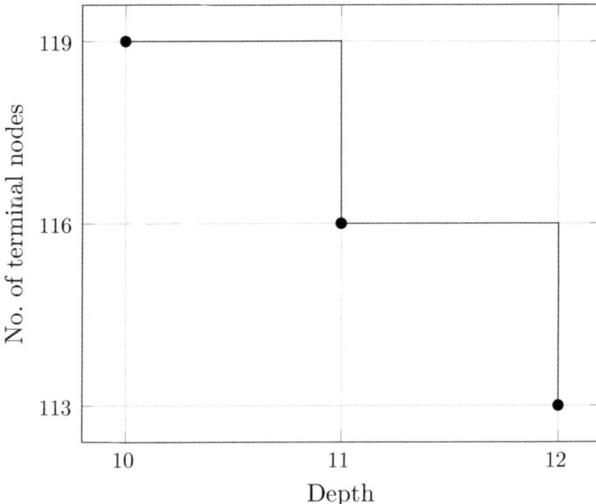

Fig. 7.13 Relationship for (h, L_t) and BOX- 50

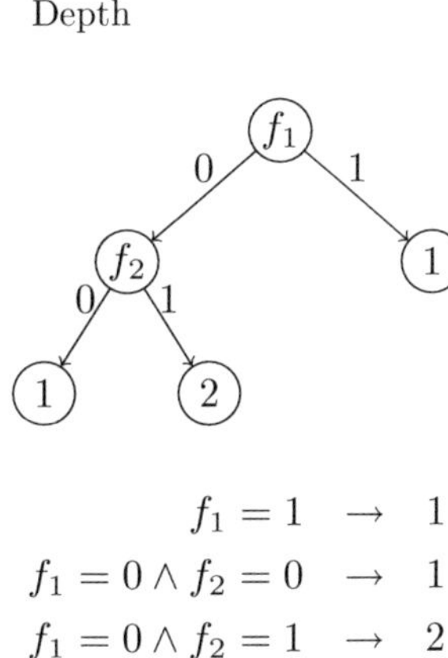

Fig. 7.14 System of decision rules for decision tree

7.5 Systems of Decision Rules Derived from Decision Trees

One of well known approaches to decision rule construction is to build a decision tree Γ and derive a decision rule from each path from the root to a terminal node of Γ [11–13] (see Fig. 7.14).

It is clear that the number of constructed rules is equal to the number $L_t(\Gamma)$ of terminal nodes in Γ, and the maximum length of the constructed rule is equal to the depth $h(\Gamma)$ of Γ. In this section, we study bi-criteria optimization of decision trees relative to h and L_t.

According to Tables 7.1 and 7.2, the best three impurity finctions for h are *s-rt*, *ws-gini*, *ws-ent*; for L_t are *ws-gini*, *ws-ent*, and *s-me*; and for (h, L_t) are *s-rt*, *ws-rt*, and *ws-me*. Figure 7.6 shows that the greedy algorithm based on *s-rt* should be used for bi-criteria optimization relative to h and L_t if $\alpha \leq 0.55$. If $\alpha > 0.55$, then we should use the greedy algorithm based on *ws-rt*.

Table 7.4 Number of Pareto optimal points for bi-criteria optimization problem for decision trees relative to (h, L_t)

Decision table	Rows	Attrs.	Alpha						
			0	0.01	0.05	0.2	0.4	0.6	0.8
ADULT-STRETCH	16	5	1	1	1	1	1	1	1
BALANCE-SCALE	8,124	22	1	1	1	1	1	1	1
BREAST-CANCER	266	10	1	1	1	1	4	3	1
CARS	1,728	6	1	1	1	1	1	1	1
FLAGS	194	26	4	4	4	2	3	3	1
HAYES-ROTH-DATA	69	4	1	1	1	1	1	1	1
HOUSE-VOTES-84	279	16	1	1	3	2	1	1	1
LENSES	11	4	1	1	1	1	1	1	1
LYMPHOGRAPHY	148	18	3	3	3	2	3	1	1
MONKS-1-TEST	432	6	1	1	1	1	1	1	1
MONKS-1-TRAIN	124	6	1	1	1	1	2	1	1
MONKS-2-TEST	432	6	1	1	1	1	1	1	1
MONKS-2-TRAIN	169	6	2	2	2	2	2	2	1
MONKS-3-TEST	432	6	1	1	1	1	1	1	1
MONKS-3-TRAIN	122	6	2	2	2	2	2	1	1
MUSHROOM	8,124	22	2	1	1	2	2	2	1
NURSERY	12,960	8	1	1	1	1	2	1	1
SHUTTLE-LAND.CONT.	15	6	1	1	1	1	1	1	1
SOYBEAN-SMALL	47	35	1	1	1	1	1	2	1
SPECT-TEST	169	22	1	1	1	1	1	2	1
TEETH	23	8	1	1	1	1	1	2	1
TIC-TAC-TOE	958	9	2	2	2	2	2	2	1
ZOO- DATA	59	16	1	1	1	1	1	1	1

The number of Pareto optimal points for bi-criteria optimization relative to h and L_t is small: see Table 7.4 which contains the number of Pareto optimal points from the set $Par(t_{h,L_t}(\Delta_{rme,\alpha}(T), T))$, where $T \in S$ (the set of 23 decision tables from UCI ML Repository) and $\alpha \in \{0, 0.01, 0.05, 0.2, 0.4, 0.6, 0.8\}$. The maximum number of Pareto optimal points in the considered examples is equal to four. Totally optimal decision trees relative to h and L_t exist in many cases (in all cases where the number of POPs is equal to one).

For $n = 5, \ldots, 13$, we randomly generate 100 partial Boolean functions with n variables and 0.7×2^n n-tuples of variable values for which functions are defined, represent these functions as decision tables and count, for each n, the average number of Pareto optimal points. Figure 7.15 shows that the average number of Pareto optimal points is at most two.

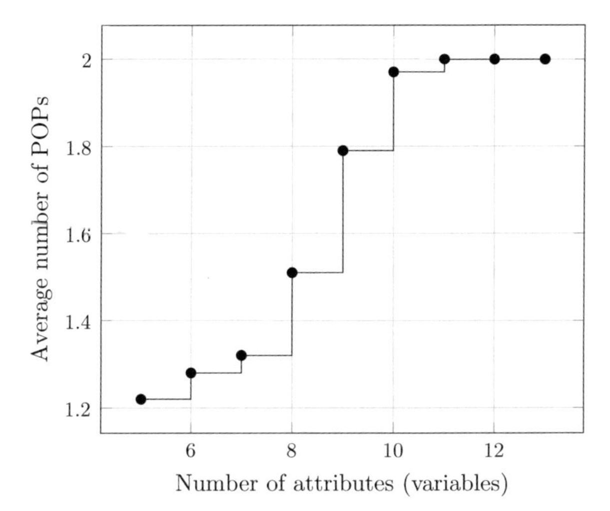

Fig. 7.15 Average number of Pareto optimal points for decision tables corresponding to partial Boolean functions

The obtained results show that the sets of POPs for (h, L_t) are not diverse. However, the construction of these sets can help in choosing the appropriate systems of rules. See, for example, Fig. 7.13 where the values for h are changing from 10 to 12 and the values for L_t are changing from 113 to 119.

References

1. Alkhalid, A., Amin, T., Chikalov, I., Hussain, S., Moshkov, M., Zielosko, B.: Dagger: A tool for analysis and optimization of decision trees and rules. In: Ficarra, F.V.C., Kratky, A., Veltman, K.H., Ficarra, M.C., Nicol, E., Brie, M. (eds.) Computational Informatics, Social Factors and New Information Technologies: Hypermedia Perspectives and Avant-Garde Experiencies in the Era of Communicability Expansion, pp. 29–39. Blue Herons (2011)
2. Alkhalid, A., Amin, T., Chikalov, I., Hussain, S., Moshkov, M., Zielosko, B.: Optimization and analysis of decision trees and rules: dynamic programming approach. Int. J. Gen. Syst. **42**(6), 614–634 (2013)
3. Alkhalid, A., Chikalov, I., Moshkov, M.: Comparison of greedy algorithms for decision tree construction. In: Filipe, J., Fred A.L.N. (eds.) International Conference on Knowledge Discovery and Information Retrieval, KDIR 2011, Paris, France, October 26–29, 2011, pp. 438–443. SciTePress (2011)
4. Alkhalid, A., Chikalov, I., Moshkov, M.: Constructing an optimal decision tree for FAST corner point detection. In: Yao, J., Ramanna, S., Wang, G., Suraj, Z. (eds.) Rough Sets and Knowledge Technology – 6th International Conference, RSKT 2011, Banff, Canada, October 9–12, 2011. Lecture Notes in Computer Science, vol. 6954, pp. 187–194. Springer, Berlin (2011)
5. Alkhalid, A., Chikalov, I., Moshkov, M.: Decision tree construction using greedy algorithms and dynamic programming–comparative study. In: Szczuka, M., Czaja, L., Skowron, A., Kacprzak, M. (eds.) 20th International Workshop on Concurrency, Specification and Programming, CS&P 2011, Pultusk, Poland, September 28–30, 2011, pp. 1–9. Białystok University of Technology (2011)
6. Chikalov, I., Hussain, S., Moshkov, M.: Bi-criteria optimization of decision trees with applications to data analysis. Eur. J. Oper. Res. **266**(2), 689–701 (2018)

7. Chikalov, I., Hussain, S., Moshkov, M.: Relationships between average depth and number of nodes for decision trees. In: Sun, F., Li, T., Li, H. (eds.) Knowledge Engineering and Management, 7th International Conference on Intelligent Systems and Knowledge Engineering, ISKE 2012, Beijing, China, December 15–17, 2012, Advances in Intelligent Systems and Computing, vol. 214, pp. 519–529. Springer, Berlin (2014)
8. Hussain, S.: Greedy heuristics for minimization of number of terminal nodes in decision trees. In: 2014 IEEE International Conference on Granular Computing, GrC 2014, Noboribetsu, Japan, October 22–24, 2014, pp. 112–115. IEEE Computer Society (2014)
9. Hussain, S.: Relationships among various parameters for decision tree optimization. In: Faucher, C., Jain, L.C. (eds.) Innovations in Intelligent Machines-4 - Recent Advances in Knowledge Engineering, Studies in Computational Intelligence, vol. 514, pp. 393–410. Springer, Berlin (2014)
10. Lichman, M.: UCI Machine Learning Repository. University of California, Irvine, School of Information and Computer Sciences (2013). http://archive.ics.uci.edu/ml
11. Michalski, R.S., Pietrzykowski, J.: iAQ: A program that discovers rules. In: 22nd AAAI Conference on Artificial Intelligence, AI Video Competition (2007). http://videolectures.net/aaai07_michalski_iaq/
12. Moshkov, M., Zielosko, B.: Combinatorial Machine Learning–A Rough Set Approach, Studies in Computational Intelligence, vol. 360. Springer, Berlin (2011)
13. Quinlan, J.R.: C4.5: Programs for Machine Learning. Morgan Kaufmann, San Francisco, CA (1993)
14. Rosten, E., Drummond, T.: Fusing points and lines for high performance tracking. In: 10th IEEE International Conference on Computer Vision, ICCV 2005, Beijing, China, October 17–20, 2005, pp. 1508–1515. IEEE Computer Society (2005)
15. Rosten, E., Drummond, T.: Machine learning for high-speed corner detection. In: Leonardis, A., Bischof, H., Pinz, A. (eds.) 9th European Conference on Computer Vision, ECCV 2006, Graz, Austria, May 7–13, 2006. Lecture Notes in Computer Science, vol. 3951, pp. 430–443. Springer, Berlin (2006)

Chapter 8
Bi-criteria Optimization Problem for Decision Trees: Cost Versus Uncertainty

In this chapter, we consider an algorithm which constructs the sets of Pareto optimal points for bi-criteria optimization problems for decision trees relative to cost and uncertainty (Sect. 8.1). We also show how the constructed set of Pareto optimal points can be transformed into the graphs of functions which describe the relationships between the considered cost function and uncertainty measure (Sect. 8.2). Some of the initial results were obtained in [3–5].

We end this chapter with illustrative examples of relationships between cost and uncertainty of decision trees (Sect. 8.3) and a discussion of a technique of multi-pruning of decision trees with applications from knowledge representation and classification [1] (Sect. 8.4). The classifiers constructed by multi-pruning process often have better accuracy than the classifiers constructed by CART [2].

8.1 Pareto Optimal Points: Cost Versus Uncertainty

Let ψ be an increasing cost function for decision trees given by the triple of functions ψ^0, F, w, U be an uncertainty measure, T be a decision table with n conditional attributes $f_1, \ldots, f_n$, and $H \in \{\max, \text{sum}\}$ where $\max = \max(x, y)$ and $\text{sum} = x + y$.

For each node Θ of the graph $\Delta(T)$, we denote

$$t_{U,H,\psi}(\Theta) = \{(U^H(\Theta, \Gamma), \psi(\Theta, \Gamma)) : \Gamma \in Tree^*(\Delta(T), \Theta)\}$$

where $Tree^*(\Delta(T), \Theta)$ is, by Proposition 4.2, the set of decision trees for Θ. We denote by $Par(t_{U,H,\psi}(\Theta))$ the set of Pareto optimal points for $t_{U,H,\psi}(\Theta)$.

H. AbouEisha et al., *Extensions of Dynamic Programming for Combinatorial Optimization and Data Mining*, Intelligent Systems Reference Library 146,
https://doi.org/10.1007/978-3-319-91839-6_8

We now describe an algorithm $\mathcal{A}_7$ which constructs the set $Par(t_{U,H,\psi}(T))$. In fact, this algorithm constructs, for each node Θ of the graph $\Delta(T)$, the set $B(\Theta) = Par(t_{U,H,\psi}(\Theta))$.

Algorithm $\mathcal{A}_7$ (construction of POPs for trees, cost versus uncertainty).

Input: Increasing cost function for decision trees ψ given by triple of functions ψ^0, F, w, an uncertainty measure U, a function $H \in \{\max, \text{sum}\}$, a decision table T with n conditional attributes $f_1, \ldots, f_n$, and the graph $\Delta(T)$.

Output: The set $Par(t_{U,H,\psi}(T))$ of Pareto optimal points for the set of pairs $t_{U,H,\psi}(\Theta) = \{(U^H(\Theta, \Gamma), \psi(\Theta, \Gamma)) : \Gamma \in Tree^*(\Delta(T), \Theta)\}$.

1. If all nodes in $\Delta(T)$ are processed, then return the set $B(T)$. Otherwise, choose a node Θ in the graph $\Delta(T)$ which is not processed yet and which is either a terminal node of $\Delta(T)$ or a nonterminal node of $\Delta(T)$ such that, for any $f_i \in E(\Theta)$ and any $a_j \in E(\Theta, f_i)$, the node $\Theta(f_i, a_j)$ is already processed, i.e., the set $B(\Theta(f_i, a_j))$ is already constructed.
2. If Θ is a terminal node, then set $B(\Theta) = \{(U(\Theta), \psi^0(\Theta))\}$. Mark the node Θ as processed and proceed to step 1.
3. If Θ is a nonterminal node then, for each $f_i \in E(\Theta)$, apply the algorithm $\mathcal{A}_2$ to the functions H, F and the sets $B(\Theta(f_i, a_1)), \ldots, B(\Theta(f_i, a_t))$, where $\{a_1, \ldots, a_t\} = E(\Theta, f_i)$. Set $C(\Theta, f_i)$ the output of the algorithm $\mathcal{A}_2$ and

$$\begin{aligned} B(\Theta, f_i) &= C(\Theta, f_i) \langle ++ \rangle \{(0, w(\Theta))\} \\ &= \{(a, b + w(\Theta)) : (a, b) \in C(\Theta, f_i)\} . \end{aligned}$$

4. Construct the multiset $A(\Theta) = \{(U(\Theta), \psi^0(\Theta))\} \cup \bigcup_{f_i \in E_G(\Theta)} B(\Theta, f_i)$ by simple transcription of elements from the sets $B(\Theta, f_i)$, $f_i \in E(\Theta)$. Apply to the obtained multiset $A(\Theta)$ the algorithm $\mathcal{A}_1$ which constructs the set $Par(A(\Theta))$. Set $B(\Theta) = Par(A(\Theta))$. Mark the node Θ as processed and proceed to step 1.

Proposition 8.1 *Let ψ be an increasing cost function for decision trees given by triple of functions ψ^0, F, w, U be an uncertainty measure, $H \in \{\max, \text{sum}\}$, and T be a decision table with n conditional attributes $f_1, \ldots, f_n$. Then, for each node Θ of the graph $\Delta(T)$, the algorithm $\mathcal{A}_7$ constructs the set $B(\Theta) = Par(t_{U,H,\psi}(\Theta))$.*

Proof We prove the considered statement by induction on nodes of $\Delta(T)$. Let Θ be a terminal node of $\Delta(T)$. Then $Tree^*(\Delta(T), \Theta) = \{tree(mcd(\Theta))\}$, $t_{U,H,\psi}(\Theta) = \{(U(\Theta), \psi^0(\Theta))\}$, and $B(\Theta) = Par(t_{U,H,\psi}(\Theta))$.

Let Θ be a nonterminal node of $\Delta(T)$ such that, for any $f_i \in E(\Theta)$ and any $a_j \in E(\Theta, f_i)$, the considered statement holds for the node $\Theta(f_i, a_j)$, i.e., $B(\Theta(f_i, a_j)) = Par(t_{U,H,\psi}(\Theta(f_i, a_j)))$.

Let $f_i \in E(\Theta)$ and $E(\Theta, f_i) = \{a_1, \ldots, a_t\}$. We denote

$$\begin{aligned} P(f_i) = \{(&H(b_1, \ldots, b_t), F(c_1, \ldots, c_t) + w(\Theta)) \\ &: (b_j, c_j) \in t_{U,H,\psi}(\Theta(f_i, a_j)), j = 1, \ldots, t\} \end{aligned}$$

and, for $j = 1, \ldots, t$, we denote $P_j = t_{U,H,\psi}(\Theta(f_i, a_j))$.

If we apply the algorithm $\mathscr{A}_2$ to the functions H, F and the sets $Par(P_1), \ldots, Par(P_t)$, we obtain the set $Par(Q_t)$ where $Q_1 = P_1$, and, for $j = 2, \ldots, t$, $Q_j = Q_{j-1} \langle HF \rangle P_j$. It is not difficult to show that $P(f_i) = Q_t \langle ++ \rangle \{(0, w(\Theta))\} = \{(a, b + w(\Theta)) : (a, b) \in Q_t\}$ and $Par(P(f_i)) = Par(Q_t) \langle ++ \rangle \{(0, w(\Theta))\}$.

According to the induction hypothesis, $B(\Theta(f_i, a_j)) = Par(P_j)$ for $j = 1, \ldots, t$. Therefore $C(\Theta, f_i) = Par(Q_t)$ and $B(\Theta, f_i) = Par(P(f_i))$.

One can show that $t_{U,H,\psi}(\Theta) = \{(U(\Theta), \psi^0(\Theta))\} \cup \bigcup_{f_i \in E(\Theta)} P(f_i)$. By Lemma 2.5,

$$
\begin{aligned}
Par(t_{U,H,\psi}(\Theta)) &= Par\left(\{(U(\Theta), \psi^0(\Theta))\} \cup \bigcup_{f_i \in E(\Theta)} P(f_i)\right) \\
&\subseteq \{(U(\Theta), \psi^0(\Theta))\} \cup \bigcup_{f_i \in E(\Theta)} Par(P(f_i)) \,.
\end{aligned}
$$

Using Lemma 2.4 we obtain

$$
Par(t_{U,H,\psi}(\Theta)) = Par\left(\{(U(\Theta), \psi^0(\Theta))\} \cup \bigcup_{f_i \in E(\Theta)} Par(P(f_i))\right) .
$$

Since $B(\Theta, f_i) = Par(P(f_i))$ for any $f_i \in E(\Theta)$, $Par(t_{U,H,\psi}(\Theta)) = Par\,(A(\Theta)) = B(\Theta)$. □

We now analyze the number of elementary operations made by the algorithm $\mathscr{A}_7$ during the construction of the set $Par(t_{U,H,\psi}(T))$ for integral cost functions, for the cases when $F(x, y) = \max(x, y)$ (h is an example of such cost function) and when $F(x, y) = x + y$ (tpl, L, L_n, and L_t are examples of such cost functions).

Let us recall that $range(T) = \max\{|E(T, f_i)| : i = 1, \ldots, n\}$,

$$
ub(\psi, T) = \max\{\psi(\Theta, \Gamma) : \Theta \in SEP(T), \Gamma \in DT(\Theta)\} ,
$$

and $\psi(\Theta, \Gamma) \in \{0, 1, \ldots, ub(\psi, T)\}$ for any separable subtable Θ of T and for any decision tree Γ for Θ. Upper bounds on the value $ub(\psi, T)$ for $\psi \in \{h, tpl, L, L_n, L_t\}$ are given in Lemma 4.2: $ub(h, T) \le n$, $ub(tpl, T) \le nN(T)$, $ub(L, T) \le 2N(T)$, $ub(L_n, T) \le N(T)$, and $ub(L_t, T) \le N(T)$.

Proposition 8.2 *Let ψ be an increasing integral cost function for decision trees given by triple of functions ψ^0, F, w, $F \in \{\max(x, y), x + y\}$, U be an uncertainty measure, $H \in \{\max, \text{sum}\}$, and T be a decision table with n conditional attributes $f_1, \ldots, f_n$. Then, to construct the set $Par(t_{U,H,\psi}(T))$, the algorithm $\mathscr{A}_7$ makes*

$$
O(L(\Delta(T))range(T)ub(\psi, T)^2 n \log(ub(\psi, T)n))
$$

elementary operations (computations of ψ^0, F, w, H, U, comparisons and additions) if $F = \max(x, y)$, and

$$O(L(\Delta(T))range(T)^2 ub(\psi, T)^2 n \log(range(T)ub(\psi, T)n))$$

elementary operations (computations of ψ^0, F, w, H, U, comparisons and additions) if $F = x + y$.

Proof It is clear that, for any node Θ of the graph $\Delta(T)$,

$$t_{U,H,\psi}(\Theta)^{(2)} = \{b : (a, b) \in t_{U,H,\psi}(\Theta)\} \subseteq \{0, \ldots, ub(\psi, T)\} .$$

From Proposition 8.1 it follows that $B(\Theta) = Par(t_{U,H,\psi}(\Theta))$ for any node Θ of the graph $\Delta(T)$. Therefore, for any node Θ of the graph $\Delta(T)$, $B(\Theta)^{(2)} = \{b : (a, b) \in B(\Theta)\} \subseteq \{0, \ldots, ub(\psi, T)\}$.

Let $F(x, y) = \max(x, y)$. To process a terminal node Θ, the algorithm $\mathcal{A}_7$ makes two elementary operations (computations of ψ^0 and U).

We now consider the processing of a nonterminal node Θ (see description of the algorithm $\mathcal{A}_7$). We know that $B(\Theta(f_i, a_j))^{(2)} \subseteq \{0, \ldots, ub(\psi, T)\}$ for $j = 1, \ldots, t$, and $t \le range(T)$. From Proposition 2.3 it follows that $|C(\Theta, f_i)| \le ub(\psi, T) + 1$, and to construct the set $C(\Theta, f_i)$, the algorithm $\mathcal{A}_2$ makes

$$O(range(T)ub(\psi, T)^2 \log ub(\psi, T))$$

elementary operations (computations of F, H and comparisons). To construct the set $B(\Theta, f_i)$ from the set $C(\Theta, f_i)$, the algorithm makes a computation of w and at most $ub(\psi, T) + 1$ additions. From here it follows that, to construct the set $B(\Theta, f_i)$, the algorithm makes

$$O(range(T)ub(\psi, T)^2 \log ub(\psi, T))$$

elementary operations. It is clear that $|E(\Theta)| \le n$. Therefore, to construct the set $B(\Theta, f_i)$ for each $f_i \in E(\Theta)$, the algorithm makes

$$O(range(T)ub(\psi, T)^2 n \log ub(\psi, T))$$

elementary operations.

Since $|C(\Theta, f_i)| \le ub(\psi, T) + 1$, $|B(\Theta, f_i)| \le ub(\psi, T) + 1$. Since $|E(\Theta)| \le n$, we have $|A(\Theta)| \le n(ub(\psi, T) + 1) + 1$ where

$$A(\Theta) = \{(U(\Theta), \psi^0(\Theta))\} \cup \bigcup_{f_i \in E(\Theta)} B(\Theta, f_i) .$$

From Proposition 2.1 it follows that, to construct the set $B(\Theta) = Par(A(\Theta))$, the algorithm $\mathcal{A}_1$ makes $O(ub(\psi, T)n \log(ub(\psi, T)n))$ comparisons. So, to process a nonterminal node Θ (to construct $B(\Theta) = Par(t_{U,H,\psi}(\Theta))$ if $B(\Theta(f_i, a_j)) =$

$Par(t_{U,H,\psi}(\Theta(f_i, a_j)))$ is known for all $f_i \in E(\Theta)$ and $a_j \in E(\Theta, f_i)$), the algorithm $\mathscr{A}_7$ makes

$$O(range(T)ub(\psi, T)^2 n \log(ub(\psi, T)n))$$

elementary operations.

To construct the set $Par(t_{U,H,\psi}(T))$ for given decision table T with n conditional attributes and the graph $\Delta(T)$, it is enough to make

$$O(L(\Delta(T))range(T)ub(\psi, T)^2 n \log(ub(\psi, T)n))$$

elementary operations (computations of ψ^0, F, w, H, U, comparisons and additions).

Let $F(x, y) = x + y$. To process a terminal node Θ, the algorithm $\mathscr{A}_7$ makes two elementary operations (computations of ψ^0 and U).

We now consider the processing of a nonterminal node Θ (see description of the algorithm $\mathscr{A}_7$). We know that $B(\Theta(f_i, a_j))^{(2)} \subseteq \{0, \dots, ub(\psi, T)\}$ for $j = 1, \dots, t$, and $t \leq range(T)$. From Proposition 2.3 it follows that $|C(\Theta, f_i)| \leq t \times ub(\psi, T) + 1 \leq range(T)ub(\psi, T) + 1$, and to construct the set $C(\Theta, f_i)$, the algorithm $\mathscr{A}_2$ makes

$$O(range(T)^2 ub(\psi, T)^2 \log(range(T)ub(\psi, T)))$$

elementary operations (computations of F, H and comparisons). To construct the set $B(\Theta, f_i)$ from the set $C(\Theta, f_i)$, the algorithm makes a computation of w and at most $range(T)ub(\psi, T) + 1$ additions. From here it follows that, to construct the set $B(\Theta, f_i)$, the algorithm makes

$$O(range(T)^2 ub(\psi, T)^2 \log(range(T)ub(\psi, T)))$$

elementary operations. It is clear that $|E(\Theta)| \leq n$. Therefore, to construct the set $B(\Theta, f_i)$ for each $f_i \in E(\Theta)$, the algorithm makes

$$O(range(T)^2 ub(\psi, T)^2 n \log(range(T)ub(\psi, T)))$$

elementary operations.

Since $|C(\Theta, f_i)| \leq range(T)ub(\psi, T) + 1$, we have

$$|B(\Theta, f_i)| \leq range(T)ub(\psi, T) + 1 \ .$$

Since $|E(\Theta)| \leq n$, we have $|A(\Theta)| \leq (range(T)ub(\psi, T) + 1)n + 1$ where $A(\Theta) = \{(U(\Theta), \psi^0(\Theta))\} \cup \bigcup_{f_i \in E(\Theta)} B(\Theta, f_i)$. From Proposition 2.1 it follows that, to construct the set $B(\Theta) = Par(A(\Theta))$, the algorithm $\mathscr{A}_1$ makes

$$O(range(T)ub(\psi, T)n \log(range(T)ub(\psi, T)n))$$

comparisons. So, to process a nonterminal node Θ (to construct

$$B(\Theta) = Par(t_{U,H,\psi}(\Theta))$$

if $B(\Theta(f_i, a_j)) = Par(t_{U,H,\psi}(\Theta(f_i, a_j)))$ is known for all $f_i \in E(\Theta)$ and $a_j \in E(\Theta, f_i)$), the algorithm $\mathscr{A}_7$ makes

$$O(range(T)^2 ub(\psi, T)^2 n \log(range(T) ub(\psi, T) n))$$

elementary operations.

To construct the set $Par(t_{U,H,\psi}(T))$ for given decision table T with n conditional attributes and the graph $\Delta(T)$, it is enough to make

$$O(L(\Delta(T)) range(T)^2 ub(\psi, T)^2 n \log(range(T) ub(\psi, T) n))$$

elementary operations (computations of ψ^0, F, w, H, U, comparisons and additions). □

Proposition 8.3 *Let ψ be a cost function for decision trees from the set $\{h, tpl, L, L_n, L_t\}$ given by triple of functions ψ^0, F, w, U be an uncertainty measure from the set $\{me, rme, ent, gini, R\}$, $H \in \{\max, \text{sum}\}$, and $\mathscr{U}$ be a restricted information system. Then the algorithm $\mathscr{A}_7$ has polynomial time complexity for decision tables from $\mathscr{T}(\mathscr{U})$ depending on the number of conditional attributes in these tables.*

Proof Since $\psi \in \{h, tpl, L, L_n, L_t\}$, ψ^0 is a constant, the function F is either $\max(x, y)$ or $x + y$, and the function w is either a constant or $N(T)$. From Proposition 8.2 and Lemma 4.2 it follows that, for the algorithm $\mathscr{A}_7$, the number of elementary operations (computations of ψ^0, F, w, H, U, comparisons and additions) is bounded from above by a polynomial depending on the size of input table T and on the number of separable subtables of T. All operations with numbers are basic ones. The computations of numerical parameters of decision tables used by the algorithm $\mathscr{A}_7$ (constants, $N(T)$, and $U(T)$) have polynomial time complexity depending on the size of decision tables.

According to Proposition 3.5, the algorithm $\mathscr{A}_7$ has polynomial time complexity for decision tables from $\mathscr{T}(\mathscr{U})$ depending on the number of conditional attributes in these tables. □

We now consider the problem of construction of the set of Pareto optimal points for the sets $t^{\max}_{U,\psi}(T) = \{(U^{\max}(T, \Gamma), \psi(T, \Gamma)) : \Gamma \in DT^{\max}_U(T)\}$ and $t^{\text{sum}}_{U,\psi}(T) = \{(U^{\text{sum}}(T, \Gamma), \psi(T, \Gamma)) : \Gamma \in DT^{\text{sum}}_U(T)\}$. Let ψ be bounded and increasing cost function for decision trees (in particular, h, tpl, L, L_n, and L_t are bounded and increasing cost functions). Then, by Lemma 2.2 and Proposition 4.13, $Par(t^{\max}_{U,\psi}(T)) = Par(t_{U,\max,\psi}(T))$. By Lemma 2.2 and Proposition 4.14,

$$Par(t^{\text{sum}}_{U,\psi}(T)) = Par(t_{U,\text{sum},\psi}(T)) .$$

So to construct $Par(t^{\max}_{U,\psi}(T)$ and $Par(t^{\text{sum}}_{U,\psi}(T))$ we can use the algorithm $\mathscr{A}_7$.

8.2 Relationships Between Cost and Uncertainty

Let ψ be bonded and increasing cost function for decision trees, U be an uncertainty measure, $H \in \{\max(x, y), \text{sum}(x, y)\}$, and T be a decision table.

To study relationships between cost function ψ and uncertainty U^H for decision trees on the set of decision trees $DT_U^H(T)$ we consider partial functions $\mathcal{T}_{T,H}^{U,\psi} : \mathbb{R} \to \mathbb{R}$ and $\mathcal{T}_{T,H}^{\psi,U} : \mathbb{R} \to \mathbb{R}$ defined as follows:

$$\mathcal{T}_{T,H}^{U,\psi}(x) = \min\{\psi(T, \Gamma) : \Gamma \in DT_U^H(T), U^H(T, \Gamma) \leq x\} ,$$
$$\mathcal{T}_{T,H}^{\psi,U}(x) = \min\{U^H(T, \Gamma) : \Gamma \in DT_U^H(T), \psi(T, \Gamma) \leq x\} .$$

Let $(a_1, b_1), \ldots, (a_k, b_k)$ be the normal representation of the set

$$Par(t_{U,\psi}^H(T)) = Par(t_{U,H,\psi}(T))$$

where $a_1 < \ldots < a_k$ and $b_1 > \ldots > b_k$. By Lemma 2.8 and Remark 2.2, for any $x \in \mathbb{R}$,

$$\mathcal{T}_{T,H}^{U,\psi}(x) = \begin{cases} \textit{undefined}, & x < a_1 \\ b_1, & a_1 \leq x < a_2 \\ \ldots & \ldots \\ b_{k-1}, & a_{k-1} \leq x < a_k \\ b_k, & a_k \leq x \end{cases} ,$$

$$\mathcal{T}_{T,H}^{\psi,U}(x) = \begin{cases} \textit{undefined}, & x < b_k \\ a_k, & b_k \leq x < b_{k-1} \\ \ldots & \ldots \\ a_2, & b_2 \leq x < b_1 \\ a_1, & b_1 \leq x \end{cases} .$$

8.3 Illustrative Examples

In this section, we consider examples which illustrate the use of tools created in the chapter. Let T be the decision table NURSERY from UCI ML Repository [6] containing 12,960 rows and 8 conditional attributes. We describe for this decision table how minimum depth, minimum average depth (recall that $h_{avg}(T, \Gamma) = tpl(T, \Gamma)/N(T)$), and minimum number of nodes of decision trees Γ for T depend on two parameters which can be interpreted as uncertainties of decision trees. The first parameter is *misclassification error rate* $me^{\text{sum}}(T, \Gamma)/N(T)$ which is equal to the number of misclassifications of Γ on T dividing by the number of rows in T. The second parameter is *local misclassification error rate* $rme^{\max}(T, \Gamma)$. For each terminal node v of Γ, we consider the number of misclassifications of Γ on rows of T for which the com-

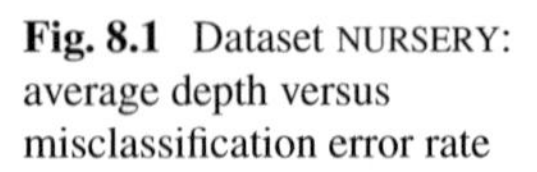

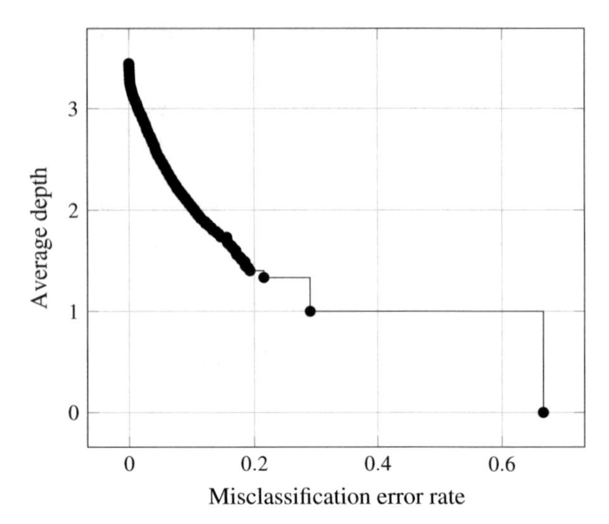

Fig. 8.1 Dataset NURSERY: average depth versus misclassification error rate

putation finishes in v divided by the number of rows of T for which the computation finishes in v, and consider maximum of this parameter among all terminal nodes of Γ.

We apply to the decision table T the algorithm $\mathscr{A}_3$ and construct the directed acyclic graph $\Delta(T)$ (see Remark 3.3). Using the algorithm $\mathscr{A}_7$ we construct the sets of Pareto optimal points for six bi-criteria optimization problems (h, me^{sum}), (h, rme^{max}), (tpl, me^{sum}), (tpl, rme^{max}),(L, me^{sum}), and (L, rme^{max}), and transform these sets to six graphs describing relationships between costs and uncertainties using technique considered in Sect. 8.2. After that, we transform some of these graphs to obtain results for average depth and misclassification error rate. The obtained graphs are depicted in Figs. 8.1, 8.2, 8.3, 8.4, 8.5, and 8.6.

Such graphs can be useful for the choice of decision trees which represent knowledge contained in the decision table: decision tree with enough small number of nodes and reasonable uncertainty can be considered as a good model of data. Similar situation is with the choice of decision trees that are considered as algorithms. In this case we are interested in finding of decision tree with small depth or average depth and reasonable uncertainty.

Similar approach also can be useful in machine learning (see Sect. 8.4).

8.4 Multi-pruning

We use bi-criteria optimization techniques (for cost versus uncertainty) similar to presented in previous sections to construct decision trees which can be used for classification and knowledge representation. This approach is applicable to decision tables with both categorical and numerical conditional attributes. We use a DAG

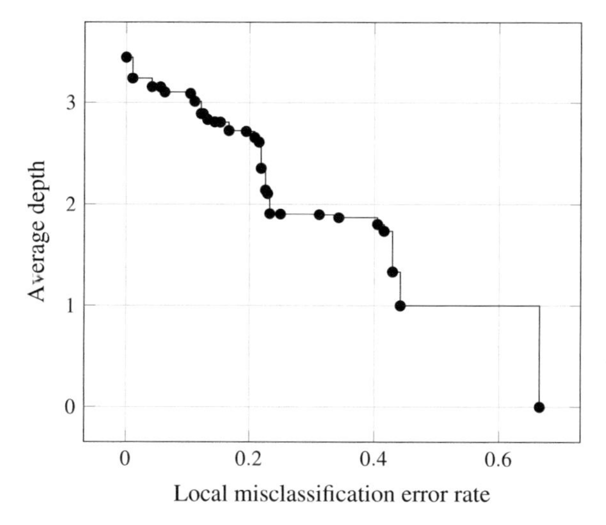

Fig. 8.2 Dataset NURSERY: average depth versus local misclassification error rate

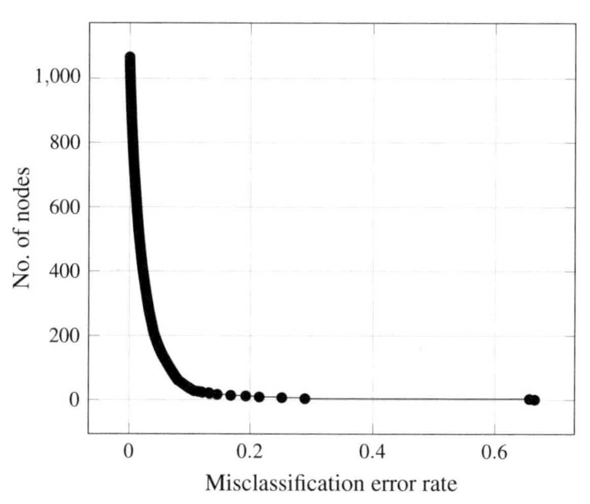

Fig. 8.3 Dataset NURSERY: no. of nodes versus misclassification error rate

to describe a large set of CART(Classification and Regression Trees)-like decision trees for the considered decision table. In particular, for the dataset BREAST- CANCER containing 266 rows and 10 attributes, the number of CART-like decision trees is equal to 1.42×10^{193}. Such trees use binary splits created on the base of conditional attributes. In contrast with standard CART [2] which use the best splits among all attributes, CART-like trees use, additionally, the best splits for each attribute. The set of CART-like decision trees is closed under the operation of usual bottom up pruning. So we call the considered approach *multi-pruning*.

We study two applications of this approach. The first application is connected with knowledge representation. We use the initial decision table to build the DAG and the set of Pareto optimal points. Then we choose a suitable Pareto optimal point

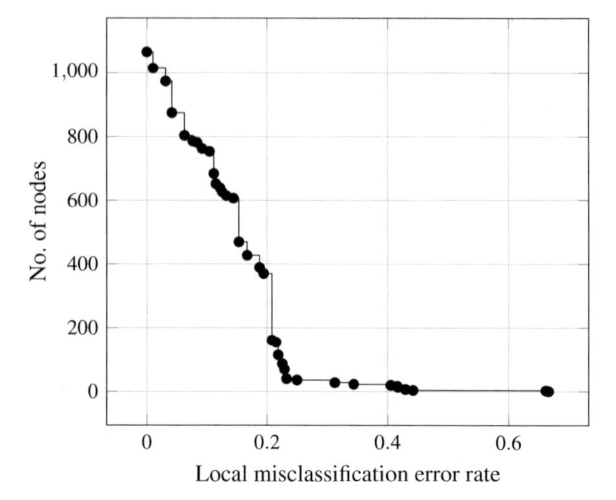

Fig. 8.4 Dataset NURSERY: no. of nodes versus local misclassification error rate

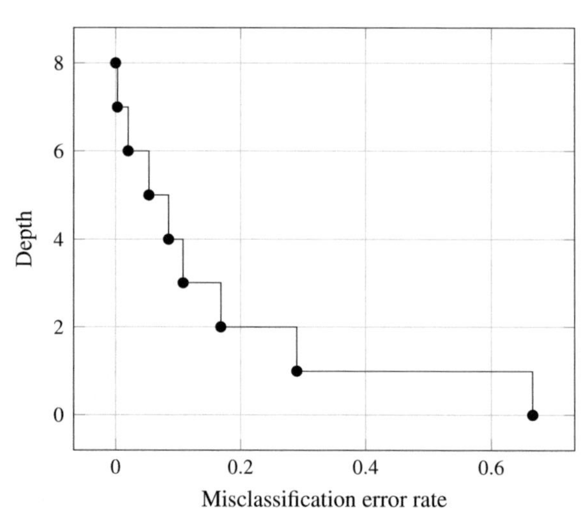

Fig. 8.5 Dataset NURSERY: depth versus misclassification error rate

and derive the corresponding decision tree. If this tree has relatively small number of nodes and relatively small number of misclassification, it can be considered as understandable and enough accurate model for the knowledge contained in the table. Results of experiments with decision tables from UCI ML Repository [6] show that we can construct such decision trees in many cases.

Another application is connected with machine learning. We divide the initial table into three subtables: training, validation, and testing. We use training subtable to build the DAG and the set of Pareto optimal points. We derive randomly a number of decision trees (five in our experiments) for each Pareto optimal point and find, based on the validation subtable, a decision tree with minimum number of misclassifications among all derived trees. We evaluate the accuracy of prediction for this tree using

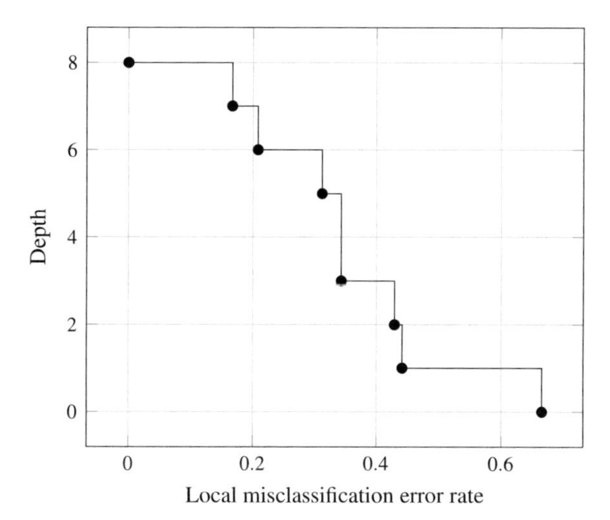

Fig. 8.6 Dataset NURSERY: depth versus local misclassification error rate

testing subtable. This process (*multi-pruning* process) is similar to the usual pruning of a decision tree but it is applied here to many decision trees since the set of CART-like decision trees is closed under the operation of usual bottom up pruning. We compare multi-pruning process with CART which continues to be one of the best algorithms for construction of classification trees. The classifiers constructed by multi-pruning process often have better accuracy than the classifiers constructed by CART.

Results presented in this section are published in [1].

8.4.1 CART-Like Decision Trees

In this section, we discuss the notion of CART-like decision trees. At each internal node of the decision tree we use an optimal split for some conditional attribute unlike CART which uses an optimal split among all the conditional attributes. As a result, we obtain a wide set of reasonable decision trees.

Let T be a decision table with n conditional attributes $f_1, \ldots, f_n$ which are categorical or numerical, and with a categorical decision attribute d. Instead of conditional attributes we use (as in CART) *binary splits* (binary attributes) each of which is based on a conditional attribute f_i. If f_i is a categorical attribute with the set of values B then we consider a partitioning of B into two nonempty subsets B_0 and B_1. The value of the corresponding binary split s is equal to 0 if the value of f_i belongs to B_0, and 1 otherwise. If f_i is a numerical attribute then we consider a real threshold α. The value of corresponding binary split s is equal to 0 if the value of f_i is less than α, and 1 otherwise. Each binary split s divides the table T into two subtables $T_{s=0}$ and $T_{s=1}$ according to the values of s on rows of T.

We use Gini index as an uncertainty measure for the decision tables. The impurity $I(T, s)$ of the split s is equal to the weighted sum of uncertainties of subtables $T_{s=0}$ and $T_{s=1}$, where the weights are proportional to the number of rows in subtables $T_{s=0}$ and $T_{s=1}$, respectively. The impurity of splits is considered as a quality measure for splits. A split s based on attribute f_i with minimum impurity $I(T, s)$ is called the *best split for T and f_i*.

We consider CART-like decision trees for T in which each leaf is labeled with a decision (a value of the decision attribute d), each internal node is labeled with a binary split corresponding to one of the conditional attributes, and two outgoing edges from this node are labeled with 0 and 1, respectively. We correspond to each node v of a decision tree Γ a subtable $T(\Gamma, v)$ of T that contains all rows of T for which the computation of Γ passes through the node v. We assume that, for each nonterminal node v, the subtable $T(\Gamma, v)$ contains rows labeled with different decisions, and the node v is labeled with a best split for $T(\Gamma, v)$ and an attribute f_i which is not constant on $T(\Gamma, v)$. We assume also that, for each terminal node v, the node v is labeled with a most common decision for $T(\Gamma, v)$ (a decision which is attached to the maximum number of rows in $T(\Gamma, v)$).

8.4.2 Directed Acyclic Graph for CART-Like Decision Trees

In this section, we study a directed acyclic graph $G(T)$ which is used to describe the set of CART-like decision trees for T and consequently using this set. Nodes of the graph $G(T)$ are some subtables of the table T. We now describe an algorithm for the construction of the directed acyclic graph $G(T)$ which can be considered as a definition of this DAG.

Algorithm $\mathscr{A}_{\mathrm{DAG}}$ (construction of DAG $G(T)$).

Input: A decision table T.
Output: The DAG $G(T)$.

1. Construct a graph which contains only one node T which is not marked as processed.
2. If all nodes of the graph are processed then return it as $G(T)$ and finish. Otherwise, choose a node (subtable) Θ which is not processed yet.
3. If all rows of Θ are labeled with the same decision mark Θ as processed and proceed to Step 2.
4. Otherwise, for each attribute f_i which is not constant in Θ, find a best split s for Θ and f_i, draw two edges from Θ to subtables $\Theta_{s=0}$ and $\Theta_{s=1}$, and label these edges with $s = 0$ and $s = 1$, respectively (this pair of edges is called an s -pair). If some of subtables $\Theta_{s=0}$ and $\Theta_{s=1}$ are not in the graph add them to the graph. Mark Θ as processed and proceed to Step 2.

One can show that the time complexity of this algorithm is bounded from above by a polynomial in the size of decision table T and the number of nodes in the graph $G(T)$.

We correspond to each node Θ of the graph $G(T)$ a set of decision trees $DT(\Theta)$. We denote by $tree(\Theta)$ the decision tree with exactly one node labeled with a most common decision for Θ. If all rows of Θ are labeled with the same decision then $DT(\Theta)$ contains only one tree $tree(\Theta)$. Otherwise, $DT(\Theta)$ contains the tree $tree(\Theta)$ and all trees of the following kind: the root of tree is labeled with a split s such that an s pair of edges starts in Θ, two edges start in the root which are labeled with 0 and 1, and enter to the roots of decision trees from $DT(\Theta_{s=0})$ and $DT(\Theta_{s=1})$, respectively. Note that the set $DT(T)$ is closed under the operation of usual bottom up pruning of decision trees. One can prove that, for each node Θ of the graph $G(T)$, the set of decision trees $DT(\Theta)$ coincides with the set of all CART-like decision trees for Θ.

8.4.3 Set of Pareto Optimal Points

Let Γ be a decision tree from $DT(T)$. We denote by $L(\Gamma)$ the number of nodes in Γ and by $mc(T, \Gamma)$ the number of misclassifications of Γ on rows of T, i.e., the number of rows of T for which the work of Γ ends in a terminal node that is labeled with a decision different from the decision attached to the row. It is clear that $mc(T, \Gamma) = me^{\text{sum}}(T, \Gamma)$. We correspond to each decision tree $\Gamma \in DT(T)$ the point $(mc(T, \Gamma), L(\Gamma))$. As a result, we obtain the set of points $\{(mc(T, \Gamma), L(\Gamma)) : \Gamma \in DT(T)\}$. Our aim is to construct for this set the set of all Pareto optimal points $POP(T) = Par(\{(mc(T, \Gamma), L(\Gamma)) : \Gamma \in DT(T)\})$.

We describe now an algorithm which attaches to each node Θ of the DAG $G(T)$ the set $POP(\Theta) = Par(\{(mc(\Theta, \Gamma), L(\Gamma)) : \Gamma \in DT(\Theta)\})$. This algorithm works in bottom-up fashion beginning with subtables in which all rows are labeled with the same decision.

Algorithm $\mathscr{A}_{\text{POPs}}$ (construction of the set $POP(T)$).

Input: A decision table T and the DAG $G(T)$.
Output: The set $POP(T)$.

1. If all nodes of $G(T)$ are processed then return the set $POP(T)$ attached to the node T and finish. Otherwise, choose a node Θ of $G(T)$ which is not processed yet and such that either all rows of Θ are labeled with the same decision or all children of Θ are already processed.
2. If all nodes of Θ are labeled with the same decision then attach to Θ the set $POP(\Theta) = \{(0, 1)\}$, mark Θ as processed, and proceed to Step 1.
3. If all children of Θ are already processed and $S(\Theta)$ is the set of splits s such that an s-pair of edges starts in Θ then attach to Θ the set

$$POP(\Theta) = Par(\{(mc(\Theta, tree(\Theta)), 1)\}$$
$$\cup \bigcup_{s \in S(\Theta)} \{(a + c, b + d + 1) : (a, b) \in POP(\Theta_{s=0}), (c, d) \in POP(\Theta_{s=1})\}) ,$$

mark Θ as processed, and proceed to Step 1.

Let Θ be a node of the graph $G(T)$ and $N(\Theta)$ be the number of rows in Θ. It is clear that all points from $POP(\Theta)$ have pairwise different first coordinates which belong to the set $\{0, 1, \ldots, N(\Theta)\}$. Therefore the cardinality of the set $POP(\Theta)$ is bounded from above by the number $N(\Theta) + 1$. From here it follows that the time complexity of the considered algorithm is bounded from above by a polynomial in the size of decision table T and the number of nodes in the graph $G(T)$.

For each Pareto optimal point from the set $POP(\Theta)$, we keep information about its construction: either it corresponds to the tree $tree(\Theta)$ or is obtained as a combination $(a + c, b + d + 1)$ of points $(a, b) \in POP(\Theta_{s=0})$ and $(c, d) \in POP(\Theta_{s=1})$ for some $s \in S(\Theta)$. In the last case, we keep all such combinations. This information allows us to derive, for each point (x, y) from $POP(T)$, decision trees from $DT(T)$, such that $(mc(T, \Gamma), L(\Gamma)) = (x, y)$ for each derived tree Γ.

8.4.4 Application to Knowledge Representation

In this section, we discuss an application of the considered approach to the problem of knowledge representation.

For a given decision table T, we can draw all points from $POP(T)$ as small filled circles (see Figs. 8.7, 8.8, 8.9, and 8.10). After that, we can choose a suitable Pareto

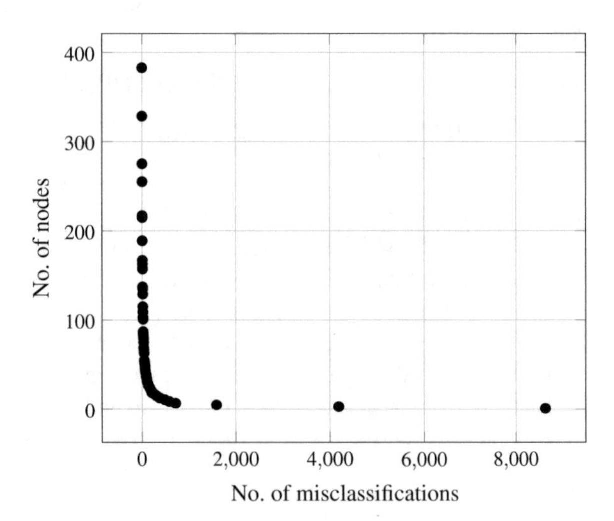

Fig. 8.7 Pareto optimal points for NURSERY dataset

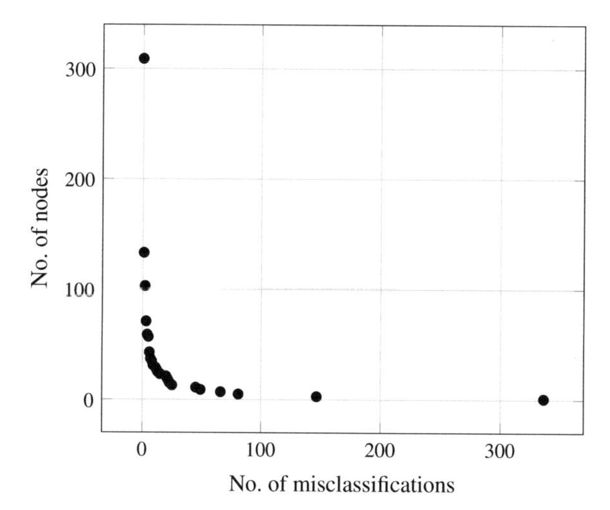

Fig. 8.8 Pareto optimal points for BALANCE- SCALE dataset

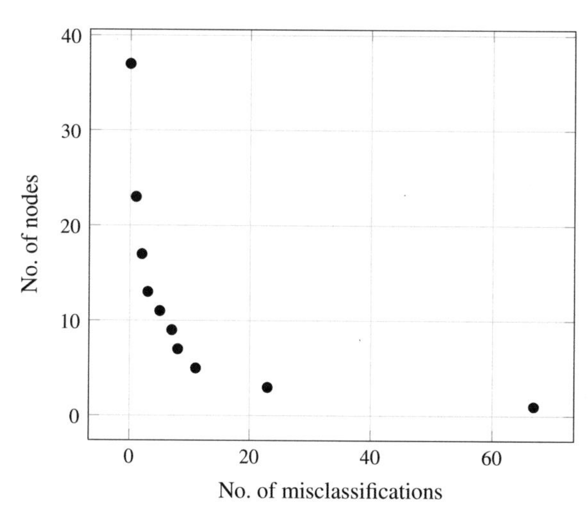

Fig. 8.9 Pareto optimal points for LYMPHOGRAPHY dataset

optimal point (a, b) and derive a decision tree Γ from $DT(\Gamma)$ such that $mc(T, \Gamma) = a$ and $L(\Gamma) = b$.

We did experiments for the first 11 of 17 decision tables from UCI ML Repository described in Table 8.1. For each decision table T, we find a Pareto optimal point and a corresponding CART-like decision tree for which the number of nodes is at most 19 (the choice for number of nodes is not arbitrary, we consider such a tree as more or less understandable; in particular, it would have at most 9 nonterminal nodes). For the considered trees, in the worst-case, the number of misclassifications is at most 12.5% of the number of misclassifications for the decision tree $tree(T)$.

Let us consider in details the decision table NURSERY which has 12,960 rows and 9 conditional attributes. For this table, the minimum number of nodes for an exact

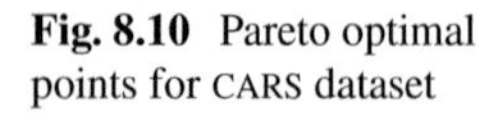

Fig. 8.10 Pareto optimal points for CARS dataset

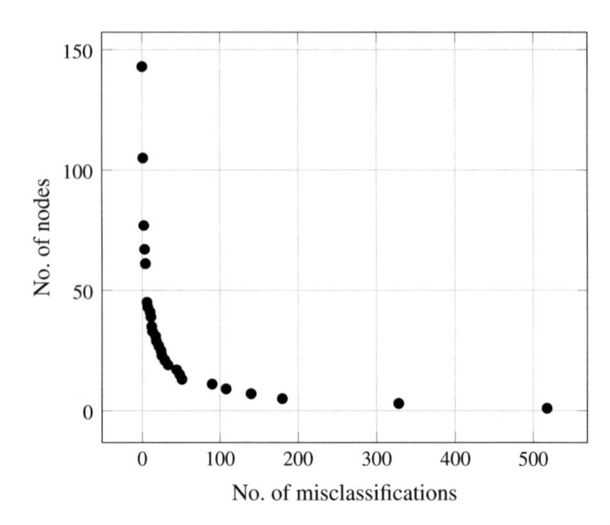

Table 8.1 Comparison of multi-pruning (MP) and CART

Dataset	Rows	Attr.	MP	CART
BALANCE- SCALE	625	5	23.26	23.75
BREAST- CANCER	266	10	29.02	29.77
CARS	1,728	7	4.69	5.23
HAYES- ROTH- DATA	69	5	24.33	34.44
HOUSE- VOTES- 84	279	17	6.88	6.60
LENSES	10	5	16.00	28.00
LYMPHOGRAPHY	148	19	25.14	27.70
NURSERY	12,960	9	1.44	1.38
SHUTTLE- LANDING	15	7	54.29	46.25
SOYBEAN- SMALL	47	36	6.74	18.75
SPECT- TEST	169	23	4.74	5.21
TIC- TAC- TOE	958	10	7.85	10.73
ZOO- DATA	59	17	21.36	22.01
BANKNOTE	1,372	5	2.05	3.38
IRIS	150	5	5.43	5.71
GLASS	214	10	38.31	39.82
WINE	178	13	8.99	11.80
Average:			16.50	18.85

decision tree (an exact decision tree is a tree without any misclassifications) is equal to 383. For the same dataset and the decision tree with one node labeled with a most common decision, the number of misclassifications is equal to 8,640. However, we can find a decision tree for this table such that the number of nodes is 19 and the number of misclassification is equal to 207 only which is less than 2.4% of 8,640 (see Fig. 8.7).

Figures 8.8, 8.9, and 8.10 contain three more examples for which there are Pareto optimal points that are not far from the origin.

8.4.5 Application to Classification

In this section, we consider an application of the considered approach to the problem of classification.

First, we describe the procedure of multi-pruning to choose a decision tree from the set of CART-like decision trees which is closed under the operation of the usual bottom up pruning. This tree will be used as a classifier.

We divide the initial decision table T into three subtables: training subtable T_{train}, validation subtable T_{val}, and testing subtable T_{test}. We construct the DAG $G(T_{\text{train}})$ and based on this DAG we construct the set of Pareto optimal points $POP(T_{\text{train}})$. For each point $(a, b) \in POP(T_{\text{train}})$, we derive randomly k decision trees (in our experiments $k = 5$) $\Gamma_1, \ldots, \Gamma_k$ from $DT(T_{\text{train}})$ such that

$$(a, b) = (mc(T_{\text{train}}, \Gamma_i), L(\Gamma_i))$$

for $i = 1, \ldots, k$. Among decision trees derived for all Pareto optimal points, we choose a decision tree Γ_0 which has minimum number of misclassifications on T_{val}. We evaluate the quality of this tree by counting the number of misclassifications on T_{test} and dividing it by the number of rows in T_{test}. As a result, we obtain the misclassification error rate of Γ_0.

We did experiments with 17 decision tables from UCI ML Repository and compared quality of classifiers (decision trees) constructed by CART and quality of classifiers constructed by multi-pruning procedure. We did some preprocessing on the datasets as some of them contained conditional attributes which were unique for the whole table. We removed such attributes. In some cases, there were identical rows, with possibly, different decisions. We replaced each group of such identical rows with a single row and the most common decision for the group of rows as the decision of this row. Similarly, missing values for an attribute were replaced with a most common value for that attribute.

We repeated 2-fold cross validation five times for each of the considered decision tables such that T_{train} contains 35% of rows, T_{val} contains 15% of rows, and T_{test} contains 50% of rows. After that we find average misclassification error rate of the constructed classifiers.

Results can be found in Table 8.1 which, for each decision table, shows the number of rows, number of conditional attributes, average misclassification error rate for multi-pruning classifiers, and average misclassification error rate for CART classifiers. One can see that, for three tables, CART classifiers outperform multi-pruning classifiers and, for the others 14 tables, multi-pruning based classifiers are better than CART classifiers.

To estimate significance of the comparison, we use (two-sided) Wilcoxon signed rank test [7], which is a non-parametric alternative to the paired Students t-test that does not make any assumption about the distribution of the measurements. The test confirmed that the difference is significant at confidence level 94.6%.

The obtained results show that multi-pruning approach (which is based on consideration of many decision trees, unlike the usual greedy approach) can be used in machine learning to improve predictive performance of individual trees. Ensembles based on such trees can have better accuracy of classification.

References

1. Azad, M., Chikalov, I., Hussain, S., Moshkov, M.: Multi-pruning of decision trees for knowledge representation and classification. In: 3rd IAPR Asian Conference on Pattern Recognition, ACPR 2015, Kuala Lumpur, Malaysia, November 3–6, 2015, pp. 604–608. IEEE (2015)
2. Breiman, L., Friedman, J.H., Olshen, R.A., Stone, C.J.: Classification and Regression Trees. Wadsworth and Brooks, Monterey (1984)
3. Chikalov, I., Hussain, S., Moshkov, M.: Average depth and number of misclassifications for decision trees. In: Popova-Zeugmann, L. (ed.) 21st International Workshop on Concurrency, Specification and Programming, CS&P 2012, Berlin, Germany, September 26–28, 2012, CEUR Workshop Proceedings, vol. 928, pp. 160–169. CEUR-WS.org (2012)
4. Chikalov, I., Hussain, S., Moshkov, M.: On cost and uncertainty of decision trees. In: Yao, J., Yang, Y., Slowinski, R., Greco, S., Li, H., Mitra, S., Polkowski L. (eds.) 8th International Conference Rough Sets and Current Trends in Computing, RSCTC 2012, Chengdu, China, August 17–20, 2012. Lecture Notes in Computer Science, vol. 7413, pp. 190–197. Springer (2012)
5. Chikalov, I., Hussain, S., Moshkov, M.: Relationships between number of nodes and number of misclassifications for decision trees. In: Yao, J., Yang, Y., Slowinski, R., Greco, S., Li, H., Mitra, S., Polkowski, L. (eds.) 8th International Conference Rough Sets and Current Trends in Computing, RSCTC 2012, Chengdu, China, August 17–20, 2012. Lecture Notes in Computer Science, vol. 7413, pp. 212–218. Springer (2012)
6. Lichman, M.: UCI Machine Learning Repository. University of California, Irvine, School of Information and Computer Sciences (2013). http://archive.ics.uci.edu/ml
7. Wilcoxon, F.: Individual comparisons by ranking methods. Biom. Bull. **1**(6), 80–83 (1945)

Part III
Decision Rules and Systems of Decision Rules

This part is devoted to the development of extensions of dynamic programming for the design and analysis of decision rules and systems of decision rules. In Chap. 9, we discuss main notions related to rules and systems of rules. In Chap. 10, we consider sequentially optimizing rules with regards to multiple criteria. In Chap. 11, we study the construction of the set of Pareto optimal points for two cost functions. Finally, in Chap. 12, we consider the construction of the set of Pareto optimal points for a cost function and an uncertainty measure.

Chapter 9
Different Kinds of Rules and Systems of Rules

A decision rule is a mapping from a set of conditions to an outcome. These rules are widely used in applications concerning data mining [9], knowledge representation and discovery, and machine learning [5]. Many problems in these areas can be solved with multiple approaches including systems of decision rules [6], decision trees [19], logical representations [16], k-nearest neighbor [8], neural networks [21], and support vector machines [7].

Although there are many comparable methods, decision rules tend to be the most meaningful and easily understood by humans [10]. This allows them to excel where knowledge representation is concerned. Even among decision rules some are better than others. For example, short decision rules are easy to remember and simple to understand while rules that cover many cases are more general and applicable.

There are many different ways to design and analyze decision rules, such as brute force, genetic algorithms [23], Boolean reasoning [18], derivation from decision trees [14, 20], sequential covering procedures [6, 10], greedy algorithms [15], and dynamic programming [2, 24]. Implementations of many of these methods can be found in programs such as LERS [12], RSES and Rseslib [3, 4] (Rseslib is available in Weka as official Weka package), RIONA [11], Rosetta [17], Weka [13], TRS Library [22], and DAGGER [1].

In this chapter, we provide definitions and basic concepts with regards to decision rules (Sect. 9.1) and systems of decision rules (Sect. 9.2). In the following chapters, we explore varied ways to constructing and analyzing rules and systems of rules based on extended dynamic programming approach.

9.1 Decision Rules

Let T be a decision table with n conditional attributes $f_1, \ldots, f_n$ and $r = (b_1, \ldots, b_n)$ be a row of T. A *decision rule over* T is an expression of the kind

H. AbouEisha et al., *Extensions of Dynamic Programming for Combinatorial Optimization and Data Mining*, Intelligent Systems Reference Library 146,
https://doi.org/10.1007/978-3-319-91839-6_9

$$f_{i_1} = a_1 \wedge \ldots \wedge f_{i_m} = a_m \rightarrow t \quad (9.1)$$

where $f_{i_1}, \ldots, f_{i_m} \in \{f_1, \ldots, f_n\}$, and $a_1, \ldots, a_m, t$ are numbers from ω. It is possible that $m = 0$. For the considered rule, we denote $T^0 = T$, and if $m > 0$ we denote $T^j = T(f_{i_1}, a_1) \ldots (f_{i_j}, a_j)$ for $j = 1, \ldots, m$. We will say that the decision rule (9.1) *covers* the row r if r belongs to T^m, i.e., $b_{i_1} = a_1, \ldots, b_{i_m} = a_m$.

A decision rule (9.1) over T is called a *decision rule for* T if $t = mcd(T^m)$, and either $m = 0$, or $m > 0$ and, for $j = 1, \ldots, m$, T^{j-1} is not degenerate, and $f_{i_j} \in E(T^{j-1})$. A decision rule (9.1) for T is called a *decision rule for* T *and* r if it covers r.

We denote by $DR(T)$ the set of decision rules for T. By $DR(T, r)$ we denote the set of decision rules for T and r.

Let U be an uncertainty measure and $\alpha \in \mathbb{R}_+$. A decision rule (9.1) for T is called a (U, α)*-decision rule for* T if $U(T^m) \leq \alpha$ and, if $m > 0$, then $U(T^j) > \alpha$ for $j = 0, \ldots, m - 1$. A (U, α)-decision rule (9.1) for T is called a (U, α)*-decision rule for* T *and* r if it covers r.

We denote by $DR_{U,\alpha}(T)$ the set of (U, α)-decision rules for T, and we denote by $DR_{U,\alpha}(T, r)$ the set of (U, α) -decision rules for T and r.

A decision rule (9.1) for T is called a U*-decision rule for* T if there exists a nonnegative real number α such that (9.1) is a (U, α)-decision rule for T. A decision rule (9.1) for T and r is called a U*-decision rule for* T *and* r if there exists a nonnegative real number α such that (9.1) is a (U, α)-decision rule for T and r.

We denote by $DR_U(T)$ the set of U-decision rules for T. By $DR_U(T, r)$ we denote the set of U-decision rules for T and r.

We define *uncertainty* $U(T, \rho)$ *of a decision rule* ρ *for* T *relative to the table* T in the following way. Let ρ be equal to (9.1). Then $U(T, \rho) = U(T^m)$.

We now consider a notion of *cost function for decision rules*. This is a function $\psi(T, \rho)$ which is defined on pairs T, ρ, where T is a nonempty decision table and ρ is a decision rule for T, and has values from the set $\mathbb{R}$ of real numbers. This cost function is given by pair of functions $\psi^0 : \mathscr{T}^+ \rightarrow \mathbb{R}$ and $F : \mathbb{R} \rightarrow \mathbb{R}$ where $\mathscr{T}^+$ is the set of nonempty decision tables. The value $\psi(T, \rho)$ is defined by induction:

- If ρ is equal to $\rightarrow mcd(T)$ then $\psi(T, \rightarrow mcd(T)) = \psi^0(T)$.
- If ρ is equal to $f_i = a \wedge \beta \rightarrow t$ then

$$\psi(T, f_i = a \wedge \beta \rightarrow t) = F(\psi(T(f_i, a), \beta \rightarrow t)) .$$

The cost function ψ is called *strictly increasing* if $F(x_1) > F(x_2)$ for any $x_1, x_2 \in \mathbb{R}$ such that $x_1 > x_2$.

Let us consider examples of strictly increasing cost functions for decision rules:

- The *length* $l(T, \rho) = l(\rho)$ for which $\psi^0(T) = 0$ and $F(x) = x + 1$. The length of the rule (9.1) is equal to m.
- The *coverage* $c(T, \rho)$ for which $\psi^0(T) = N_{mcd(T)}(T)$ and $F(x) = x$. The coverage of the rule (9.1) for table T is equal to $N_{mcd(T^m)}(T^m)$.

- The *relative coverage* $rc(T, \rho)$ for which $\psi^0(T) = N_{mcd(T)}(T)/N(T)$ and $F(x) = x$. The relative coverage of the rule (9.1) for table T is equal to

$$N_{mcd(T^m)}(T^m)/N(T^m) .$$

- The *modified coverage* $c_M(T, \rho)$ for which $\psi^0(T) = N^M(T)$ and $F(x) = x$. Here M is a set of rows of T and, for any subtable Θ of T, $N^M(\Theta)$ is the number of rows of Θ which do not belong to M. The modified coverage of the rule (9.1) for table T is equal to $N^M(T^m)$.
- The *miscoverage* $mc(T, \rho)$ for which $\psi^0(T) = N(T) - N_{mcd(T)}(T)$ and $F(x) = x$. The miscoverage of the rule (9.1) for table T is equal to $N(T^m) - N_{mcd(T^m)}(T^m)$.
- The *relative miscoverage* $rmc(T, \rho)$ for which

$$\psi^0(T) = (N(T) - N_{mcd(T)}(T))/N(T)$$

and $F(x) = x$. The relative miscoverage of the rule (9.1) for table T is equal to $(N(T^m) - N_{mcd(T^m)}(T^m))/N(T^m)$.

We need to minimize length, miscoverage and relative miscoverage, and maximize coverage, relative coverage, and modified coverage. However, we will consider only algorithms for the minimization of cost functions. Therefore, instead of maximization of coverage c we will minimize the negation of coverage $-c$ given by pair of functions $\psi^0(T) = -N_{mcd(T)}(T)$ and $F(x) = x$. Similarly, instead of maximization of relative coverage rc we will minimize the negation of relative coverage $-rc$ given by pair of functions $\psi^0(T) = -N_{mcd(T)}(T)/N(T)$ and $F(x) = x$. Instead of maximization of modified coverage c_M we will minimize the negation of modified coverage $-c_M$ given by pair of functions $\psi^0(T) = -N^M(T)$ and $F(x) = x$. The cost functions $-c$, $-rc$, and $-c_M$ are strictly increasing cost functions.

For a given cost function ψ and decision table T, we denote $Range_\psi(T) = \{\psi(\Theta, \rho) : \Theta \in SEP(T), \rho \in DR(\Theta)\}$. By $q_\psi(T)$ we denote the cardinality of the set $Range_\psi(T)$. It is easy to prove the following statement:

Lemma 9.1 *Let T be a decision table with n conditional attributes. Then*

- $Range_l(T) \subseteq \{0, 1, \ldots, n\}$, $q_l(T) \leq n + 1$.
- $Range_{-c}(T) \subseteq \{0, -1, \ldots, -N(T)\}$, $q_{-c}(T) \leq N(T) + 1$.
- $Range_{mc}(T) \subseteq \{0, 1, \ldots, N(T)\}$, $q_{mc}(T) \leq N(T) + 1$.
- $Range_{-rc}(T) \subseteq \{-a/b : a, b \in \{0, 1, \ldots, N(T)\}, b > 0\}$, $q_{-rc}(T) \leq N(T)(N(T) + 1)$.
- $Range_{-c_M}(T) \subseteq \{0, -1, \ldots, -N(T)\}$, $q_{-c_M}(T) \leq N(T) + 1$.
- $Range_{rmc}(T) \subseteq \{a/b : a, b \in \{0, 1, \ldots, N(T)\}, b > 0\}$, $q_{rmc}(T) \leq N(T)(N(T) + 1)$.

9.2 Systems of Decision Rules

Let T be a nonempty decision table with n conditional attributes $f_1, \ldots, f_n$ and $N(T)$ rows $r_1, \ldots, r_{N(T)}$, and U be an uncertainty measure.

A *system of decision rules for* T is an $N(T)$-tuple $S = (\rho_1, \ldots, \rho_{N(T)})$ where $\rho_1 \in DR(T, r_1), \ldots, \rho_{N(T)} \in DR(T, r_{N(T)})$. Let $\alpha \in \mathbb{R}_+$. The considered system is called a (U, α)-*system of decision rules for* T if $\rho_i \in DR_{U,\alpha}(T, r_i)$ for $i = 1, \ldots, N(T)$. This system is called a U-*system of decision rules for* T if $\rho_i \in DR_U(T, r_i)$ for $i = 1, \ldots, N(T)$.

We now consider a notion of *cost function for systems of decision rules*. This is a function $\psi_f(T, S)$ which is defined on pairs T, S, where T is a nonempty decision table and $S = (\rho_1, \ldots, \rho_{N(T)})$ is a system of decision rules for T, and has values from the set $\mathbb{R}$. This function is given by cost function for decision rules ψ and a function $f : \mathbb{R}^2 \to \mathbb{R}$. The value $\psi_f(T, S)$ is equal to $f(\psi(T, \rho_1), \ldots, \psi(T, \rho_{N(T)}))$ where the value $f(x_1, \ldots, x_k)$, for any natural k, is defined by induction: $f(x_1) = x_1$ and, for $k > 2$, $f(x_1, \ldots, x_k) = f(f(x_1, \ldots, x_{k-1}), x_k)$.

The cost function for systems of decision rules ψ_f is called *strictly increasing* if ψ is strictly increasing cost function for decision rules and f is an increasing function from $\mathbb{R}^2$ to $\mathbb{R}$, i.e., $f(x_1, y_1) \leq f(x_2, y_2)$ for any $x_1, x_2, y_1, y_2 \in \mathbb{R}$ such that $x_1 \leq x_2$ and $y_1 \leq y_2$.

For example, if $\psi \in \{l, -c, -rc, -c_M, mc, rmc\}$ and

$$f \in \{\text{sum}(x, y) = x + y, \max(x, y)\}$$

then ψ_f is a strictly increasing cost function for systems of decision rules.

We now consider a notion of *uncertainty for systems of decision rules*. This is a function $U_g(T, S)$ which is defined on pairs T, S, where T is a nonempty decision table and $S = (\rho_1, \ldots, \rho_{N(T)})$ is a system of decision rules for T, and has values from the set $\mathbb{R}$. This function is given by uncertainty measure U and a function $g : \mathbb{R}^2 \to \mathbb{R}$. The value $U_g(T, S)$ is equal to $g(U(T, \rho_1), \ldots, U(T, \rho_{N(T)}))$ where the value $g(x_1, \ldots, x_k)$, for any natural k, is defined by induction: $g(x_1) = x_1$ and, for $k > 2$, $g(x_1, \ldots, x_k) = g(g(x_1, \ldots, x_{k-1}), x_k)$.

References

1. Alkhalid, A., Amin, T., Chikalov, I., Hussain, S., Moshkov, M., Zielosko, B.: Dagger: A tool for analysis and optimization of decision trees and rules. In: Ficarra, F.V.C., Kratky, A., Veltman, K.H., Ficarra, M.C., Nicol, E., Brie, M. (eds.) Computational Informatics, Social Factors and New Information Technologies: Hypermedia Perspectives and Avant-Garde Experiencies in the Era of Communicability Expansion, pp. 29–39. Blue Herons (2011)
2. Amin, T., Chikalov, I., Moshkov, M., Zielosko, B.: Dynamic programming approach for exact decision rule optimization. In: Skowron, A., Suraj, Z. (eds.) Rough Sets and Intelligent Systems – Professor Zdzisław Pawlak in Memoriam – Volume 1. Intelligent Systems Reference Library, vol. 42, pp. 211–228. Springer, Berlin (2013)

3. Bazan, J.G., Szczuka, M.S.: RSES and RSESlib - a collection of tools for rough set computations. In: Ziarko, W., Yao, Y.Y. (eds.) Rough Sets and Current Trends in Computing, Second International Conference, RSCTC 2000, Banff, Canada, October 16–19, 2000, Revised Papers. Lecture Notes in Computer Science, vol. 2005, pp. 106–113. Springer, Berlin (2001)
4. Bazan, J.G., Szczuka, M.S., Wojna, A., Wojnarski, M.: On the evolution of rough set exploration system. In: Tsumoto, S., Słowinski, R., Komorowski, H.J., Grzymała-Busse, J.W. (eds.) Rough Sets and Current Trends in Computing - 4th International Conference, RSCTC 2004, Uppsala, Sweden, June 1–5, 2004. Lecture Notes in Computer Science, vol. 3066, pp. 592–601. Springer, Berlin (2004)
5. Carbonell, J.G., Michalski, R.S., Mitchell, T.M.: An overview of machine learning. In: Michalski, R.S., Carbonell, J.G., Mitchell, T.M. (eds.) Machine Learning, An Artificial Intelligence Approach, pp. 1–23. Tioga Publishing, Palo Alto (1983)
6. Clark, P., Niblett, T.: The CN2 induction algoirthm. Mach. Learn. **3**(4), 261–283 (1989)
7. Cortes, C., Vapnik, V.: Support-vector networks. Mach. Learn. **20**(3), 273–297 (1995)
8. Dasarathy, B.V. (ed.): Nearest Neighbor(NN) Norms: NN Pattern Classification Techniques. IEEE Computer Society Press, Los Alamitos (1990)
9. Fayyad, U., Piatetsky-Shapiro, G., Smyth, P.: From data mining to knowledge discovery in databases. AI Mag. **17**, 37–54 (1996)
10. Fürnkranz, J.: Separate-and-conquer rule learning. Artif. Intell. Rev. **13**(1), 3–54 (1999)
11. Góra, G., Wojna, A.: RIONA: A new classification system combining rule induction and instance-based learning. Fundam. Inform. **51**(4), 369–390 (2002)
12. Grzymała-Busse, J.W.: LERS – a system for learning from examples based on rough sets. In: Słowiński, R. (ed.) Intelligent Decision Support. Handbook of Applications and Advances of the Rough Sets Theory, pp. 3–18. Kluwer Academic Publishers, Dordrecht (1992)
13. Hall, M., Frank, E., Holmes, G., Pfahringer, B., Reutemann, P., Witten, I.H.: The WEKA data mining software: an update. SIGKDD Explor. **11**(1), 10–18 (2009)
14. Moshkov, M., Zielosko, B.: Combinatorial Machine Learning - A Rough Set Approach, Studies in Computational Intelligence, vol. 360. Springer, Heidelberg (2011)
15. Moshkov, M., Piliszczuk, M., Zielosko, B.: Partial Covers, Reducts and Decision Rules in Rough Sets - Theory and Applications, Studies in Computational Intelligence, vol. 145. Springer, Heidelberg (2008)
16. Muggleton, S.: Learning stochastic logic programs. Electron. Trans. Artif. Intell. **4**(B), 141–153 (2000)
17. Øhrn, A., Komorowski, J., Skowron, A., Synak, P.: The design and implementation of a knowledge discovery toolkit based on rough sets: The ROSETTA system. In: Polkowski, L., Skowron, A. (eds.) Rough Sets in Knowledge Discovery 1: Methodology and Applications. Studies in Fuzziness and Soft Computing, vol. 18, pp. 376–399. Physica-Verlag (1998)
18. Pawlak, Z., Skowron, A.: Rough sets and boolean reasoning. Inf. Sci. **177**(1), 41–73 (2007)
19. Quinlan, J.R.: Induction of decision trees. Mach. Learn. **1**(1), 81–106 (1986)
20. Quinlan, J.R.: Simplifying decision trees. Int. J. Man. Mach. Stud. **27**(3), 221–234 (1987)
21. Rumelhart, D.E., McClelland, J.L., CORPORATE PDP Research Group (ed.): Parallel Distributed Processing: Explorations in the Microstructure of Cognition, Vol. 1: Foundations. MIT Press, Cambridge (1986)
22. Sikora, M.: Decision rule-based data models using TRS and NetTRS - methods and algorithms. In: Peters, J.F., Skowron, A. (eds.) Trans. Rough Sets XI. Lecture Notes in Computer Science, vol. 5946, pp. 130–160. Springer, Berlin (2010)
23. Ślęzak, D., Wróblewski, J.: Order based genetic algorithms for the search of approximate entropy reducts. In: Wang, G., Liu, Q., Yao, Y., Skowron, A. (eds.) Rough Sets, Fuzzy Sets, Data Mining, and Granular Computing – 9th International Conference, RSFDGrC 2003, Chongqing, China, May 26–29, 2003. Lecture Notes in Computer Science, vol. 2639, pp. 308–311. Springer, Berlin (2003)
24. Zielosko, B., Chikalov, I., Moshkov, M., Amin, T.: Optimization of decision rules based on dynamic programming approach. In: Faucher, C., Jain, L.C. (eds.) Innovations in Intelligent Machines-4 – Recent Advances in Knowledge Engineering. Studies in Computational Intelligence, vol. 514, pp. 369–392. Springer, Berlin (2014)

Chapter 10
Multi-stage Optimization of Decision Rules

Decision rules can be characterized by many different parameters, such as the cost functions described in Sect. 9.1. Unfortunately, it is often very difficult to find decision rules with optimal cost. For example, finding an exact decision rule for a given row that has minimum length is an NP-hard problem (see, for example, [6]). The same situation applies to finding an exact decision rule for a given row that has maximum coverage [4].

In this chapter, we present a dynamic programming algorithm that can perform multi-stage optimization relative to multiple cost functions. The approach used by this algorithm is based on the construction of a DAG $\Delta_{U,\alpha}(T)$ as per Algorithm $\mathscr{A}_3$ described in Sect. 3.3. The DAG is reduced to a subgraph by eliminating edges corresponding to suboptimal paths for a particular cost function. Eventually we are left with a subgraph from which we can extract the set of all (U, α)-decision rules that are optimal with respect to our chosen cost. Furthermore, we can reapply the algorithm to the resultant subgraph in order to apply another stage of optimization, this time with respect to a different cost function.

When we apply multi-stage optimization with respect to length and coverage, we can observe an interesting phenomenon for certain decision tables: the existence of decision rules that have the minimum length and the maximum coverage simultaneously (totally optimal decision rules relative to the length and coverage). In such cases, the order in which we apply each stage of optimization does not affect the resulting subgraph or the set of rules derived from it. This situation occurs for a significant number of decision tables, but also does not occur in many others [1]. We explore this phenomenon in detail and even describe some situations where it is guaranteed.

This chapter also contains a simulation of a greedy algorithm for constructing relatively small sets of (U, α)-decision rules. This algorithm is similar to the greedy algorithm for construction of partial covers for the set cover problem [8, 9]. Furthermore, we adapt a nontrivial lower bound on the minimum cardinality for partial covers considered in [6] to our algorithm.

H. AbouEisha et al., *Extensions of Dynamic Programming for Combinatorial Optimization and Data Mining*, Intelligent Systems Reference Library 146,
https://doi.org/10.1007/978-3-319-91839-6_10

Section 10.1 describes how to represent the set of all (U, α)-decision rules, and Sect. 10.2 shows how to apply stages of optimization to this set. In Sect. 10.3 we present an algorithm for counting the number of decision rules. Following that, Sect. 10.4 describes a simulation of greedy algorithm for constructing relatively small sets of (U, α)-decision rules and presents a lower bound on the minimum cardinality of such sets based on the work of the algorithm, and Sect. 10.5 describes experimental results for simulation of greedy algorithm. Section 10.6 shows our work concerning totally optimal decision rules. The work in this section is similar to the paper [2]. Finally, Sect. 10.7 compares minimum depth of deterministic and nondeterministic decision trees when dealing with total Boolean functions.

10.1 Representation of the Set of (U, α)-Decision Rules

Let T be a nonempty decision table with n conditional attributes $f_1, \ldots, f_n$, U be an uncertainty measure, $\alpha \in \mathbb{R}_+$, and G be a proper subgraph of $\Delta_{U,\alpha}(T)$ (it is possible that $G = \Delta_{U,\alpha}(T)$).

Let τ be a directed path from a node Θ of G to a terminal node Θ' in which edges (in the order from Θ to Θ') are labeled with pairs $(f_{i_1}, c_{i_1}), \ldots, (f_{i_m}, c_{i_m})$, and $t = mcd(\Theta')$. We denote by $rule(\tau)$ the decision rule over T

$$f_{i_1} = c_{i_1} \wedge \ldots \wedge f_{i_m} = c_{i_m} \rightarrow t \ .$$

If $m = 0$ (if $\Theta = \Theta'$) then the rule $rule(\tau)$ is equal to $\rightarrow t$.

Let $r = (b_1, \ldots, b_n)$ be a row of T, and Θ be a node of G (subtable of T) containing the row r. We denote by $Rule(G, \Theta, r)$ the set of rules $rule(\tau)$ corresponding to all directed paths τ from Θ to terminal nodes Θ' containing r.

Proposition 10.1 *Let T be a nonempty decision table with n conditional attributes $f_1, \ldots, f_n$, $r = (b_1, \ldots, b_n)$ be a row of T, U be an uncertainty measure, $\alpha \in \mathbb{R}_+$, and Θ be a node of the graph $\Delta_{U,\alpha}(T)$ containing r. Then the set $Rule(\Delta_{U,\alpha}(T), \Theta, r)$ coincides with the set of all (U, α)-decision rules for Θ and r, i.e.,*

$$Rule(\Delta_{U,\alpha}(T), \Theta, r) = DR_{U,\alpha}(\Theta, r) \ .$$

Proof From the definition of the graph $\Delta_{U,\alpha}(T)$ it follows that each rule from $Rule(\Delta_{U,\alpha}, \Theta, r)$ is a (U, α) -decision rule for Θ and r.

Let us consider an arbitrary (U, α)-decision rule ρ for Θ and r:

$$f_{i_1} = b_{i_1} \wedge \ldots \wedge f_{i_m} = b_{i_m} \rightarrow t.$$

It is easy to show that there is a directed path $\Theta_0 = \Theta, \Theta_1, \ldots, \Theta_m$ in $\Delta_{U,\alpha}(T)$ such that, for $j = 1, \ldots, m$, $\Theta_j = \Theta(f_{i_1}, b_{i_1}) \ldots (f_{i_j}, b_{i_j})$, there is an edge from Θ_{j-1} to Θ_j labeled with (f_{i_j}, b_{i_j}), and Θ_m is a terminal node in $\Delta_{U,\alpha}(T)$. Therefore $\rho \in Rule(\Delta_{U,\alpha}, \Theta, r)$. □

10.2 Procedure of Optimization

We describe now a procedure of optimization (minimization of cost) of rules for row $r = (b_1, \ldots, b_n)$ relative to a strictly increasing cost function ψ given by pair of functions ψ^0 and F. We will move from terminal nodes of the graph G to the node T. We will attach to each node Θ of the graph G containing r the minimum cost $c(\Theta, r)$ of a rule from $Rule(G, \Theta, r)$ and, probably, we will remove some bundles of edges starting in nonterminal nodes. As a result we obtain a proper subgraph $G^\psi = G^\psi(r)$ of the graph G.

Algorithm $\mathscr{A}_8$ (procedure of decision rule optimization).

Input: A proper subgraph G of the graph $\Delta_{U,\alpha}(T)$ for some decision table T with n conditional attributes $f_1, \ldots, f_n$, uncertainty measure U, a number $\alpha \in \mathbb{R}_+$, a row $r = (b_1, \ldots, b_n)$ of T, and a strictly increasing cost function ψ for decision rules given by pair of functions ψ^0 and F.
Output: The proper subgraph $G^\psi = G^\psi(r)$ of the graph G.

1. If all nodes of the graph G containing r are processed then return the obtained graph as G^ψ and finish the work of the algorithm. Otherwise, choose a node Θ of the graph G containing r which is not processed yet and which is either a terminal node of G or a nonterminal node of G for which all children containing r are processed.
2. If Θ is a terminal node then set $c(\Theta, r) = \psi^0(\Theta)$, mark node Θ as processed and proceed to step 1.
3. If Θ is a nonterminal node then, for each $f_i \in E_G(\Theta)$, compute the value $c(\Theta, r, f_i) = F(c(\Theta(f_i, b_i), r))$ and set $c(\Theta, r) = \min\{c(\Theta, r, f_i) : f_i \in E_G(\Theta)\}$. Remove all f_i-bundles of edges starting from Θ for which $c(\Theta, r) < c(\Theta, r, f_i)$. Mark the node Θ as processed and proceed to step 1.

Proposition 10.2 *Let G be a proper subgraph of the graph $\Delta_{U,\alpha}(T)$ for some decision table T with n conditional attributes $f_1, \ldots, f_n$, U be an uncertainty measure, $\alpha \in \mathbb{R}_+$, $r = (b_1, \ldots, b_n)$ be a row of T, and ψ be a strictly increasing cost function for decision rules given by pair of functions ψ^0 and F. Then, to construct the graph $G^\psi = G^\psi(r)$, the algorithm $\mathscr{A}_8$ makes*

$$O(nL(G))$$

elementary operations (computations of ψ^0, F, and comparisons).

Proof In each terminal node of the graph G, the algorithm $\mathcal{A}_8$ computes the value of ψ^0. In each nonterminal node of G, the algorithm $\mathcal{A}_8$ computes the value of F at most n times and makes at most $2n$ comparisons. Therefore the algorithm $\mathcal{A}_8$ makes

$$O(nL(G))$$

elementary operations. □

Proposition 10.3 *Let $\mathcal{U}$ be a restricted information system and $\psi \in \{l, -c, -rc, -c_M, mc, rmc\}$. Then the algorithm $\mathcal{A}_8$ has polynomial time complexity for decision tables from $\mathcal{T}(\mathcal{U})$ depending on the number of conditional attributes in these tables.*

Proof Since $\psi \in \{l, -c, -rc, -c_M, mc, rmc\}$,

$$\psi^0 \in \{0, -N_{mcd(T)}(T), -N_{mcd(T)}(T)/N(T), -N^M(T), N(T) - N_{mcd(T)}(T), (N(T) - N_{mcd(T)}(T))/N(T)\} ,$$

and F is either x or $x + 1$. Therefore the elementary operations used by the algorithm $\mathcal{A}_8$ are either basic numerical operations or computations of numerical parameters of decision tables which have polynomial time complexity depending on the size of decision tables. From Proposition 10.2 it follows that the number of elementary operations is bounded from above by a polynomial depending on the size of input table T and on the number of separable subtables of T.

According to Proposition 3.5, the algorithm $\mathcal{A}_8$ has polynomial time complexity for decision tables from $\mathcal{T}(\mathcal{U})$ depending on the number of conditional attributes in these tables. □

Theorem 10.1 *Let T be a nonempty decision table with n conditional attributes $f_1, \ldots, f_n$, $r = (b_1, \ldots, b_n)$ be a row of T, U be an uncertainty measure, $\alpha \in \mathbb{R}_+$, G be a proper subgraph of the graph $\Delta_{U,\alpha}(T)$, and ψ be a strictly increasing cost function given by pair of functions ψ^0 and F. Then, for any node Θ of the graph $G^\psi = G^\psi(r)$ containing the row r, $c(\Theta, r) = \min\{\psi(\Theta, \rho) : \rho \in Rule(G, \Theta, r)\}$ and the set $Rule(G^\psi, \Theta, r)$ coincides with set of rules from $Rule(G, \Theta, r)$ that have minimum cost relative to ψ.*

Proof We prove this theorem by induction on nodes of G^ψ containing r. If Θ is a terminal node containing r then $Rule(G^\psi, \Theta, r) = Rule(G, \Theta, r) = \{\rightarrow mcd(\Theta)\}$ and $c(\Theta, r) = \psi^0(\Theta) = \psi(\Theta, \rightarrow mcd(\Theta))$. Therefore the statement of theorem holds for Θ.

Let Θ be a nonterminal node containing r such that, for each child of Θ containing r, the statement of theorem holds. It is clear that

$$Rule(G, \Theta, r) = \bigcup_{f_i \in E_G(\Theta)} Rule(G, \Theta, r, f_i)$$

where, for $f_i \in E_G(\Theta)$,

$$Rule(G, \Theta, r, f_i) = \{f_i = b_i \wedge \gamma \to t : \gamma \to t \in Rule(G, \Theta(f_i, b_i), r)\} .$$

By inductive hypothesis, for any $f_i \in E_G(\Theta)$, the minimum cost of rule from $Rule(G, \Theta(f_i, b_i), r)$ is equal to $c(\Theta(f_i, b_i), r)$. Since ψ is a strictly increasing cost function, the minimum cost of a rule from the set $Rule(G, \Theta, r, f_{i_j})$ is equal to $F(c(\Theta(f_i, b_i), r) = c(\Theta, r, f_i)$. Therefore $c(\Theta, r) = \min\{c(\Theta, r, f_i) : f_i \in E_G(\Theta)\}$ is the minimum cost of rule from $Rule(G, \Theta, r)$. Set $q = c(\Theta, r)$.

Let $f_i = b_i \wedge \gamma \to t$ be a rule from $Rule(G, \Theta, r)$ which cost is equal to q. It is clear that G contains the node $\Theta(f_i, b_i)$ and the edge e which starts in Θ, enters $\Theta(f_i, b_i)$, and is labeled with (f_i, b_i). Let p be the minimum cost of a rule from $Rule(G, \Theta(f_i, b_i), r)$, i.e., $p = c(\Theta(f_i, b_i), r)$. The rule $\gamma \to t$ belongs to the set $Rule(G, \Theta(f_i, b_i), r)$ and, since ψ is strictly increasing, the cost of $\gamma \to t$ is equal to p (otherwise, the minimum cost of a rule from $Rule(G, \Theta, r)$ is less than q). Therefore $F(p) = q$, and the edge e belongs to the graph G^{ψ}. By the induction hypothesis, the set $Rule(G^{\psi}, \Theta(f_i, b_i), r)$ coincides with the set of rules from $Rule(G, \Theta(f_i, b_i), r)$ which cost is equal to p. From here it follows that $\gamma \to t$ belongs to $Rule(G^{\psi}, \Theta(f_i, b_i), r)$ and $f_i = b_i \wedge \gamma \to t$ belongs to $Rule(G^{\psi}, \Theta, r)$.

Let $f_i = b_i \wedge \gamma \to t$ belong to $Rule(G^{\psi}, \Theta, r)$. Then $\Theta(f_i, b_i)$ is a child of Θ in the graph G^{ψ}, $\gamma \to t$ belongs to $Rule(G^{\psi}, \Theta(f_i, b_i), r)$ and, by the induction hypothesis, the set $Rule(G^{\psi}, \Theta(f_i, b_i), r)$ coincides with the set of rules from $Rule(G, \Theta(f_i, b_i), r)$ which cost is equal to p – the minimum cost of a rule from $Rule(G, \Theta(f_i, b_i), r)$. From the description of the procedure of optimization and from the fact that $\Theta(f_i, b_i)$ is a child of Θ in the graph G^{ψ} it follows that $F(p) = q$. Therefore, the cost of rule $f_i = b_i \wedge \gamma \to t$ is equal to q. □

We can make sequential optimization of (U, α)-rules for T and r relative to a sequence of strictly increasing cost functions $\psi_1, \psi_2, \ldots$ for decision rules. We begin from the graph $G = \Delta_{U,\alpha}(T)$ and apply to it the procedure of optimization (algorithm $\mathcal{A}_8$) relative to the cost function ψ_1. As a result, we obtain a proper subgraph G^{ψ_1} of the graph G. By Proposition 10.1, the set $Rule(G, T, r)$ is equal to the set of all (U, α) -rules for T and r. From here and from Theorem 10.1 it follows that the set $Rule(G^{\psi_1}, T, r)$ is equal to the set of all (U, α) -rules for T and r which have minimum cost relative to ψ_1. If we apply to the graph G^{ψ_1} the procedure of optimization relative to the cost function ψ_2 we obtain a proper subgraph G^{ψ_1,ψ_2} of the graph G^{ψ_1}. The set $Rule(G^{\psi_1,\psi_2}, T, r)$ is equal to the set of all rules from the set $Rule(G^{\psi_1}, T, r)$ which have minimum cost relative to ψ_2, etc.

We described the work of optimization procedure for one row. If we would like to work with all rows in parallel, then instead of removal of bundles of edges we will change the list of bundles attached to row. We begin from the graph $G = \Delta_{U,\alpha}(T)$. In this graph, for each nonterminal node Θ, each row r of Θ is labeled with the set of attributes $E(G, \Theta, r) = E(\Theta)$. It means that, for the row r, we consider only f_i-bundles of edges starting from Θ such that $f_i \in E(G, \Theta, r)$. During the work of the procedure of optimization relative to a cost function ψ we will not change the "topology" of the graph G but will change sets $E(G, \Theta, r)$ attached to rows r of nonterminal nodes Θ. In a new graph G^{ψ} (we will say about this graph as about *labeled proper subgraph* of the graph G), for each nonterminal node Θ, each row r of Θ is labeled with a subset $E(G^{\psi}, \Theta, r)$ of the set $E(G, \Theta, r)$ containing only attributes for which corresponding bundles were not removed during the optimization relative to ψ for the row r.

We can study also totally optimal decision rules relative to various combinations of cost functions. For a cost function ψ, we denote $\psi^{U,\alpha}(T, r) = \min\{\psi(T, \rho) : \rho \in DR_{U,\alpha}(T, r)\}$, i.e., $\psi^{U,\alpha}(T, r)$ is the minimum cost of a (U, α) -decision rule for T and r relative to the cost function ψ. Let $\psi_1, \ldots, \psi_m$ be cost functions and $m \geq 2$. A (U, α) -decision rule ρ for T and r is called a *totally optimal* (U, α)*-decision rule for* T *and* r *relative to the cost functions* $\psi_1, \ldots, \psi_m$ if $\psi_1(T, \rho) = \psi_1^{U,\alpha}(T, r), \ldots, \psi_m(T, \rho) = \psi_m^{U,\alpha}(T, r)$, i.e., ρ is optimal relative to $\psi_1, \ldots, \psi_m$ simultaneously.

Assume that $\psi_1, \ldots, \psi_m$ are strictly increasing cost functions for decision rules. We now describe how to recognize the existence of a (U, α)-decision rule for T and r which is a totally optimal (U, α)-decision rule for T and r relative to the cost functions $\psi_1, \ldots, \psi_m$.

First, we construct the graph $G = \Delta_{U,\alpha}(T)$ using the Algorithm $\mathcal{A}_3$. For $i = 1, \ldots, m$, we apply to G and r the procedure of optimization relative to ψ_i (the Algorithm $\mathcal{A}_8$). As a result, we obtain, for $i = 1, \ldots, m$, the graph $G^{\psi_i}(r)$ and the number $\psi_i^{U,\alpha}(T, r)$ attached to the node T of $G^{\psi_i}(r)$. Next, we apply to G sequentially the procedures of optimization relative to the cost functions $\psi_1, \ldots, \psi_m$. As a result, we obtain graphs $G^{\psi_1}(r), G^{\psi_1,\psi_2}(r), \ldots, G^{\psi_1,\ldots,\psi_m}(r)$ and numbers $\varphi_1, \varphi_2, \ldots, \varphi_m$ attached to the node T of these graphs. One can show that a totally optimal (U, α)-decision rule for T and r relative to the cost functions $\psi_1, \ldots, \psi_m$ exists if and only if $\varphi_i = \psi_i^{U,\alpha}(T, r)$ for $i = 1, \ldots, m$.

10.3 Number of Rules in $Rule(G, \Theta, r)$

Let T be a nonempty decision table with n conditional attributes $f_1, \ldots, f_n$, $r = (b_1, \ldots, b_n)$ be a row of T, U be an uncertainty measure, $\alpha \in \mathbb{R}_+$, and G be a proper subgraph of the graph $\Delta_{U,\alpha}(T)$. We describe now an algorithm which counts, for each node Θ of the graph G containing r, the cardinality $C(\Theta, r)$ of the set $Rule(G, \Theta, r)$, and returns the number $C(T, r) = |Rule(G, T, r)|$.

Algorithm $\mathcal{A}_9$ (counting the number of decision rules).

Input: A proper subgraph G of the graph $\Delta_{U,\alpha}(T)$ for some decision table T with n conditional attributes $f_1, \ldots, f_n$, uncertainty measure U, number $\alpha \in \mathbb{R}_+$, and row $r = (b_1, \ldots, b_n)$ of the table T.
Output: The number $|Rule(G, T, r)|$.

1. If all nodes of the graph G containing r are processed then return the number $C(T, r)$ and finish the work of the algorithm. Otherwise, choose a node Θ of the graph G containing r which is not processed yet and which is either a terminal

node of G or a nonterminal node of G such that, for each $f_i \in E_G(T)$ and $a_j \in E(\Theta, f_i)$, the node $\Theta(f_i, a_j)$ is processed.
2. If Θ is a terminal node then set $C(\Theta, r) = 1$, mark the node Θ as processed, and proceed to step 1.
3. If Θ is a nonterminal node then set

$$C(\Theta, r) = \sum_{f_i \in E_G(\Theta)} C(\Theta(f_i, b_i), r) ,$$

mark the node Θ as processed, and proceed to step 1.

Proposition 10.4 *Let U be an uncertainty measure, $\alpha \in \mathbb{R}_+$, T be a decision table with n attributes $f_1, \ldots, f_n$, G be a proper subgraph of the graph $\Delta_{U,\alpha}(T)$, and $r = (b_1, \ldots, b_n)$ be a row of the table T. Then the algorithm $\mathscr{A}_9$ returns the number $|Rule(G, T, r)|$ and makes at most $nL(G)$ operations of addition.*

Proof We prove by induction on the nodes of G that $C(\Theta, r) = |Rule(G, \Theta, r)|$ for each node Θ of G containing r. Let Θ be a terminal node of G. Then $Rule(G, \Theta, r) = \{\to mcd(\Theta)\}$ and $|Rule(G, \Theta, r)| = 1$. Therefore the considered statement holds for Θ. Let now Θ be a nonterminal node of G such that the considered statement holds for its children containing r. It is clear that

$$Rule(G, \Theta, r) = \bigcup_{f_i \in E_G(\Theta)} Rule(G, \Theta, r, f_i)$$

where, for each $f_i \in E_G(\Theta)$,

$$Rule(G, \Theta, r, f_i) = \{f_i = b_i \wedge \gamma \to t : \gamma \to t \in Rule(G, \Theta(f_i, b_i), r)\}$$

and $\left|Rule(G, \Theta, r, f_{i_j})\right| = \left|Rule(G, \Theta(f_{i_j}, b_{i_j}), r)\right|$. Therefore

$$|Rule(G, \Theta, r)| = \sum_{f_i \in E_G(\Theta)} |Rule(G, \Theta(f_i, b_i), r)| .$$

By the induction hypothesis, $C(\Theta(f_i, b_i), r) = |Rule(G, \Theta(f_i, b_i), r)|$ for any $f_i \in E_G(\Theta)$. Therefore $C(\Theta, r) = |Rule(G, \Theta, r)|$. Hence, the considered statement holds. From here it follows that $C(T, r) = |Rule(G, T, r)|$, and the algorithm $\mathscr{A}_9$ returns the cardinality of the set $Rule(G, T, r)$.

It is easy to see that the considered algorithm makes at most $nL(G)$ operations of addition where $L(G)$ is the number of nodes in the graph G. □

Proposition 10.5 *Let $\mathscr{U}$ be a restricted information system. Then the algorithm $\mathscr{A}_9$ has polynomial time complexity for decision tables from $\mathscr{T}(\mathscr{U})$ depending on the number of conditional attributes in these tables.*

Proof All operations made by the algorithm $\mathscr{A}_9$ are basic numerical operations (additions). From Proposition 10.4 it follows that the number of these operations is bounded from above by a polynomial depending on the size of input table T and on the number of separable subtables of T.

According to Proposition 3.5, the algorithm $\mathscr{A}_9$ has polynomial time complexity for decision tables from $\mathscr{T}(\mathscr{U})$ depending on the number of conditional attributes in these tables. □

10.4 Simulation of Greedy Algorithm for Construction of Decision Rule Set

Let U be an uncertainty measure, $\alpha, \beta \in \mathbb{R}_+$, $0 \leq \beta \leq 1$, and T be a decision table with n conditional attributes $f_1, \ldots, f_n$. We would like to construct a set S of (U, α)-decision rules for T with minimum cardinality such that rules from S cover at least $(1 - \beta)N(T)$ rows of T. Unfortunately, our approach does not allow us to do this. However, we can simulate the work of a greedy algorithm for the set cover problem (algorithm $\mathscr{A}_{10}$).

The algorithm $\mathscr{A}_{10}$ works with the graph $G = \Delta_{U,\alpha}(T)$. During each step, this algorithm constructs (based on algorithm $\mathscr{A}_8$) a (U, α)-rule for T with minimum length among all (U, α)-rules for T which cover maximum number of uncovered previously rows. The algorithm finishes the work when at least $(1 - \beta)N(T)$ rows of T are covered. We will call the constructed set a *(U, α, β)-greedy set of decision rules for T*. Using results of Slavik for the set cover problem [9] we obtain that the cardinality of the constructed set of rules S is less than

$$C_{\min}(U, \alpha, \beta, T)(\ln \lceil (1 - \beta)N(T) \rceil - \ln \ln \lceil (1 - \beta)N(T) \rceil + 0.78)$$

where $C_{\min}(U, \alpha, \beta, T)$ is the minimum cardinality of a set of (U, α)-rules for T which cover at least $(1 - \beta)N(T)$ rows of T.

Let us recall that the cost function $-c_M$ is given by the pair of functions $\psi^0(T) = -N^M(T)$ and $F(x) = x$, and the cost function l is given by the pair of functions $\varphi^0(T) = 0$ and $H(x) = x + 1$.

Algorithm $\mathscr{A}_{10}$ (construction of (U, α, β)-greedy set of decision rules for decision table).

Input: A decision table T with n conditional attributes $f_1, \ldots, f_n$, uncertainty measure U, and numbers $\alpha, \beta \in \mathbb{R}_+$, $0 \leq \beta \leq 1$.
Output: A (U, α, β)-greedy set of decision rules for T.

1. Set $M = \emptyset$ and $S = \emptyset$.
2. If $|M| \geq (1 - \beta)N(T)$ then return S and finish the algorithm.

3. Apply the algorithm $\mathscr{A}_8$ to each row of T two times: first, as the procedure of optimization of rules relative to the cost function $-c_M$, and after that, as the procedure of optimization of rules relative to the cost function l.
4. As a result, for each row r of T, we obtain two numbers $-c_M(r)$ which is the minimum cost of a rule from $Rule(G, T, r)$ relative to the cost function $-c_M$, and $l(r)$ which is the minimum length among rules from $Rule(G, T, r)$ which have minimum cost relative to the cost function $-c_M$.
5. Choose a row r of T for which the value of $-c_M(r)$ is minimum among all rows of T, and the value of $l(r)$ is minimum among all rows of T with minimum value of $-c_M(r)$. In the graph $G^{-c_M,l}$, which is the result of the bi-stage procedure of optimization of rules for the row r relative to the cost functions $-c_M$ and l, choose a directed path τ from T to a terminal node of $G^{-c_M,l}$ containing r.
6. Add the rule $rule(\tau)$ to the set S, and add all rows covered by $rule(\tau)$, which do not belong to M, to the set M. Proceed to step 2.

Proposition 10.6 *Let T be a decision table with n conditional attributes, U be an uncertainty measure, and $\alpha, \beta \in \mathbb{R}_+$, $0 \le \beta \le 1$. Then the algorithm $\mathscr{A}_{10}$ returns a (U, α, β)-greedy set of decision rules for T and makes*

$$O(N(T)^2 nL(G))$$

elementary operations (computations of ψ^0, F, φ^0, H, and comparisons).

Proof Let us analyze one iteration of the algorithm $\mathscr{A}_{10}$ (steps 3–6) and rule $rule(\tau)$ added at the step 6 to the set S. Using Proposition 10.1 and Theorem 10.1 we obtain that the rule $rule(\tau)$ is a (U, α)-rule for T which covers the maximum number of uncovered rows (rows which does not belong to M) and has minimum length among such rules. Therefore the algorithm $\mathscr{A}_{10}$ returns a (U, α, β) -greedy set of decision rules for T.

Let us analyze the number of elementary operations (computations of ψ^0, F, φ^0, H, and comparisons) which algorithm $\mathscr{A}_{10}$ makes during one iteration. We know that the algorithm $\mathscr{A}_8$, under the bi-stage optimization of rules for one row, makes

$$O(nL(G))$$

elementary operations (computations of ψ^0, F, φ^0, H, and comparisons). The number of rows is equal to $N(T)$. To choose a row r of T for which the value of $-c_M(r)$ is minimum among all rows of T, and the value of $l(r)$ is minimum among all rows of T with minimum value of $-c_M(r)$, the algorithm $\mathscr{A}_{10}$ makes at most $2N(T)$ comparisons. Therefore the number of elementary operations which algorithm $\mathscr{A}_{10}$ makes during one iteration is $O(N(T)nL(G))$.

The number of iterations is at most $N(T)$. Therefore, during the construction of a (U, α, β)-greedy set of decision rules for T, the algorithm $\mathscr{A}_{10}$ makes

$$O(N(T)^2 nL(G))$$

elementary operations (computations of ψ^0, F, φ^0, H, and comparisons). □

Proposition 10.7 *Let $\mathscr{U}$ be a restricted information system. Then the algorithm $\mathscr{A}_{10}$ has polynomial time complexity for decision tables from $\mathscr{T}(\mathscr{U})$ depending on the number of conditional attributes in these tables.*

Proof From Proposition 10.6 it follows that, for the algorithm $\mathscr{A}_{10}$, the number of elementary operations (computations of ψ^0, F, φ^0, H, and comparisons) is bounded from above by a polynomial depending on the size of input table T and on the number of separable subtables of T. The computations of numerical parameters of decision tables used by the algorithm $\mathscr{A}_{10}$ (constant 0 and $-N^M(T)$) have polynomial time complexity depending on the size of decision tables. All operations with numbers are basic ones (x, $x + 1$, comparisons).

According to Proposition 3.5, the algorithm $\mathscr{A}_{10}$ has polynomial time complexity for decision tables from $\mathscr{T}(\mathscr{U})$ depending on the number of conditional attributes in these tables. □

Using information based on the work of algorithm $\mathscr{A}_{10}$, we can obtain lower bound on the parameter $C_{\min}(U, \alpha, \beta, T)$ that is the minimum cardinality of a set of (U, α)-decision rules for T which cover at least $(1 - \beta)N(T)$ rows of T. During the construction of (U, α, β)-greedy set of decision rules for T, let the algorithm $\mathscr{A}_{10}$ choose consequently rules $\rho_1, \ldots, \rho_t$. Let $B_1, \ldots, B_t$ be sets of rows covered by rules $\rho_1, \ldots, \rho_t$, respectively. Set $B_0 = \emptyset$, $\delta_0 = 0$ and, for $i = 1, \ldots, t$, set $\delta_i = |B_i \setminus (B_0 \cup \ldots \cup B_{i-1})|$. The information derived from the algorithm's $\mathscr{A}_{10}$ work consists of the tuple $(\delta_1, \ldots, \delta_t)$ and the numbers $N(T)$ and β.

From the results obtained in [6] regarding a greedy algorithm for the set cover problem it follows that

$$C_{\min}(U, \alpha, \beta, T) \geq \max \left\{ \left\lceil \frac{\lceil (1-\beta)N(T) \rceil - (\delta_0 + \ldots + \delta_i)}{\delta_{i+1}} \right\rceil : i = 0, \ldots, t-1 \right\}. \quad (10.1)$$

10.5 Experimental Results for Simulation of Greedy Algorithm

Here we describe the results of our experiments with the simulated greedy algorithm. We applied the algorithm $\mathscr{A}_{10}$ on 22 decision tables from the UCI Machine Learning

Repository [5] using relative misclassification error rme as uncertainty measure and values from $\{0, 0.1, 0.3, 0.5\}$ and $\{0, 0.01, 0.05, 0.1, 0.15, 0.2\}$ for α and β, respectively. The work and results of the algorithm provided us with an upper bound (cardinality of the set of rules constructed by $\mathscr{A}_{10}$) and a lower bound (10.1) for the minimum cardinality of a set of (rme, α)-decision rules for T which cover at least $(1 - \beta)N(T)$ rows of T for the considered tables T.

The values for the upper and lower bounds when $\alpha = 0$ and $\beta = 0$ can be found in Table 10.1. The column "Table name" displays the name of the considered decision table while "Rows" and "Attr" present the size of the table in terms of rows and columns, respectively. The following two columns, "Upper bound" and "Lower bound" present both bounds.

A system of decision rules needs to consist of few rules in order to be used as an understandable representation of knowledge. As such, we have chosen 40 rules as a threshold for the number of rules in a reasonably small system of decision rules. In Table 10.1, we have marked all upper bounds in bold where this threshold is exceeded.

Table 10.1 Upper and lower bounds on the minimum cardinality of a set of $(rme, 0)$-decision rules covering all rows for 22 decision tables from UCI ML Repository

Table name	Rows	Attr	Upper bound	Lower bound
Adult-stretch	16	4	3	2
Agaricus-lepiota	8124	22	13	4
Balance-scale	625	4	**303**	**177**
Breast-cancer	266	9	**79**	33
Cars	1728	6	**246**	**159**
Hayes-roth-data	69	5	25	13
House-votes-84	279	16	23	9
Lenses	10	4	9	6
Lymphography	148	18	24	9
Monks-1-test	432	6	22	18
Monks-1-train	124	6	23	9
Monks-2-test	432	6	**254**	**226**
Monks-2-train	169	6	**66**	32
Monks-3-test	432	6	12	9
Monks-3-train	122	6	23	09
Nursery	12960	8	**555**	**175**
Shuttle-landing	15	6	10	7
Soybean-small	47	35	4	3
Spect-test	169	22	14	5
Teeth	23	8	23	23
Tic-tac-toe	958	9	24	11
Zoo-data	59	16	9	4

Table 10.2 Bounds for "Balance-scale"

	α							
β	0		0.1		0.3		0.5	
0	303	177	229	135	126	74	54	22
0.01	297	171	223	129	120	68	48	16
0.05	272	146	198	104	95	43	29	15
0.1	241	115	167	76	64	23	21	13
0.15	210	107	136	68	43	22	16	11
0.2	178	100	104	60	**34**	20	13	9

Table 10.3 Bounds for "Cars"

	α							
β	0		0.1		0.3		0.5	
0	246	159	246	159	206	129	206	129
0.01	229	142	229	142	189	112	189	112
0.05	160	83	160	83	120	55	120	49
0.1	82	61	82	61	65	34	62	29
0.15	61	40	61	40	43	25	39	18
0.2	**39**	18	39	18	27	17	23	10

Table 10.4 Bounds for "Monks-2-test"

	α							
β	0		0.1		0.3		0.5	
0	254	226	254	226	210	178	132	96
0.01	250	222	250	222	206	174	128	92
0.05	233	205	233	205	189	157	111	75
0.1	211	183	211	183	167	135	89	53
0.15	190	162	190	162	146	114	68	32
0.2	168	140	168	140	124	92	46	23

Furthermore, we have also marked in bold the lower bounds so as to identify tables where it is impossible to construct a reasonably small system of $(rme, 0)$-decision rules which cover all rows. For these four cases ("balance-scale", "cars", "monks-2-test", and "nursery"), we conducted further experiments to identify values of α and β such that the number of rules can be reduced to within our threshold.

In Tables 10.2, 10.3, 10.4, and 10.5 we can see the upper and lower bounds for our four tables for various values of α and β. The first value in each cell is the upper bound and the second value is the lower bound. We have marked in bold cells where the upper bound is within our threshold of 40 to show that it is possible to

Table 10.5 Bounds for "Nursery"

	α							
β	0		0.1		0.3		0.5	
0	555	175	384	124	157	46	20	11
0.01	449	153	288	102	85	24	18	11
0.05	266	98	138	52	44	14	15	9
0.1	166	69	79	36	28	12	13	7
0.15	115	47	50	29	21	10	11	5
0.2	79	35	**38**	22	16	8	8	5

significantly reduce the size of systems of decision rules by changing α and β. With the exception of "Monks-2-test" (Table 10.4) we were able to find good values of α and β for reasonably small systems of decision rules.

10.6 Totally Optimal Decision Rules for Complete Decision Tables

In this section, we study the existence of decision rules that are totally optimal with respect to cost functions length l and coverage c. This study allows us to partially explain a phenomenon observed experimentally in many cases: the existence of decision rules that have the minimum length and the maximum coverage simultaneously.

Let T be a nonempty decision table, r be a row from T, and $DR_{rme,0}(T, r)$ be the set of $(rme, 0)$-decision rules for T and r. A *totally optimal decision rule* ρ *for* T *and* r is a decision rule $\rho \in DR_{rme,0}(T, r)$ such that $l(T, \rho) = \min\{l(T, \rho') : \rho' \in DR_{rme,0}(T, r)\}$ and $c(T, \rho) = \max\{c(T, \rho') : \rho' \in DR_{rme,0}(T, r)\}$.

We now define the concept of a *complete* decision table as a representation of a total function in which different variables can have domains of differing values. Given nonempty finite subsets of the set of nonnegative integers $B_1, \ldots, B_n$, $I = B_1 \times \ldots \times B_n$, and $\nu : I \to \omega$, we can define $T = (I, \nu)$ as a complete decision table T with conditional attributes $f_1, \ldots, f_n$ that have values from the sets $B_1, \ldots, B_n$, respectively. n-Tuples from I are rows of T with the decision $\nu(b_1, \ldots, b_n)$ attached to a row $(b_1, \ldots, b_n) \in I$. The table T is nondegenerate if and only if ν is a non-constant function (which has at least two different values).

We now consider the existence of totally optimal decision rules for a complete decision table $T = (I, \nu)$ with conditional attributes $f_1, \ldots, f_n$ that have values from the sets $B_1, \ldots, B_n$, respectively. Let $v_i = |B_i|$ for $i = 1, \ldots, n$, $2 \le v_1 \le \ldots \le v_n$, and $f_{j_1} = b_{j_1} \wedge \ldots \wedge f_{j_k} = b_{j_k} \to \nu(r)$ be a $(rme, 0)$-decision rule for T and the row $r = (b_1, ..., b_n)$ from T. We denote this rule as ρ. Because T is a complete table, $c(T, \rho) = \prod_{i \in \{1,...,n\} \setminus \{j_1,...,j_k\}} v_i$. Furthermore, it follows that $c(T, \rho) \le v_{k+1} \times \ldots \times v_n$, and $c(T, \rho) \ge v_1 \times \ldots \times v_{n-k}$.

Let us consider the system of inequalities

$$\{\nu_{m+2} \times \ldots \times \nu_n \leq \nu_1 \times \ldots \times \nu_{n-m} : m = 1, \ldots, n-1\} . \qquad (10.2)$$

If $m = n - 2$ then we have the inequality $\nu_n \leq \nu_1 \times \nu_2$. If $m = n - 1$ then we have the inequality $1 \leq \nu_1$ which holds since $\nu_1 \geq 2$.

Theorem 10.2 *Let $B_1, \ldots, B_n$ be nonempty finite subsets of the set ω, $\nu_i = |B_i|$ for $i = 1, \ldots, n$, $2 \leq \nu_1 \leq \ldots \leq \nu_n$, and $I = B_1 \times \ldots \times B_n$.*

If each inequality from the system (10.2) holds then, for each non-constant function $\nu : I \to \omega$ and for each row r of the complete decision table $T = (I, \nu)$, there exists a totally optimal decision rule for T and r.

Otherwise, for each row r from I there exists a non-constant function $\nu_r : I \to \omega$ such that there is no totally optimal decision rule for $T = (I, \nu_r)$ and r.

Proof Let each equality from the system (10.2) hold, $\nu : I \to \omega$ be a non-constant function, and $r = (b_1, \ldots, b_n)$ be a row of the decision table $T = (I, \nu)$. We will show that there exists a totally optimal rule for T and r. Let us assume the contrary. In this case there exist two rules ρ_1 and ρ_2 for T and r such that $l(T, \rho_2) > l(T, \rho_1)$ and $c(T, \rho_2) > c(T, \rho_1)$. Let $m = l(T, \rho_1)$. Then $l(T, \rho_2) \geq m + 1$, and $m > 0$ since ν is a non-constant function.

Since $\nu_1 \leq \ldots \leq \nu_n$, we have $c(T, \rho_1) \geq \nu_1 \times \ldots \times \nu_{n-m}$ and $c(T, \rho_2) \leq \nu_{m+2} \times \ldots \times \nu_n$. We know that $c(T, \rho_2) > c(T, \rho_1)$. Therefore $\nu_{m+2} \times \ldots \times \nu_n > \nu_1 \times \ldots \times \nu_{n-m}$, but this is impossible. Hence a totally optimal decision rule for T and r exists.

Let there exist $m \in \{1, \ldots, n-2\}$ such that $\nu_{m+2} \times \ldots \times \nu_n > \nu_1 \times \ldots \times \nu_{n-m}$. For an arbitrary $r = (b_1, \ldots, b_n) \in I$, we define a function $\nu_r : I \to \omega$. For $x = (x_1, \ldots, x_n) \in I$, $\nu_r(x) = 1$ if $x_1 = b_1, \ldots, x_{m+1} = b_{m+1}$ or $x_{n-m+1} = b_{n-m+1}, \ldots, x_n = b_n$. Otherwise, $\nu_r(x) = 0$. The function ν_r is non-constant: $\nu_r(r) = 1$ and $\nu_r(x) = 0$ if, in particular, $x_1 \neq b_1$ and $x_n \neq b_n$.

Let $T = (I, \nu_r)$, $p_1 = \nu_1 \times \ldots \times \nu_{n-m}$ and $p_2 = \nu_{m+2} \times \ldots \times \nu_n$. We now show that $p_1 < p_2$ since it is obvious that $m < m + 1$. We denote by ρ_1 the rule

$$f_{n-m+1} = b_{n-m+1} \wedge \ldots \wedge f_n = b_n \to 1 ,$$

and by ρ_2 we denote the rule

$$f_1 = b_1 \wedge \ldots \wedge f_{m+1} = b_{m+1} \to 1 .$$

It is not difficult to show that ρ_1 and ρ_2 are decision rules for T and r such that $l(T, \rho_1) = m$, $c(T, \rho_1) = p_1$, $l(T, \rho_2) = m + 1$, and $c(T, \rho_2) = p_2$. Therefore $l(T, \rho_1) < l(T, \rho_2)$ and $c(T, \rho_1) < c(T, \rho_2)$ meaning that neither is a totally optimal decision rule.

Let ρ be an $(rme, 0)$-decision rule for T and r, and ρ be equal to

$$f_{i_1} = b_{i_1} \wedge \ldots \wedge f_{i_k} = b_{i_k} \to 1 .$$

Let $A = \{i_1, \ldots, i_k\}$, $B = \{n - m + 1, \ldots, n\}$, and $C = \{1, \ldots, m + 1\}$. We now show that $B \subseteq A$ or $C \subseteq A$. Let us assume the contrary: there is $s \in B$ such that $s \notin A$, and there is $t \in C$ such that $t \notin A$. We denote by x a n-tuple $(x_1, \ldots, x_n)$ from I such that $x_i = b_i$ if $i \in \{1, \ldots, n\} \setminus \{s, t\}$, $x_s \neq b_s$, and $x_t \neq b_t$. It is clear that x is covered by the rule ρ and $\nu_r(x) = 0$, but this is impossible.

If $B \subseteq A$ then $l(T, \rho) \geq l(T, \rho_1)$ and $c(T, \rho) \leq c(T, \rho_1)$. If $C \subseteq A$ then $l(T, \rho) \geq l(T, \rho_2)$ and $c(T, \rho) \leq c(T, \rho_2)$. Therefore either

$$(c(T, \rho), l(T, \rho)) \in \{(p_1, m), (p_2, m + 1)\}$$

or the values $(c(T, \rho), l(T, \rho))$ are worse (from the point of view of length and coverage) than either (p_1, m) or $(p_2, m + 1)$. Furthermore, $p_1 < p_2$ and $m < m + 1$. From here it follows that there is no totally optimal decision rule for T and r. □

Example 10.1 Let $v_1 = 3$, $v_2 = 4$, $v_3 = 4$, and $v_4 = 5$. The system of inequalities (10.2) in this case result in the following:

$$\begin{array}{rcl} v_3 \times v_4 \leq v_1 \times v_2 \times v_3 & 4 \times 5 \leq 3 \times 4 \times 4 & : m = 1 \\ v_4 \leq v_1 \times v_2 & 5 \leq 3 \times 4 & : m = 2 \\ 1 \leq v_1 & 1 \leq 3 & : m = 3 \end{array}$$

We see that all inequalities hold.

Example 10.2 Let $v_1 = 2$, $v_2 = 3$, $v_3 = 5$, and $v_4 = 8$. The system of inequalities (10.2) in this case result in the following:

$$\begin{array}{rcl} v_3 \times v_4 \leq v_1 \times v_2 \times v_3 & 5 \times 8 \leq 2 \times 3 \times 5 & : m = 1 \\ v_4 \leq v_1 \times v_2 & 8 \leq 2 \times 3 & : m = 2 \\ 1 \leq v_1 & 1 \leq 2 & : m = 3 \end{array}$$

Here the first and second inequalities and clearly false.

We consider now a special case when all $v_1, \ldots, v_n$ are equal. It is easy to prove the following statement.

Proposition 10.8 *Let $v_1 = \ldots = v_n = v \geq 2$. Then the system of inequalities (10.2) is equal to the system $\{v^{n-m-1} \leq v^{n-m} : m = 1, \ldots, n - 1\}$ in which each inequality holds.*

Let n, k be natural numbers, $k \geq 2$, and $E_k = \{0, \ldots, k - 1\}$. A function $f : E_k^n \to E_k$ is called a *function of k-valued logic* (if $k = 2$ then f is a *Boolean* function). We correspond to this function a complete decision table $T_f = (E_k^n, f)$ with conditional attributes $x_1, \ldots, x_n$. From Theorem 10.2 and Proposition 10.8 it follows that, for each row r of the decision table T_f, there exists a totally optimal decision rule for T_f and r relative to c and l.

We consider now another special case when among $v_1, \ldots, v_n$ there are only two different numbers.

Proposition 10.9 *Let* $v_1 = \ldots = v_{n_a} = v_a$ *and* $v_{n_a+1} = \ldots = v_{n_a+n_b} = v_b$, *where* $n = n_a + n_b$, $n_a > 0$, $n_b > 0$, *and* $2 \le v_a < v_b$. *If* $n_a > n_b$ *then each inequality from the system of inequalities (10.2) holds if and only if the inequality* $v_b^{n_b} \le v_a^{n_b+1}$ *holds. If* $n_a \le n_b$ *then each inequality from the system of inequalities (10.2) holds if and only if the inequality* $v_b^{n_a-1} \le v_a^{n_a}$ *holds.*

Proof For $m \in \{1, \ldots, n-1\}$, we denote by $ineq(m)$ the inequality $v_{m+2} \times \ldots \times v_n \le v_1 \times \ldots \times v_{n-m}$ from the system of inequalities (10.2). We will say that two inequalities are *equivalent* if they either both hold or both do not hold.

Let $n_a > n_b$. We consider three cases. If $1 \le m \le n_b$ then $ineq(m)$ is equal to $v_a^{n_a-m-1} \times v_b^{n_b} \le v_a^{n_a} \times v_b^{n_b-m}$, and this inequality is equivalent to $v_b^m \le v_a^{m+1}$. If $n_b < m < n_a$ then $ineq(m)$ is equal to $v_a^{n_a-m-1} \times v_b^{n_b} \le v_a^{n_a-(m-n_b)}$, and this inequality is equivalent to $v_b^{n_b} \le v_a^{n_b+1}$. If $n_a \le m \le n-1$ then $ineq(m)$ is equal to $v_b^{n-m-1} \le v_a^{n-m}$ (note that $0 \le n-m-1 \le n_b-1$).

The inequality $v_b^{n_b} \le v_a^{n_b+1}$ is equivalent to the inequality $ineq(n_b)$. If each inequality from the system (10.2) holds then the inequality $v_b^{n_b} \le v_a^{n_b+1}$ holds. Let now the inequality $v_b^{n_b} \le v_a^{n_b+1}$ hold. If $n_b < m < n_a$ then the inequality $ineq(m)$ holds (it is equivalent to $v_b^{n_b} \le v_a^{n_b+1}$). If $1 \le m \le n_b$ or $n_a \le m \le n-1$ then the inequality $ineq(m)$ is equivalent to the inequality $v_b^k \le v_a^{k+1}$ for some $k \in \{0, \ldots, n_b\}$. Let $t = n_b - k$. If we divide the left-hand side of the inequality $v_b^{n_b} \le v_a^{n_b+1}$ by v_b^t and the right-hand side by v_a^t we obtain the inequality $v_b^k \le v_a^{k+1}$ which holds also. Therefore each inequality from the system (10.2) holds.

Let $n_a \le n_b$. We consider three cases. If $1 \le m < n_a$ then $ineq(m)$ is equal to $v_a^{n_a-m-1} \times v_b^{n_b} \le v_a^{n_a} \times v_b^{n_b-m}$, and this inequality is equivalent to $v_b^m \le v_a^{m+1}$. If $n_a \le m \le n_b$ then $ineq(m)$ is equal to $v_b^{n_b-(m+1-n_a)} \le v_a^{n_a} \times v_b^{n_b-m}$, and this inequality is equivalent to $v_b^{n_a-1} \le v_a^{n_a}$. If $n_b < m \le n-1$ then $ineq(m)$ is equal to $v_b^{n-m-1} \le v_a^{n-m}$ (note that $0 \le n-m-1 \le n_a-2$).

The inequality $v_b^{n_a-1} \le v_a^{n_a}$ is equivalent to the inequality $ineq(n_a)$. If each inequality from the system (10.2) holds then the inequality $v_b^{n_a-1} \le v_a^{n_a}$ holds. Let now the inequality $v_b^{n_a-1} \le v_a^{n_a}$ hold. If $n_a \le m \le n_b$ then the inequality $ineq(m)$ holds (it is equivalent to $v_b^{n_a-1} \le v_a^{n_a}$). If $1 \le m < n_a$ or $n_b < m \le n-1$ then the inequality $ineq(m)$ is equivalent to the inequality $v_b^k \le v_a^{k+1}$ for some $k \in \{0, \ldots, n_a-1\}$. Let $t = n_a - k - 1$. If we divide the left-hand side of the inequality $v_b^{n_a-1} \le v_a^{n_a}$ by v_b^t and the right-hand side by v_a^t we obtain the inequality $v_b^k \le v_a^{k+1}$ which also holds. Therefore each inequality from the system (10.2) holds. □

Corollary 10.1 *Let* $v_1 = v_a$ *and* $v_2 = \ldots = v_n = v_b$, *where* $2 \le v_a < v_b$. *Then each inequality from the system of inequalities (10.2) holds.*

Table 10.6 Details regarding eight complete decision tables from UCI ML Repository

Table name	Attr	Rows	Theorem 10.2	Tot opt
Adult-stretch	4	16	Yes	Yes
Balance-scale	4	625	Yes	Yes
Cars	6	1728	Yes	Yes
Lenses	4	24	Yes	Yes
Monks-1-test	6	432	Yes	Yes
Monks-2-test	6	432	Yes	Yes
Monks-3-test	6	432	Yes	Yes
Nursery	8	12960	No	Yes

10.6.1 Experimental Results for Totally Optimal Decision Rules

In this section, we present our findings with regards to totally optimal decision rules when working with eight complete decision tables from the UCI Machine Learning Repository [5].

In Table 10.6, we can find details from our experimentation. The columns "Attr" and "Rows" contain the number of attributes and the number of rows in each decision table, respectively. The column "Theorem 10.2" is labeled with "yes" for decision tables where the system of equalities (10.2) studied in Theorem 10.2 holds and "no" where they do not. Finally, the column "Tot opt" states whether or not totally optimal decision rules exist for every row in the decision table. It is interesting to note that the "Nursery" decision table, despite failing Theorem 10.2 still has totally optimal decision rules for each row.

10.7 Deterministic Versus Nondeterministic Decision Trees for Total Boolean Functions

In this section, we compare minimum depth of deterministic and nondeterministic decision trees for randomly generated Boolean functions.

A total Boolean function $f(x_1, \ldots, x_n)$ is a function of the kind $f : \{0, 1\}^n \rightarrow \{0, 1\}$. We consider a decision table representation of the function f which is a rectangular table T_f with n columns filled with numbers from the set $\{0, 1\}$. Columns of the table are labeled with variables $x_1, \ldots, x_n$ which are considered as attributes. Rows of the table are pairwise different, and the set of rows coincides with the set of n-tuples from $\{0, 1\}^n$. Each row is labeled with the value of function f on this row.

We use relative misclassification error rme as uncertainty measure and study both exact decision trees for T_f ((rme, 0)-decision trees for T_f) and exact systems of decision rules for T_f ($(rme, 0)$-systems of decision rules for T_f) which can be considered as nondeterministic decision trees for T_f.

We denote by $h(f)$ the minimum depth of $(rme, 0)$-decision tree for T_f. For a row r of T_f, we denote by $l(T_f, r)$ the minimum length of $(rme, 0)$-decision rule for T_f and r. Let $l(f) = \max\{l(T_f, r) : r \in \{0, 1\}^n\}$. The parameter $l(f)$ is known as the minimum depth of a nondeterministic decision tree for the function f and as the certificate complexity of f. The following inequalities hold for any total Boolean function f: $l(f) \leq h(f) \leq l(f)^2$ (see [3]).

We performed experiments using total Boolean functions of n variables with n ranging from 7 to 13. For each n, we randomly generated 2^{n+1} total Boolean functions with n variables. For each function f, we constructed the decision table T_f. For this table, we constructed the directed acyclic graph $\Delta_{rme,0}(T_f)$ by the algorithm $\mathscr{A}_3$. Using algorithms $\mathscr{A}_5$ and $\mathscr{A}_8$ we computed the parameters $h(f)$ and $l(f)$.

We found the results to be split into two cases:

1. Case 1: $h(f) = n$ and $l(f) = n - 1$,
2. Case 2: $h(f) = n$ and $l(f) = n$.

Note that $h(f) = n$ in both cases.

In Table 10.7 we can see the results of our experiments. Column "n" is simply the number of variables in the Boolean functions under consideration, and column "# Functions" is how many randomly generated Boolean functions of that size were tested. Columns "Case 1" and "Case 2" show for how many of the test cases $l(f) = n - 1$ and $l(f) = n$, respectively. Finally, the last column is the percentage of test cases for which $l(f) = n - 1$.

Firstly, the fact that $h(f) = n$ for all cases is as expected: from [7] we know that, for most total Boolean functions f with n variables, $h(f) = n$, especially as n grows. What is interesting is that $l(f) \geq n - 1$ for each case and, in particular $l(f) = n - 1$ for just over a third of the Boolean functions displayed this situation. Furthermore, these results were consistent when increasing n.

Table 10.7 Comparison of deterministic and nondeterministic decision trees

n	# Functions	Case 1	Case 2	Case 1 (%)
7	256	101	155	39
8	512	186	326	36
9	1024	361	663	35
10	2048	721	1327	35
11	4096	1463	2633	36
12	8192	3004	5188	37
13	16384	5873	10511	36

References

1. Amin, T., Chikalov, I., Moshkov, M., Zielosko, B.: Dynamic programming approach for exact decision rule optimization. In: Skowron, A., Suraj, Z. (eds.) Rough Sets and Intelligent Systems – Professor Zdzisław Pawlak in Memoriam – Volume 1. Intelligent Systems Reference Library, vol. 42, pp. 211–228. Springer, Berlin (2013)
2. Amin, T., Moshkov, M.: Totally optimal decision rules. Discret. Appl. Math. **236**, 453–458 (2018)
3. Blum, M., Impagliazzo, R.: Generic oracles and oracle classes (extended abstract). In: 28th Annual Symposium on Foundations of Computer Science, Los Angeles, California, USA, October 27–29, 1987, pp. 118–126. IEEE Computer Society (1987)
4. Bonates, T.O., Hammer, P.L., Kogan, A.: Maximum patterns in datasets. Discret. Appl. Math. **156**(6), 846–861 (2008)
5. Lichman, M.: UCI Machine Learning Repository. University of California, Irvine, School of Information and Computer Sciences (2013). http://archive.ics.uci.edu/ml
6. Moshkov, M., Piliszczuk, M., Zielosko, B.: Partial Covers, Reducts and Decision Rules in Rough Sets - Theory and Applications. Studies in Computational Intelligence, vol. 145. Springer, Heidelberg (2008)
7. Rivest, R.L., Vuillemin, J.: On recognizing graph properties from adjacency matrices. Theor. Comput. Sci. **3**(3), 371–384 (1976)
8. Slavìk, P.: A tight analysis of the greedy algorithm for set cover. J. Algorithm **25**(2), 237–254 (1997)
9. Slavìk, P.: Approximation algorithms for set cover and related problems. Ph.D. thesis, University of New York at Buffalo (1998)

Chapter 11
Bi-criteria Optimization Problem for Decision Rules and Systems of Rules: Cost Versus Cost

A decision rule can be optimized for many beneficial qualities: a shorter length leads to easier understandability, a larger coverage grants more generality, and a smaller miscoverage means higher accuracy. However, these and other such qualities cannot all be optimized for at the same time. Often, a rule that is better with respect to one criterion will be worse with respect to another. In order to understand the trade-offs available and to find the best possible values for multiple criteria, we present algorithms for the construction of the set of Pareto optimal points for decision rules and systems of decision rules with respect to two criteria. We further transform these sets of Pareto optimal points into graphs of functions describing relationships between the considered criteria.

The construction of all Pareto optimal patterns (the equivalent of decision rules for logical analysis of data) was considered in [1] as being "intractable" due to the large quantity of them. While we do not construct all Pareto optimal rules or Pareto optimal rule systems, we do construct the entire set of Pareto optimal points. The cardinality of the set of Pareto optimal points is usually much less than the cardinality of the set of Pareto optimal rule systems. In particular, for the "Balance scale" dataset [2] with 4 attributes and 625 rows we have one Pareto optimal point and approximately 2.64×10^{653} Pareto optimal systems of rules.

This chapter also uses Pareto optimal points to measure the performance of heuristics with regards to two criteria at once. To evaluate a heuristic with respect to a single criterion, it is possible to simply compare the value derived from the heuristic with the optimal value which can be obtained via dynamic programming. When considering two criteria at once, it is often the case that two heuristics are incomparable (one is better with regards to one criterion, but worse with regards to another). In order to measure the performance of a heuristic with respect to both criteria, we can construct the set of Pareto optimal points for systems of decision rules as well as a system of decision rules using the greedy heuristic. We can then consider the normalized distance between the point corresponding to the heuristically created

H. AbouEisha et al., *Extensions of Dynamic Programming for Combinatorial Optimization and Data Mining*, Intelligent Systems Reference Library 146,
https://doi.org/10.1007/978-3-319-91839-6_11

system and the nearest Pareto optimal point. This way, both criteria are considered at once.

In Sect. 11.1, we present a method of deriving the set of Pareto optimal points for decision rules for a given row of a decision table with respect to two strictly increasing cost functions. Then, we use this set to construct functions describing the relationships between the considered cost functions. This theory is then expanded to work with systems of decision rules in Sect. 11.2. Finally, in Sect. 11.3, we present an experiment where we construct systems of decision rules using greedy heuristics and compare their performance using the set of Pareto optimal points.

11.1 Pareto Optimal Points for Decision Rules: Cost Versus Cost

Let ψ and φ be strictly increasing cost functions for decision rules given by pairs of functions ψ^0, F and φ^0, H, respectively. Let T be a nonempty decision table with n conditional attributes $f_1, \ldots, f_n$, $r = (b_1, \ldots, b_n)$ be a row of T, U be an uncertainty measure, $\alpha \in \mathbb{R}_+$, and G be a proper subgraph of the graph $\Delta_{U,\alpha}(T)$ (it is possible that $G = \Delta_{U,\alpha}(T)$).

For each node Θ of the graph G containing r, we denote $p_{\psi,\varphi}(G, \Theta, r) = \{(\psi(\Theta, \rho), \varphi(\Theta, \rho)) : \rho \in Rule(G, \Theta, r)\}$. We denote by $Par(p_{\psi,\varphi}(G, \Theta, r))$ the set of Pareto optimal points for $p_{\psi,\varphi}(G, \Theta, r)$. Note that, by Proposition 10.1, if $G = \Delta_{U,\alpha}(T)$ then the set $Rule(G, \Theta, r)$ is equal to the set of (U, α)-decision rules for Θ and r. Another interesting case is when G is the result of application of procedure of optimization of rules for r (algorithm $\mathcal{A}_8$) relative to cost functions different from ψ and φ to the graph $\Delta_{U,\alpha}(T)$.

We now describe an algorithm $\mathcal{A}_{11}$ constructing the set $Par(p_{\psi,\varphi}(G, T, r))$. In fact, this algorithm constructs, for each node Θ of the graph G, the set $B(\Theta, r) = Par(p_{\psi,\varphi}(G, \Theta, r))$.

Algorithm $\mathcal{A}_{11}$ (construction of POPs for decision rules, cost versus cost).

Input: Strictly increasing cost functions ψ and φ for decision rules given by pairs of functions ψ^0, F and φ^0, H, respectively, a nonempty decision table T with n conditional attributes $f_1, \ldots, f_n$, a row $r = (b_1, \ldots, b_n)$ of T, and a proper subgraph G of the graph $\Delta_{U,\alpha}(T)$ where U is an uncertainty measure and $\alpha \in \mathbb{R}_+$.

Output: The set $Par(p_{\psi,\varphi}(G, T, r))$ of Pareto optimal points for the set of pairs $p_{\psi,\varphi}(G, T, r) = \{(\psi(T, \rho), \varphi(T, \rho)) : \rho \in Rule(G, T, r)\}$.

1. If all nodes in G containing r are processed, then return the set $B(T, r)$. Otherwise, choose in the graph G a node Θ containing r which is not processed yet and which is either a terminal node of G or a nonterminal node of G such that, for any $f_i \in E_G(\Theta)$, the node $\Theta(f_i, b_i)$ is already processed, i.e., the set $B(\Theta(f_i, b_i), r)$ is already constructed.
2. If Θ is a terminal node, then set $B(\Theta, r) = \{(\psi^0(\Theta), \varphi^0(\Theta))\}$. Mark the node Θ as processed and proceed to step 1.

3. If Θ is a nonterminal node then, for each $f_i \in E_G(\Theta)$, construct the set

$$B(\Theta(f_i, b_i), r)^{FH} ,$$

and construct the multiset

$$A(\Theta, r) = \bigcup_{f_i \in E_G(\Theta)} B(\Theta(f_i, b_i), r)^{FH}$$

by simple transcription of elements from the sets $B(\Theta(f_i, b_i), r)^{FH}$, $f_i \in E_G(\Theta)$.
4. Apply to the multiset $A(\Theta, r)$ the algorithm $\mathscr{A}_1$ which constructs the set

$$Par(A(\Theta, r)) .$$

Set $B(\Theta, r) = Par(A(\Theta, r))$. Mark the node Θ as processed and proceed to step 1.

Proposition 11.1 *Let ψ and φ be strictly increasing cost functions for decision rules given by pairs of functions ψ^0, F and φ^0, H, respectively, T be a nonempty decision table with n conditional attributes $f_1, \ldots, f_n$, $r = (b_1, \ldots, b_n)$ be a row of T, U be an uncertainty measure, $\alpha \in \mathbb{R}_+$, and G be a proper subgraph of the graph $\Delta_{U,\alpha}(T)$. Then, for each node Θ of the graph G containing r, the algorithm $\mathscr{A}_{11}$ constructs the set $B(\Theta, r) = Par(p_{\psi,\varphi}(G, \Theta, r))$.*

Proof We prove the considered statement by induction on nodes of G. Let Θ be a terminal node of G containing r. Then $Rule(G, \Theta, r) = \{\to mcd(\Theta)\}$,

$$p_{\psi,\varphi}(G, \Theta, r) = Par(p_{\psi,\varphi}(G, \Theta, r)) = \{(\psi^0(\Theta), \varphi^0(\Theta))\} ,$$

and $B(\Theta, r) = Par(p_{\psi,\varphi}(G, \Theta, r))$.

Let Θ be a nonterminal node of G containing r such that, for any $f_i \in E_G(\Theta)$, the considered statement holds for the node $\Theta(f_i, b_i)$, i.e.,

$$B(\Theta(f_i, b_i), r) = Par(p_{\psi,\varphi}(G, \Theta(f_i, b_i), r)) .$$

It is clear that

$$p_{\psi,\varphi}(G, \Theta, r) = \bigcup_{f_i \in E_G(\Theta)} p_{\psi,\varphi}(G, \Theta(f_i, b_i), r)^{FH} .$$

From Lemma 2.5 it follows that

$$Par(p_{\psi,\varphi}(G, \Theta, r)) \subseteq \bigcup_{f_i \in E_G(\Theta)} Par(p_{\psi,\varphi}(G, \Theta(f_i, b_i), r)^{FH}) .$$

By Lemma 2.7, $Par(p_{\psi,\varphi}(G, \Theta(f_i, b_i), r)^{FH}) = Par(p_{\psi,\varphi}(G, \Theta(f_i, b_i), r))^{FH}$ for any $f_i \in E_G(\Theta)$. Therefore

$$Par(p_{\psi,\varphi}(G, \Theta, r)) \subseteq \bigcup_{f_i \in E_G(\Theta)} Par(p_{\psi,\varphi}(G, \Theta(f_i, b_i), r))^{FH} \subseteq p_{\psi,\varphi}(G, \Theta, r) .$$

Using Lemma 2.4 we obtain

$$Par(p_{\psi,\varphi}(G, \Theta, r)) = Par\left(\bigcup_{f_i \in E_G(\Theta)} Par(p_{\psi,\varphi}(G, \Theta(f_i, b_i), r))^{FH}\right) .$$

Since $B(\Theta, r) = Par\left(\bigcup_{f_i \in E_G(\Theta)} B(\Theta(f_i, b_i), r)^{FH}\right)$ and

$$B(\Theta(f_i, b_i), r) = Par(p_{\psi,\varphi}(G, \Theta(f_i, b_i), r))$$

for any $f_i \in E_G(\Theta)$, we have $B(\Theta, r) = Par(p_{\psi,\varphi}(G, \Theta, r))$. □

We now evaluate the number of elementary operations (computations of F, H, ψ^0, φ^0, and comparisons) made by the algorithm $\mathscr{A}_{11}$. Let us recall that, for a given cost function ψ for decision rules and decision table T,

$$q_{\psi}(T) = |\{\psi(\Theta, \rho) : \Theta \in SEP(T), \rho \in DR(\Theta)\}| .$$

In particular, by Lemma 9.1, $q_l(T) \leq n+1$, $q_{-c}(T) \leq N(T)+1$, $q_{-c_M}(T) \leq N(T)+1$, $q_{-rc}(T) \leq N(T)(N(T)+1)$, $q_{mc}(T) \leq N(T)+1$, and $q_{rmc}(T) \leq N(T)(N(T)+1)$.

Proposition 11.2 *Let ψ and φ be strictly increasing cost functions for decision rules given by pairs of functions ψ^0, F and φ^0, H, respectively, T be a nonempty decision table with n conditional attributes $f_1, \ldots, f_n$, $r = (b_1, \ldots, b_n)$ be a row of T, U be an uncertainty measure, $\alpha \in \mathbb{R}_+$, and G be a proper subgraph of the graph $\Delta_{U,\alpha}(T)$. Then, to construct the set $Par(p_{\psi,\varphi}(G, T, r))$, the algorithm $\mathscr{A}_{11}$ makes*

$$O(L(G) \min(q_{\psi}(T), q_{\varphi}(T))n \log(\min(q_{\psi}(T), q_{\varphi}(T))n))$$

elementary operations (computations of F, H, ψ^0, φ^0, and comparisons).

Proof To process a terminal node, the algorithm $\mathscr{A}_{11}$ makes two elementary operations – computes ψ^0 and φ^0. We now evaluate the number of elementary operations under the processing of a nonterminal node Θ. From Lemma 2.3 it follows that $|Par(p_{\psi,\varphi}(G, \Theta(f_i, b_i), r))| \leq \min(q_{\psi}(T), q_{\varphi}(T))$ for any $f_i \in E_G(\Theta)$. It is clear that $|E_G(\Theta)| \leq n$, $|Par(p_{\psi,\varphi}(G, \Theta(f_i, b_i), r))^{FH}| = |Par(p_{\psi,\varphi}(G, \Theta(f_i, b_i), r))|$ for any $f_i \in E_G(\Theta)$. From Proposition 11.1 it follows that

$$B(\Theta(f_i, b_i), r) = Par(p_{\psi,\varphi}(G, \Theta(f_i, b_i), r))$$

and $B(\Theta(f_i, b_i), r)^{FH} = Par(p_{\psi,\varphi}(G, \Theta(f_i, b_i), r))^{FH}$ for any $f_i \in E_G(\Theta)$. Hence

$$|A(\Theta, r)| \leq \min(q_\psi(T), q_\varphi(T))n \ .$$

Therefore to construct the sets $B(\Theta(f_i, b_i), r)^{FH}$, $f_i \in E_G(\Theta)$, from the sets

$$B(\Theta(f_i, b_i), r) \ ,$$

$f_i \in E_G(\Theta)$, the algorithm $\mathscr{A}_{11}$ makes $O(\min(q_\psi(T), q_\varphi(T))n)$ computations of F and H, and to construct the set

$$Par(A(\Theta, r)) = Par(p_{\psi,\varphi}(G, \Theta, r))$$

from the set $A(\Theta, r)$, the algorithm $\mathscr{A}_{11}$ makes

$$O(\min(q_\psi(T), q_\varphi(T))n \log(\min(q_\psi(T), q_\varphi(T))n))$$

comparisons (see Proposition 2.1). Hence, to treat a nonterminal node Θ, the algorithm makes

$$O(\min(q_\psi(T), q_\varphi(T))n \log(\min(q_\psi(T), q_\varphi(T))n))$$

computations of F, H, and comparisons.

To construct the set $Par(p_{\psi,\varphi}(G, T, r))$ the algorithm $\mathscr{A}_{11}$ makes

$$O(L(G) \min(q_\psi(T), q_\varphi(T))n \log(\min(q_\psi(T), q_\varphi(T))n))$$

elementary operations (computations of F, H, ψ^0, φ^0, and comparisons). □

Proposition 11.3 *Let ψ and φ be strictly increasing cost functions for decision rules given by pairs of functions ψ^0, F and φ^0, H, respectively,*

$$\psi, \varphi \in \{l, -c, -rc, -c_M, mc, rmc\} \ ,$$

and $\mathscr{U}$ be a restricted information system. Then the algorithm $\mathscr{A}_{11}$ has polynomial time complexity for decision tables from $\mathscr{T}(\mathscr{U})$ depending on the number of conditional attributes in these tables.

Proof Since $\psi, \varphi \in \{l, -c, -rc, -c_M, mc, rmc\}$,

$$\begin{aligned}\psi^0, \varphi^0 \in \{&0, -N_{mcd(T)}(T), -N_{mcd(T)}(T)/N(T), -N^M(T),\\ &N(T) - N_{mcd(T)}(T), (N(T) - N_{mcd(T)}(T))/N(T)\} \ ,\end{aligned}$$

and $F, H \in \{x, x+1\}$. From Lemma 9.1 and Proposition 11.2 it follows that, for the algorithm $\mathscr{A}_{11}$, the number of elementary operations (computations of F, H,

ψ^0, φ^0, and comparisons) is bounded from above by a polynomial depending on the size of input table T and on the number of separable subtables of T. All operations with numbers are basic ones. The computations of numerical parameters of decision tables used by the algorithm $\mathscr{A}_{11}$ (0, $-N_{mcd(T)}(T)$, $-N_{mcd(T)}(T)/N(T)$, $-N^M(T)$, $N(T) - N_{mcd(T)}(T)$, and $(N(T) - N_{mcd(T)}(T))/N(T)$) have polynomial time complexity depending on the size of decision tables.

According to Proposition 3.5, the algorithm $\mathscr{A}_{11}$ has polynomial time complexity for decision tables from $\mathscr{T}(\mathscr{U})$ depending on the number of conditional attributes in these tables. □

11.1.1 Relationships for Decision Rules: Cost Versus Cost

Let ψ and φ be strictly increasing cost functions for decision rules, T be a nonempty decision table with n conditional attributes $f_1, \ldots, f_n$, $r = (b_1, \ldots, b_n)$ be a row of T, U be an uncertainty measure, $\alpha \in \mathbb{R}_+$, and G be a proper subgraph of the graph $\Delta_{U,\alpha}(T)$ (it is possible that $G = \Delta_{U,\alpha}(T)$).

To study relationships between cost functions ψ and φ on the set of rules $Rule(G, T, r)$, we consider partial functions $\mathscr{R}^{\psi,\varphi}_{G,T,r} : \mathbb{R} \to \mathbb{R}$ and $\mathscr{R}^{\varphi,\psi}_{G,T,r} : \mathbb{R} \to \mathbb{R}$ defined in the following way:

$$\mathscr{R}^{\psi,\varphi}_{G,T,r}(x) = \min\{\varphi(T, \rho) : \rho \in Rule(G, T, r), \psi(T, \rho) \le x\} ,$$
$$\mathscr{R}^{\varphi,\psi}_{G,T,r}(x) = \min\{\psi(T, \rho) : \rho \in Rule(G, T, r), \varphi(T, \rho) \le x\} .$$

Let $p_{\psi,\varphi}(G, T, r) = \{(\psi(T, \rho), \varphi(T, \rho)) : \rho \in Rule(G, T, r)\}$ and

$$(a_1, b_1), \ldots, (a_k, b_k)$$

be the normal representation of the set $Par(p_{\psi,\varphi}(G, T, r))$ where $a_1 < \cdots < a_k$ and $b_1 > \ldots > b_k$. By Lemma 2.8 and Remark 2.2, for any $x \in \mathbb{R}$,

$$\mathscr{R}^{\psi,\varphi}_{G,T,r}(x) = \begin{cases} undefined, & x < a_1 \\ b_1, & a_1 \le x < a_2 \\ \ldots & \ldots \\ b_{k-1}, & a_{k-1} \le x < a_k \\ b_k, & a_k \le x \end{cases} ,$$

$$\mathscr{R}^{\varphi,\psi}_{G,T,r}(x) = \begin{cases} undefined, & x < b_k \\ a_k, & b_k \le x < b_{k-1} \\ \ldots & \ldots \\ a_2, & b_2 \le x < b_1 \\ a_1, & b_1 \le x \end{cases} .$$

11.2 Pareto Optimal Points for Systems of Decision Rules: Cost Versus Cost

Let T be a nonempty decision table with n conditional attributes $f_1, \ldots, f_n$ and $N(T)$ rows $r_1, \ldots, r_{N(T)}$, U be an uncertainty measure, $\alpha \in \mathbb{R}_+$, and $\mathbf{G} = (G_1, \ldots, G_{N(T)})$ be an $N(T)$-tuple of proper subgraphs of the graph $\Delta_{U,\alpha}(T)$. Let $G = \Delta_{U,\alpha}(T)$ and ξ be a strictly increasing cost function for decision rules. Then $(G, \ldots, G)$ and $(G^{\xi}(r_1), \ldots, G^{\xi}(r_{N(T)}))$ are examples of interesting for study $N(T)$-tuples of proper subgraphs of the graph $\Delta_{U,\alpha}(T)$.

We denote by $\mathscr{S}(\mathbf{G}, T)$ the set $Rule(G_1, T, r_1) \times \cdots \times Rule(G_{N(T)}, T, r_{N(T)})$ of (U, α)-systems of decision rules for T. Let ψ, φ be strictly increasing cost functions for decision rules given by pairs of functions ψ^0, F and φ^0, H, respectively, and f, g be increasing functions from $\mathbb{R}^2$ to $\mathbb{R}$. It is clear that ψ_f and φ_g are strictly increasing cost functions for systems of decision rules.

We describe now an algorithm which constructs the set of Pareto optimal points for the set of pairs $p^{f,g}_{\psi,\varphi}(\mathbf{G}, T) = \{(\psi_f(T, S), \varphi_g(T, S)) : S \in \mathscr{S}(\mathbf{G}, T)\}$.

Algorithm $\mathscr{A}_{12}$ (construction of POPs for decision rule systems, cost versus cost).

Input: Strictly increasing cost functions for decision rules ψ and φ given by pairs of functions ψ^0, F and φ^0, H, respectively, increasing functions f, g from $\mathbb{R}^2$ to $\mathbb{R}$, a nonempty decision table T with n conditional attributes $f_1, \ldots, f_n$ and $N(T)$ rows $r_1, \ldots, r_{N(T)}$, and an $N(T)$-tuple $\mathbf{G} = (G_1, \ldots, G_{N(T)})$ of proper subgraphs of the graph $\Delta_{U,\alpha}(T)$ where U is an uncertainty measure and $\alpha \in \mathbb{R}_+$.

Output: The set $Par(p^{f,g}_{\psi,\varphi}(\mathbf{G}, T))$ of Pareto optimal points for the set of pairs $p^{f,g}_{\psi,\varphi}(\mathbf{G}, T) = \{((\psi_f(T, S), (\varphi_g(T, S)) : S \in \mathscr{S}(\mathbf{G}, T)\}$.

1. Using the algorithm $\mathscr{A}_{11}$ construct, for $i = 1, \ldots, N(T)$, the set $Par(P_i)$ where

$$P_i = p_{\psi,\varphi}(G_i, T, r_i) = \{(\psi(T, \rho), \varphi(T, \rho)) : \rho \in Rule(G_i, T, r_i)\} .$$

2. Apply the algorithm $\mathscr{A}_2$ to the functions f, g and the sets

$$Par(P_1), \ldots, Par(P_{N(T)}) .$$

Set $C(\mathbf{G}, T)$ the output of the algorithm $\mathscr{A}_2$ and return it.

Proposition 11.4 *Let ψ, φ be strictly increasing cost functions for decision rules given by pairs of functions ψ^0, F and φ^0, H, respectively, f, g be increasing functions from $\mathbb{R}^2$ to $\mathbb{R}$, U be an uncertainty measure, $\alpha \in \mathbb{R}_+$, T be a decision table with n conditional attributes $f_1, \ldots, f_n$ and $N(T)$ rows $r_1, \ldots, r_{N(T)}$, and $\mathbf{G} = (G_1, \ldots, G_{N(T)})$ be an $N(T)$-tuple of proper subgraphs of the graph $\Delta_{U,\alpha}(T)$. Then the algorithm $\mathscr{A}_{12}$ constructs the set $C(\mathbf{G}, T) = Par(p^{f,g}_{\psi,\varphi}(\mathbf{G}, T))$.*

Proof For $i = 1, \ldots, N(T)$, denote $P_i = p_{\psi,\varphi}(G_i, T, r_i)$. During the first step, the algorithm $\mathscr{A}_{12}$ constructs (using the algorithm $\mathscr{A}_{11}$) the sets $Par(P_1), \ldots, Par(P_{N(T)})$ (see Proposition 11.1). During the second step, the algorithm $\mathscr{A}_{12}$ constructs (using the algorithm $\mathscr{A}_2$) the set $C(\mathbf{G}, T) = Par(Q_{N(T)})$ where $Q_1 = P_1$, and, for $i = 2, \ldots, N(T)$, $Q_i = Q_{i-1} \langle fg \rangle P_i$ (see Proposition 2.2). One can show that $Q_{N(T)} = p^{f,g}_{\psi,\varphi}(\mathbf{G}, T)$. Therefore $C(\mathbf{G}, T) = Par(p^{f,g}_{\psi,\varphi}(\mathbf{G}, T))$. □

Let us recall that, for a given cost function ψ and a decision table T, $q_\psi(T) = |\{\psi(\Theta, \rho) : \Theta \in SEP(T), \rho \in DR(\Theta)\}|$. In particular, by Lemma 9.1, $q_l(T) \leq n + 1$, $q_{-c}(T) \leq N(T) + 1$, $q_{-rc}(T) \leq N(T)(N(T) + 1)$, $q_{-c_M}(T) \leq N(T) + 1$, $q_{mc}(T) \leq N(T) + 1$, and $q_{rmc}(T) \leq N(T)(N(T) + 1)$.

Let us recall also that, for a given cost function ψ for decision rules and a decision table T, $Range_\psi(T) = \{\psi(\Theta, \rho) : \Theta \in SEP(T), \rho \in DR(\Theta)\}$. By Lemma 9.1, $Range_l(T) \subseteq \{0, 1, \ldots, n\}$, $Range_{-c}(T) \subseteq \{0, -1, \ldots, -N(T)\}$, $Range_{-c_M}(T) \subseteq \{0, -1, \ldots, -N(T)\}$, and $Range_{mc}(T) \subseteq \{0, 1, \ldots, N(T)\}$. Let $t_l(T) = n$, $t_{-c}(T) = N(T)$, $t_{-c_M}(T) = N(T)$, and $t_{mc}(T) = N(T)$.

Proposition 11.5 *Let ψ, φ be strictly increasing cost functions for decision rules given by pairs of functions ψ^0, F and φ^0, H, respectively, $\psi \in \{l, -c, -c_M, mc\}$, f, g be increasing functions from $\mathbb{R}^2$ to $\mathbb{R}$, $f \in \{x + y, \max(x, y)\}$, U be an uncertainty measure, $\alpha \in \mathbb{R}_+$, T be a decision table with n conditional attributes $f_1, \ldots, f_n$ and $N(T)$ rows $r_1, \ldots, r_{N(T)}$, $\mathbf{G} = (G_1, \ldots, G_{N(T)})$ be an $N(T)$-tuple of proper subgraphs of the graph $\Delta_{U,\alpha}(T)$, and $L(\mathbf{G}) = L(\Delta_{U,\alpha}(T))$. Then, to construct the set $Par(p^{f,g}_{\psi,\varphi}(\mathbf{G}, T))$, the algorithm $\mathscr{A}_{12}$ makes*

$$O(N(T)L(\mathbf{G}) \min(q_\psi(T), q_\varphi(T))n \log(\min(q_\psi(T), q_\varphi(T))n))$$
$$+O(N(T)t_\psi(T)^2 \log(t_\psi(T)))$$

elementary operations (computations of F, H, ψ^0, φ^0, f, g and comparisons) if $f = \max(x, y)$, and

$$O(N(T)L(\mathbf{G}) \min(q_\psi(T), q_\varphi(T))n \log(\min(q_\psi(T), q_\varphi(T))n))$$
$$+O(N(T)^2 t_\psi(T)^2 \log(N(T)t_\psi(T)))$$

elementary operations (computations of F, H, ψ^0, φ^0, f, g and comparisons) if $f = x + y$.

Proof To construct the sets $Par(P_i) = Par(p_{\psi,\varphi}(G_i, T, r_i))$, $i = 1, \ldots, N(T)$, the algorithm $\mathscr{A}_{11}$ makes

$$O(N(T)L(\mathbf{G}) \min(q_\psi(T), q_\varphi(T))n \log(\min(q_\psi(T), q_\varphi(T))n))$$

elementary operations (computations of F, H, ψ^0, φ^0, and comparisons) – see Proposition 11.2.

We now evaluate the number of elementary operations (computations of f, g, and comparisons) made by the algorithm $\mathscr{A}_2$ during the construction of the set $C(\mathbf{G}, T) = Par(Q_{N(T)}) = Par(p^{f,g}_{\psi,\varphi}(\mathbf{G}, T))$. We know that $\psi \in \{l, -c, -c_M, mc\}$ and $f \in \{x + y, \max(x, y)\}$.

For $i = 1, \ldots, N(T)$, let $P^1_i = \{a : (a, b) \in P_i\}$. Since $\psi \in \{l, -c, -c_M, mc\}$, we have $P^1_i \subseteq \{0, 1, \ldots, t_\psi(T)\}$ for $i = 1, \ldots, N(T)$ or $P^1_i \subseteq \{0, -1, \ldots, -t_\psi(T)\}$ for $i = 1, \ldots, N(T)$.

Using Proposition 2.3 we obtain the following.

If $f = x + y$, then to construct the set $Par(Q_{N(T)})$ the algorithm $\mathscr{A}_2$ makes

$$O(N(T)^2 t_\psi(T)^2 \log(N(T) t_\psi(T)))$$

elementary operations (computations of f, g, and comparisons).

If $f = \max(x, y)$, then to construct the sets $Par(Q_{N(T)})$ the algorithm $\mathscr{A}_2$ makes

$$O(N(T) t_\psi(T)^2 \log(t_\psi(T)))$$

elementary operations (computations of f, g, and comparisons). □

Similar analysis can be done for the case when $\varphi \in \{l, -c, -c_M, mc\}$ and $g \in \{x + y, \max(x, y)\}$.

Proposition 11.6 *Let ψ, φ be strictly increasing cost functions for decision rules given by pairs of functions ψ^0, F and φ^0, H, respectively, f, g be increasing functions from $\mathbb{R}^2$ to $\mathbb{R}$, $\psi \in \{l, -c, -c_M, mc\}$, $\varphi \in \{l, -c, -rc, -c_M, mc, rmc\}$, $f, g \in \{x + y, \max(x, y)\}$, and $\mathscr{U}$ be a restricted information system. Then the algorithm $\mathscr{A}_{12}$ has polynomial time complexity for decision tables from $\mathscr{T}(\mathscr{U})$ depending on the number of conditional attributes in these tables.*

Proof Since $\psi \in \{l, -c, -c_M, mc\}$ and $\varphi \in \{l, -c, -rc, -c_M, mc, rmc\}$,

$$\psi^0, \varphi^0 \in \{0, -N_{mcd(T)}(T), -N_{mcd(T)}(T)/N(T), -N^M(T), \\ N(T) - N_{mcd(T)}(T), (N(T) - N_{mcd(T)}(T))/N(T)\},$$

and $F, H \in \{x, x + 1\}$. From Lemma 9.1 and Proposition 11.5 it follows that, for the algorithm $\mathscr{A}_{12}$, the number of elementary operations (computations of F, H, ψ^0, φ^0, f, g, and comparisons) is bounded from above by a polynomial depending on the size of input table T and on the number of separable subtables of T. All operations with numbers are basic ones. The computations of numerical parameters of decision tables used by the algorithm $\mathscr{A}_{12}$ ($0, -N_{mcd(T)}(T), -N_{mcd(T)}(T)/N(T), -N^M(T), N(T) - N_{mcd(T)}(T)$, and $(N(T) - N_{mcd(T)}(T))/N(T)$) have polynomial time complexity depending on the size of decision tables.

According to Proposition 3.5, the algorithm $\mathscr{A}_{12}$ has polynomial time complexity for decision tables from $\mathscr{T}(\mathscr{U})$ depending on the number of conditional attributes in these tables. □

11.2.1 Relationships for Systems of Decision Rules: Cost Versus Cost

Let ψ and φ be strictly increasing cost functions for decision rules, and f, g be increasing functions from $\mathbb{R}^2$ to $\mathbb{R}$. Let T be a nonempty decision table, U be an uncertainty measure, $\alpha \in \mathbb{R}_+$, and $\mathbf{G} = (G_1, \ldots, G_{N(T)})$ be an $N(T)$-tuple of proper subgraphs of the graph $\Delta_{U,\alpha}(T)$.

To study relationships between cost functions ψ_f and φ_g on the set of systems of rules $\mathscr{S}(\mathbf{G}, T)$, we consider two partial functions $\mathscr{R}_{\mathbf{G},T}^{\psi,f,\varphi,g} : \mathbb{R} \to \mathbb{R}$ and $\mathscr{R}_{\mathbf{G},T}^{\varphi,g,\psi,f} : \mathbb{R} \to \mathbb{R}$ defined in the following way:

$$\mathscr{R}_{\mathbf{G},T}^{\psi,f,\varphi,g}(x) = \min\{\varphi_g(T, S) : S \in \mathscr{S}(\mathbf{G}, T), \psi_f(T, S) \leq x\},$$
$$\mathscr{R}_{\mathbf{G},T}^{\varphi,g,\psi,f}(x) = \min\{\psi_f(T, S) : S \in \mathscr{S}(\mathbf{G}, T), \varphi_g(T, S) \leq x\}.$$

Let $p_{\psi,\varphi}^{f,g}(\mathbf{G}, T)=\{(\psi_f(T, S), \varphi_g(T, S)) : S \in \mathscr{S}(\mathbf{G}, T)\}$, and $(a_1, b_1), \ldots, (a_k, b_k)$ be the normal representation of the set $Par(p_{\psi,\varphi}^{f,g}(\mathbf{G}, T))$ where $a_1 < \cdots < a_k$ and $b_1 > \cdots > b_k$. By Lemma 2.8 and Remark 2.2, for any $x \in \mathbb{R}$,

$$\mathscr{R}_{\mathbf{G},T}^{\psi,f,\varphi,g}(x) = \begin{cases} undefined, & x < a_1 \\ b_1, & a_1 \leq x < a_2 \\ \ldots & \ldots \\ b_{k-1}, & a_{k-1} \leq x < a_k \\ b_k, & a_k \leq x \end{cases},$$

$$\mathscr{R}_{\mathbf{G},T}^{\varphi,g,\psi,f}(x) = \begin{cases} undefined, & x < b_k \\ a_k, & b_k \leq x < b_{k-1} \\ \ldots & \ldots \\ a_2, & b_2 \leq x < b_1 \\ a_1, & b_1 \leq x \end{cases}.$$

11.3 Comparison of Heuristics

One application that POPs can be used for is to compare the performance of heuristics using bi-criteria optimization as a standard. Normally, comparisons are performed with regards to either a single criteria or with weighted averages and similar methods for multiple criteria. However, with the knowledge of POPs, we can simply compare how far from the nearest Pareto optimal point a heuristic is. By utilizing this methodology we can easily discern which heuristics perform closely to POPs despite having shortcomings when considered from the perspective of a single criterion.

We consider ten greedy heuristics that operate in a similar fashion. Each heuristic constructs a decision rule for a given row by incrementally adding terms to the left hand side starting with an empty rule until the uncertainty value of the rule is within a certain threshold. We only considered exact rules, but our approach can easily be extended to approximate rules. Exact rule is a $(U, 0)$-decision rule where U is an uncertainty measure. This notion does not depend on the considered uncertainty measure. Instead of exact rules we will write often about $(rme, 0)$-decision rules.

We define a *greedy heuristic* as an algorithm that inductively creates a decision rule ρ for a given row $r = (b_1, \ldots, b_n)$ of decision table T based on a heuristic function $F(T, r, t)$ that selects a conditional attribute from $E(T)$. To create a system of decision rules, we construct one rule for each row and combine them together as a system. The algorithm used to construct a rule for a given row is as follows:

Algorithm $\mathscr{A}_{\text{greedy}}$ (greedy heuristic for decision rule construction).

Input: A decision table T with n conditional attributes $f_1, \ldots, f_n$, a row $r = (b_1, \ldots, b_n)$ from T labeled with decision t, and a heuristic function $F(T, r, t)$.
Output: An exact decision rule for T and r.

1. Create a rule ρ_0 of the form $\to t$.
2. For $k \geq 0$, assume ρ_k is already defined as $f_{i_1} = b_{i_1} \wedge \cdots \wedge f_{i_k} = b_{i_k} \to t$ (step 1 defines ρ_k for $k = 0$).
3. Define a subtable T^k such that $T^k = T(f_{i_1}, b_{i_1}) \ldots (f_{i_k}, b_{i_k})$.
4. If T^k is degenerate then end the algorithm with ρ_k as the constructed rule.
5. Select an attribute $f_{i_{k+1}} \in E(T^k)$ such that $F(T^k, r, t) = f_{i_{k+1}}$.
6. Define ρ_{k+1} by adding the term $f_{i_{k+1}} = b_{i_{k+1}}$ to the left-hand side of ρ_k.
7. Repeat from step 2 with $k = k + 1$.

Now we define the heuristic functions $F(T, r, t)$ used by our algorithm. Given that T is a decision table with conditional attributes $f_1, \ldots, f_n$, $r = (b_1, \ldots, b_n)$ is a row T labeled with decision t, let $N_t(T)$ be the number of rows in T labeled with decision t. Let $M(T, t) = N(T) - N_t(T)$ and $U(T, t) = N_t(T)/N(T)$. For any attribute $f_j \in E(T)$, we define $a(T, r, f_j, t) = N_t(T) - N_t(T(f_j, b_j))$ and $b(T, r, f_j, t) = M(T, t) - M(T(f_j, b_j), t)$. By $mcd(T)$, we denote the *most common decision* for T which is the minimum decision t_0 from $D(T)$ such that $N_{t_0}(T) = \max\{N_t(T) : t \in D(T)\}$.

In this case we can define the following heuristic functions and corresponding heuristics that were used in our experiments:

1. *Div:* $F(T, r, t) = f_j$ such that $f_j \in E(T)$ and $\frac{b(T,r,f_j,t)}{a(T,r,f_j,t)+1}$ is maximized.
2. *Log.* $F(T, r, t) = f_j$ such that $f_j \in E(T)$ and $\frac{b(T,r,f_j,t)}{\log_2(a(T,r,f_j,t)+2)}$ is maximized.
3. *MC:* $F(T, r, t) = f_j$ such that $f_j \in E(T)$, $b(T, r, f_j, t) > 0$ and $a(T, r, f_j, t)$ is minimized.
4. *rme:* $F(T, r, t) = f_j$ such that $f_j \in E(T)$, $T' = T(f_j, b_j)$ and $M(T', t)$ is minimized.
5. *rmeR:* $F(T, r, t) = f_j$ such that $f_j \in E(T)$, $T' = T(f_j, b_j)$ and $\frac{M(T',t)}{N(T')}$ is minimized.
6. *me:* $F(T, r, t) = f_j$ such that $f_j \in E(T)$, $T' = T(f_j, b_j)$ and $M(T', mcd(T'))$ is minimized.

7. *meR:* $F(T, r, t) = f_j$ such that $f_j \in E(T)$, $T' = T(f_j, b_j)$ and $\frac{M(T', mcd(T'))}{N(T')}$ is minimized.
8. *ent:* $F(T, r, t) = f_j$ such that $f_j \in E(T)$, $T' = T(f_j, b_j)$ and

$$\sum_{t \in D(T')} U(T', t) \log_2 U(T', t)$$

is maximized.
9. *gin:* $F(T, r, t) = f_j$ such that $f_j \in E(T)$, $T' = T(f_j, b_j)$ and $1 - \sum_{t \in D(T')} U(T', t)^2$ is minimized.
10. *RT:* $F(T, r, t) = f_j$ such that $f_j \in E(T)$, $T' = T(f_j, b_j)$ and the number of unordered pairs of rows of T' labeled with different decisions is minimized.

When comparing heuristics against optimal rules created via dynamic programming, we need to first determine the criteria being applied. When considering a single criterion (cost function $\psi \in \{l, -c\}$), we can simply compare the *average relative difference* (ARD) between the greedy solution and the optimal dynamic programming solution.

In order to compute the ARD between a system of decision rules S_g constructed by a greedy heuristic and an optimal system S_d constructed by dynamic programming, we first calculate the value of our selected criteria (cost function ψ) for both. For a given table T, this gives us $q_g = \psi_{\text{sum}}(T, S_g)$ and $q_d = \psi_{\text{sum}}(T, S_d)$. At this point we can define the average relative difference as follows:

$$\left| \frac{q_g - q_d}{q_d} \right| .$$

Using ARD to as a measurement allows us to avoid bias caused by the varying ranges of different cost functions. For example, l varies from 0 to $n = dim(T)$ and $-c$ from 0 to $-N(T)$. ARD serves to normalize these differences.

The benefit of ARD is to measure concretely how well a given heuristic performs when comparing a single criterion at a time. However, it does not help us when we compare multiple criteria at once. In the case of bi-criteria optimization relative to l and $-c$, we can utilize Pareto optimal points. We can measure the distance between a greedy solution and the nearest Pareto optimal point and get a good idea of its performance. In order to remove any bias due to differing values for different cost functions (length only varies from 0 to n while minus coverage varies from 0 to $-N(T)$) we normalized the distance along each axis first before calculating the distances.

Let the sequence $(a_1, b_1), \ldots, (a_k, b_k)$ be the normal representation for the set $Par(p_{l,-c}^{\text{sum,sum}}(\mathbf{G}, T))$ where $\mathbf{G} = (\Delta(T), \ldots, \Delta(T))$ and $\Delta(T) = \Delta_{rme,0}(T)$. Let S_g be a system of decision rules constructed by a greedy heuristic, $a = l_{\text{sum}}(T, S_g)$, and $b = -c_{\text{sum}}(T, S_g)$. We define the normalized distance from the point (a, b) corresponding to the heuristic (*heuristic point*) to the set of Pareto optimal points $Par(p_{l,-c}^{\text{sum,sum}}(\mathbf{G}, T))$ as follows:

$$\min_{1\leq i\leq k}\sqrt{\left(\frac{a_i-a}{a_k}\right)^2+\left(\frac{b_i-b}{b_k}\right)^2}.$$

In order to reduce bias from table sizes and to avoid penalizing heuristics for doing badly on a few datasets, we can consider rankings instead of exact values. Lower ranks are better, with 1 being the best rank. Furthermore, all heuristics that are tied will receive the same, averaged rank. For example, if two heuristics tie with second and third rank, then they will both be ranked 2.5 instead. This helps obtain reasonable results when considering few but greatly varying datasets such as those from the UCI ML Repository.

Our first set of experiments involved the UCI ML Repository. In Table 11.1 we present details regarding these tables. Columns "Rows" and "Attr" specify the dimensions while column "# POPs" displays the number of Pareto optimal points when considering average length and average coverage of rules in rule systems. One can show that the number of Pareto optimal points is equal to one if and only if each row

Table 11.1 Details regarding decision tables from UCI ML repository

Table name	Rows	Attr	# POPs
Adult-stretch	16	4	1
Agaricus-lepiota	8124	22	10,822
Balance-scale	625	4	1
Breast-cancer	266	9	204
Cars	1728	6	1
Flags	194	26	423
Hayes-roth-data	69	4	1
House-votes-84	279	16	275
Lenses	24	4	1
Lymphography	148	18	140
Monks-1-test	432	6	1
Monks-1-train	124	6	5
Monks-2-test	432	6	1
Monks-2-train	169	6	13
Monks-3-test	432	6	1
Monks-3-train	122	6	1
Nursery	12,960	8	1
Shuttle-landing-control	15	6	4
Soybean-small	47	35	11
Spect-test	169	22	83
Teeth	23	8	1
Tic-tac-toe	958	9	17
Zoo-data	59	16	20

Table 11.2 ARD for top five heuristics: length

Table name	meR	rmeR	RT	me	Log
Adult-stretch	0.00	0.00	0.40	0.40	0.00
Agaricus-lepiota	0.00	0.00	0.00	0.00	0.27
Balance-scale	0.00	0.00	0.04	0.04	0.00
Breast-cancer	0.02	0.21	0.03	0.03	0.11
Cars	0.04	0.02	0.12	0.11	0.02
Flags	0.02	0.10	0.02	0.02	0.53
Hayes-roth-data	0.00	0.01	0.05	0.08	0.01
House-votes-84	0.01	0.07	0.04	0.05	0.11
Lenses	0.06	0.00	0.09	0.19	0.00
Lymphography	0.07	0.15	0.05	0.09	0.39
Monks-1-test	0.00	0.00	0.11	0.11	0.00
Monks-1-train	0.09	0.18	0.15	0.14	0.12
Monks-2-test	0.08	0.05	0.12	0.08	0.02
Monks-2-train	0.05	0.08	0.10	0.08	0.05
Monks-3-test	0.00	0.00	0.03	0.06	0.05
Monks-3-train	0.01	0.02	0.02	0.03	0.01
Nursery	0.05	0.03	0.23	0.11	0.04
Shuttle-landing-control	0.00	0.09	0.00	0.00	0.00
Soybean-small	0.00	0.00	0.00	0.00	0.21
Spect-test	0.02	0.03	0.03	0.10	0.05
Teeth	0.00	0.00	0.00	0.00	0.00
Tic-tac-toe	0.25	0.19	0.32	0.39	0.23
Zoo-data	0.03	0.03	0.03	0.01	0.21
Average	0.04	0.05	0.09	0.09	0.10

of the table has a totally optimal rule. The comparison with Table 10.6 shows that even decision tables that are not complete can have totally optimal rules for each row (see, for example, tables “Hayes-roth-data”, “Monks-3-train”, and “Teeth”).

The considered experiments show the performance of our heuristics when working with datasets created from real-life information. In Tables 11.2 and 11.3 we can see the ARD for the top five heuristics ordered from best to worst for length and coverage, respectively.

One thing that we can notice is that heuristics that perform well for one cost function don’t necessarily perform well for another. For example, Div, the leading heuristic for coverage, was not in the top five for length. Similarly, meR performed best for length but did not do so well for coverage. Of particular interest are Log and rmeR which came up in the top five for both cost functions. We can also note that the best heuristics for each cost function performed very close to the optimal dynamic programming approach: the ARD was only 4% and 7% for length and coverage, respectively.

Table 11.3 ARD for top five heuristics: coverage

Table name	Div	Log	rmeR	gin	ent
Adult-stretch	0.00	0.00	0.00	0.11	0.11
Agaricus-lepiota	0.08	0.18	0.55	0.56	0.56
Balance-scale	0.00	0.00	0.00	0.12	0.12
Breast-cancer	0.28	0.24	0.53	0.55	0.55
Cars	0.00	0.00	0.00	0.00	0.00
Flags	0.19	0.20	0.59	0.62	0.63
Hayes-roth-data	0.00	0.01	0.05	0.05	0.04
House-votes-84	0.06	0.07	0.23	0.27	0.27
Lenses	0.00	0.00	0.02	0.06	0.05
Lymphography	0.15	0.24	0.50	0.56	0.56
Monks-1-test	0.00	0.00	0.00	0.20	0.20
Monks-1-train	0.14	0.08	0.13	0.14	0.14
Monks-2-test	0.03	0.15	0.17	0.17	0.17
Monks-2-train	0.11	0.23	0.34	0.39	0.39
Monks-3-test	0.10	0.10	0.00	0.02	0.02
Monks-3-train	0.03	0.01	0.16	0.17	0.17
Nursery	0.01	0.00	0.01	0.01	0.01
Shuttle-landing-control	0.03	0.12	0.12	0.12	0.12
Soybean-small	0.00	0.00	0.29	0.29	0.29
Spect-test	0.07	0.08	0.21	0.21	0.21
Teeth	0.00	0.00	0.00	0.00	0.00
Tic-tac-toe	0.27	0.66	0.55	0.69	0.69
Zoo-data	0.00	0.00	0.08	0.09	0.09
Average	0.07	0.10	0.20	0.23	0.23

In Tables 11.4 and 11.5 we can see the ranked values ordered from best to worst for the top five heuristics for coverage and length, respectively. Of particular interest are Log and rmeR which came up in the top three for both cost functions.

In order to further study the performance of heuristics in comparison to each other, we also performed experiments involving bi-criteria optimization. The results for these experiments for the five best heuristics can be seen in Table 11.6. Also, the graphs in Figs. 11.1, 11.2, and 11.3 show plots for three datasets in more detail. For the convenience, instead of the sum of lengths of rules in a system we consider average length of rules in the system, and instead of minus sum of coverage of rules in the system we consider average coverage of rules in the system.

From Table 11.6 we can conclude that the Log heuristic outperforms all the other heuristics for most tables. The next best heuristics are Div and rmeR. There is a major gap between these and the other heuristics. Furthermore, this hierarchy is emphasized

Table 11.4 Heuristic comparison: coverage ranking

Table Name	Div	Log	rmeR	meR	MC
Adult-stretch	3	3	3	3	3
Agaricus-lepiota	1	2	4	8	3
Balance-scale	3	3	3	3	3
Breast-cancer	2	1	3	7	10
Cars	3	1	2	4	10
Flags	1	2	4	8	3
Hayes-roth-data	1	2	6	7	3
House-votes-84	1	2	4	8	3
Lenses	2	2	4	5.5	2
Lymphography	1	2	4	8	3
Monks-1-test	2.5	2.5	2.5	2.5	10
Monks-1-train	4	1	3	2	10
Monks-2-test	1	2	4.5	8.5	7
Monks-2-train	1	2	4	8	3
Monks-3-test	8.5	8.5	1.5	1.5	10
Monks-3-train	2	1	3	4	10
Nursery	2	1	3	5	10
Shuttle-landing	1	5.5	5.5	5.5	10
Soybean-small	2	2	7	7	2
Spect-test	1	2	3	7	10
Teeth	5.5	5.5	5.5	5.5	5.5
Tic-tac-toe	1	4	3	8	2
Zoo-data	1.5	3	4	5	1.5
Average	2.22	2.61	3.76	5.70	5.83

when we take a look at the graphs for some of the decision tables. The examples in Figs. 11.1, 11.2, and 11.3 show that heuristics closest to Pareto optimality are not necessarily very close to being optimal for either criteria on their own. The overall rankings for all heuristics can be found in Table 11.7.

Our second set of experiments involved determining the behavior of heuristics when dealing with randomly generated partial Boolean functions (PBFs).

Our experiments involved considering tables of varying sizes created by PBFs of ten variables. When dealing with how many inputs to define, we considered ten values spaced evenly from 0 to 2^{10} (total Boolean function). For each size we ran 100 experiments and took average results. We used length and coverage as our target cost functions.

Table 11.5 Heuristic comparison: length ranking

Table name	meR	rmeR	Log	RT	me
Adult-stretch	3	3	3	8	8
Agaricus-lepiota	4	4	8	4	4
Balance-scale	3	3	3	8	8
Breast-cancer	1	5	4	2	3
Cars	4	1.5	1.5	8.5	6.5
Flags	2.5	7	8	1	2.5
Hayes-roth-data	1	3	3	6.5	8.5
House-votes-84	1	4	5	2	3
Lenses	5	2.5	2.5	6.5	10
Lymphography	2	4	5.5	1	3
Monks-1-test	2.5	2.5	2.5	7	7
Monks-1-train	1	8	2	4	3
Monks-2-test	7.5	4	1	10	7.5
Monks-2-train	2	3	1	5	4
Monks-3-test	1.5	1.5	4.5	3	7.5
Monks-3-train	1.5	4	1.5	3	5
Nursery	3.5	1	2	9	7
Shuttle-landing	4	8	4	4	4
Soybean-small	4	4	8.5	4	4
Spect-test	1	2	4	3	5
Teeth	5	5	5	5	5
Tic-tac-toe	4	1	3	5	6
Zoo-data	4.5	4.5	8	4.5	1.5
Average	2.98	3.72	3.93	4.96	5.35

In Figs. 11.4 and 11.5, we can see the ARD for each heuristic for coverage and length, respectively. Lower values are closer to optimal. The values shown for bi-criteria optimization in Fig. 11.6 are the average normalized distances from heuristic points to POPs as described earlier. When considering coverage, we can see that Div and Log both perform well. Furthermore, all heuristics are clearly separated and show distinction in their performance. For length, we mostly see a similar display, this time with meR, Log, and rmeR being the leading heuristics. There are some interesting things to note here. Firstly, the heuristic MC shows great variance depending on how sparsely defined the PBF is. Also, the ARD values for length tend to be much closer to optimal than the ARD values for coverage. The variance in performance of heuristics also seems to be greater for coverage than for length.

For bi-criteria optimization, we can see that Log performs best, although Div matches its performance for more densely defined PBFs. We see once again that MC does far better with total Boolean functions than it does with sparse PBFs.

Table 11.6 Heuristic comparison: bi-criteria ranking

Table name	Log	Div	rmeR	meR	ent
Adult-stretch	3	3	3	3	8
Agaricus-lepiota	1	2	4	8	6
Balance-scale	3	3	3	3	8
Breast-cancer	1	2	3	4	6
Cars	1	3	2	4	5
Flags	1	9	2	6	5
Hayes-roth-data	2	1	5	7	4
House-votes-84	1	2	4	8	6
Lenses	2	2	4	5	6
Lymphography	1	2	3	8	6
Monks-1-test	2.5	2.5	2.5	2.5	7
Monks-1-train	1	4	3	2	6
Monks-2-test	2	1	4.5	8.5	4.5
Monks-2-train	2	1	4	8	6
Monks-3-test	8.5	8.5	1.5	1.5	5.5
Monks-3-train	1	2	3	4	6
Nursery	1	2	3	5	4
Shuttle-landing	4	9	8	4	4
Soybean-small	1.5	1.5	7	7	7
Spect-test	1	2	3	7	5
Teeth	5	5	5	5	5
Tic-tac-toe	4	1	3	8	6
Zoo-data	1	2	3	4	5.5
Average	2.20	3.07	3.63	5.33	5.72

Table 11.7 Heuristic overall rankings: UCI ML repository datasets

Heuristic	Div	ent	gin	Log	MC	me	meR	rme	rmeR	RT
Coverage	**1**	6	7	**2**	5	9	4	8	**3**	10
Length	6	7	8	**3**	10	5	**1**	9	**2**	4
Bi-criteria	**2**	5	6	**1**	8	9	4	7	**3**	10

With the exception of MC, we can see that there is a clear distinction between two classes of heuristics. One group tends to perform better and another clearly performs worse. This distinction exists in all three graphs. It is most clear when looking at total Boolean functions as all heuristics converge on one of two points. Also, the rankings for heuristics acquired during the UCI ML Repository experiments match very closely to the order seen in the PBF tests. Both of these facts help to confirm the optimality of certain heuristics over others.

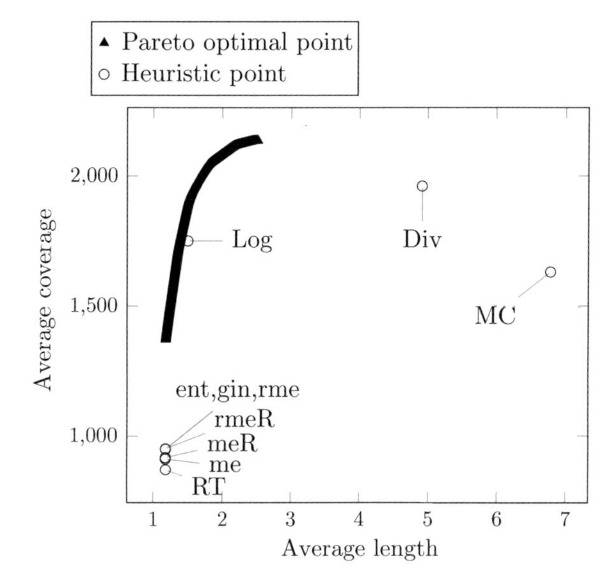

Fig. 11.1 Pareto optimal points and heuristic points for "Agaricus-lepiota"

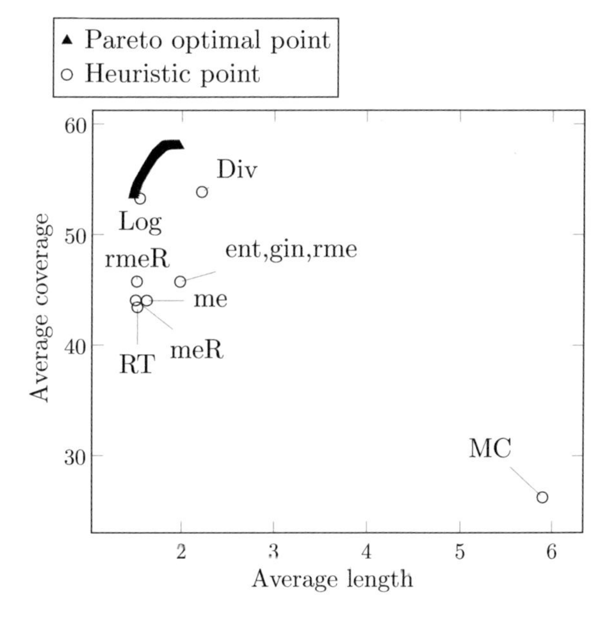

Fig. 11.2 Pareto optimal points and heuristic points for "Spect-test"

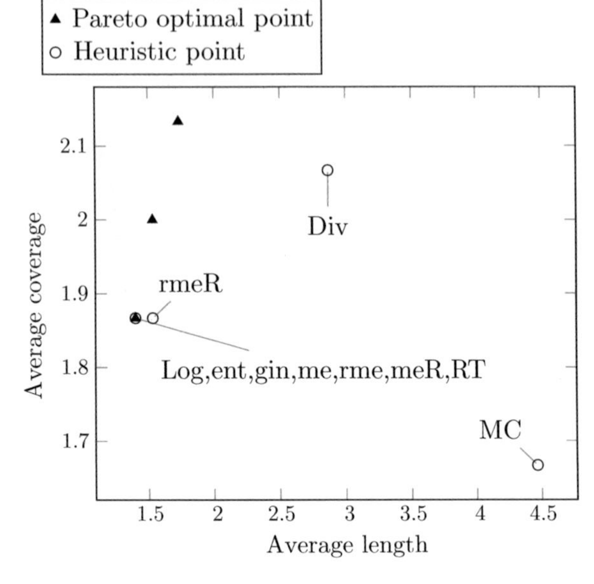

Fig. 11.3 Pareto optimal points and heuristic points for "Shuttle-landing-control"

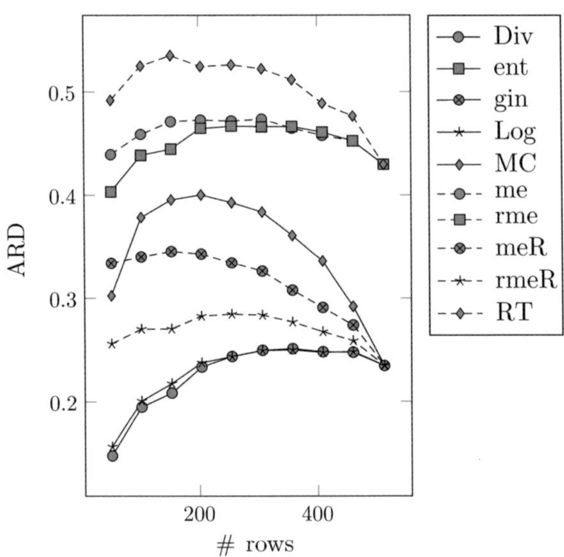

Fig. 11.4 Comparison of heuristics relative to coverage: 10 variable PBFs

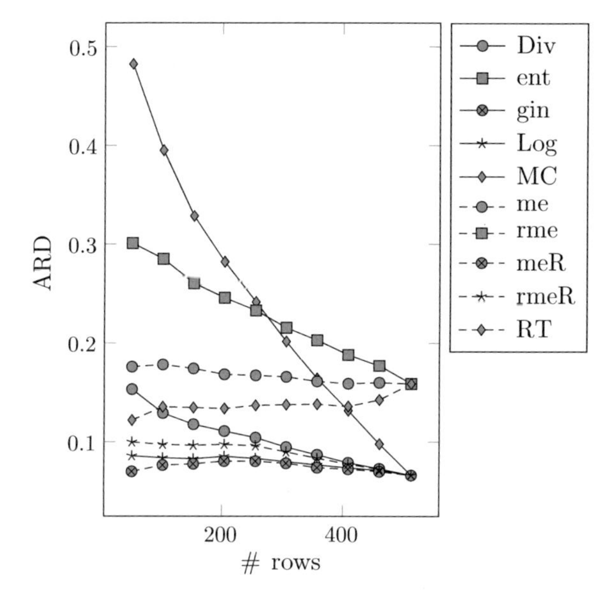

Fig. 11.5 Comparison of heuristics relative to length: 10 variable PBFs

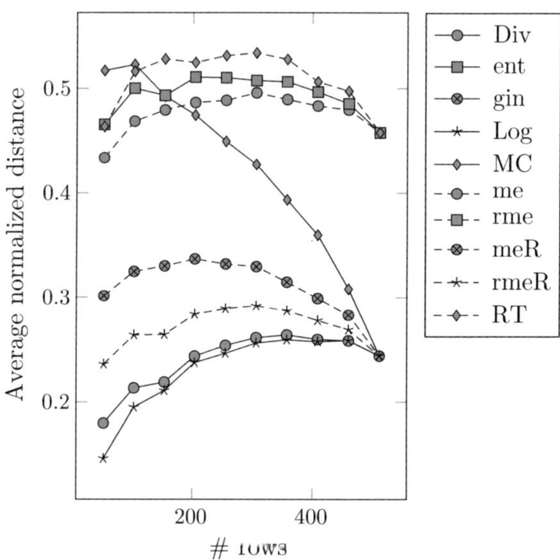

Fig. 11.6 Comparison of heuristics relative to bi-criteria optimization: 10 variable PBFs

References

1. Hammer, P.L., Kogan, A., Simeone, B., Szedmák, S.: Pareto-optimal patterns in logical analysis of data. Discret. Appl. Math. **144**(1–2), 79–102 (2004)
2. Lichman, M.: UCI Machine Learning Repository. University of California, Irvine, School of Information and Computer Sciences (2013). http://archive.ics.uci.edu/ml

Chapter 12
Bi-criteria Optimization Problem for Decision Rules and Systems of Rules: Cost Versus Uncertainty

The trade-off between accuracy and simplicity of a knowledge model is a complex problem. This problem has been studied for both, decision trees (see, for example [1]) and decision rules (see, for example [2]). In this chapter, we consider the same problem with regards to decision rules and systems of decision rules. In order to understand the trade-off and to present all possible optimal solutions, we construct the set of Pareto optimal points comparing a measure of accuracy (uncertainty measure) with a cost function.

In Sect. 12.1, we present a method of deriving the set of Pareto optimal points for decision rules for a given row of a decision table with respect to a cost function and an uncertainty measure. Then, we use this set to construct functions describing the relationships between the considered cost function and uncertainty measure. This theory is then expanded to work with systems of decision rules in Sect. 12.2. Finally, in Sect. 12.3, we present some illustrative examples displaying trade-offs that exist between cost and uncertainty for systems of decision rules in tables consisting of real-world information.

12.1 Pareto Optimal Points for Decision Rules: Cost Versus Uncertainty

Let ψ be a cost function for decision rules given by pair of functions ψ^0, F, U be an uncertainty measure, T be a nondegenerate decision table with n conditional attributes $f_1, \ldots, f_n$, and $r = (b_1, \ldots, b_n)$ be a row of T.

Let Θ be a node of the graph $\Delta(T) = \Delta_{U,0}(T)$ containing r. Let us recall that a rule over Θ is called a U-decision rule for Θ and r if there exists a nonnegative real number α such that the considered rule is a (U, α)-decision rule for Θ and r. Let $p_{\psi,U}(\Theta, r) = \{((\psi(\Theta, \rho), U(\Theta, \rho)) : \rho \in DR_U(\Theta, r)\}$ where $DR_U(\Theta, r)$ is the set of U-decision rules for Θ and r. Our aim is to find the set $Par(p_{\psi,U}(T, r))$.

H. AbouEisha et al., *Extensions of Dynamic Programming for Combinatorial Optimization and Data Mining*, Intelligent Systems Reference Library 146,
https://doi.org/10.1007/978-3-319-91839-6_12

In fact, we will find the set $Par(p_{\psi,U}(\Theta, r))$ for each node Θ of the graph $\Delta(T)$ containing r.

To this end we consider an auxiliary set of rules $Path(\Theta, r)$ for each node Θ of $\Delta(T)$ containing r. Let τ be a directed path from the node Θ to a node Θ' of $\Delta(T)$ containing r in which edges are labeled with pairs $(f_{i_1}, b_{i_1}), \ldots, (f_{i_m}, b_{i_m})$. We denote by $rule(\tau)$ the decision rule over Θ

$$f_{i_1} = b_{i_1} \wedge \ldots \wedge f_{i_m} = b_{i_m} \rightarrow mcd(\Theta') .$$

If $m = 0$ (if $\Theta = \Theta'$) then the rule $rule(\tau)$ is equal to $\rightarrow mcd(\Theta)$. We denote by $Path(\Theta, r)$ the set of rules $rule(\tau)$ corresponding to all directed paths τ from Θ to a node Θ' of $\Delta(T)$ containing r (we consider also the case when $\Theta = \Theta'$). We correspond to the set of rules $Path(\Theta, r)$ the set of pairs $p^{path}_{\psi,U}(\Theta, r) = \{((\psi(\Theta, \rho), U(\Theta, \rho)) : \rho \in Path(\Theta, r)\}$.

Lemma 12.1 *Let U be an uncertainty measure, T be a nondegenerate decision table, r be a row of T, and Θ be a node of $\Delta(T)$ containing r. Then $DR_U(\Theta, r) \subseteq Path(\Theta, r)$.*

Proof Let $\alpha \in \mathbb{R}_+$. Then either $\Delta_{U,\alpha}(T) = \Delta_{U,0}(T) = \Delta(T)$ or the graph $\Delta_{U,\alpha}(T)$ is obtained from the graph $\Delta(T)$ by removal some nodes and edges. From here and from the definition of the set of rules $Rule(\Delta_{U,\alpha}(T), \Theta, r)$ it follows that this set is a subset of the set $Path(\Theta, r)$. From Proposition 10.1 it follows that the set $Rule(\Delta_{U,\alpha}(T), \Theta, r)$ coincides with the set of all (U, α)-decision rules for Θ and r. Since α is an arbitrary nonnegative real number, we have $DR_U(\Theta, r) \subseteq Path(\Theta, r)$. □

Lemma 12.2 *Let U be a strictly increasing uncertainty measure, T be a nondegenerate decision table with n conditional attributes $f_1, \ldots, f_n$, $r = (b_1, \ldots, b_n)$ be a row of T, and Θ be a node of $\Delta(T)$ containing r. Then $DR_U(\Theta, r) = Path(\Theta, r)$.*

Proof From Lemma 12.1 it follows that $DR_U(\Theta, r) \subseteq Path(\Theta, r)$. Let us show that $Path(\Theta, r) \subseteq DR_U(\Theta, r)$. Let $rule(\tau) \in Path(\Theta, r)$ and τ be a directed path from Θ to a node Θ' containing r in which edges are labeled with pairs

$$(f_{i_1}, b_{i_1}), \ldots, (f_{i_m}, b_{i_m}) .$$

Then $rule(\tau)$ is equal to $f_{i_1} = b_{i_1} \wedge \ldots \wedge f_{i_m} = b_{i_m} \rightarrow mcd(\Theta')$. Let $\Theta^0 = \Theta$, $\Theta^1 = \Theta(f_{i_1}, b_{i_1}), \ldots, \Theta^m = \Theta(f_{i_1}, b_{i_1}), \ldots, (f_{i_m}, b_{i_m}) = \Theta'$.

If $m = 0$ then $rule(\tau)$ is, evidently, a $(U, U(\Theta))$-decision rule for Θ and r. Therefore $rule(\tau) \in DR_U(\Theta, r)$.

Let $m > 0$. By definition of the graph $\Delta(T)$, $f_{i_j} \in E(\Theta^{j-1})$ and $b_{i_j} \in E(\Theta^{j-1}, f_{i_j})$ for $j = 1, \ldots, m$. Since U is a strictly increasing uncertainty measure, $U(\Theta^0) > U(\Theta^1) > \ldots > U(\Theta^m)$. Therefore $rule(\tau)$ is a $(U, U(\Theta'))$-decision rule for Θ and r, and $rule(\tau) \in DR_U(\Theta, r)$. Hence $DR_U(\Theta, r) = Path(\Theta, r)$. □

Corollary 12.1 *Let U be a strictly increasing uncertainty measure, ψ be a cost function for decision rules, T be a nondegenerate decision table, r be a row of T, and Θ be a node of $\Delta(T)$ containing r. Then $p^{path}_{\psi,U}(\Theta, r) = p_{\psi,U}(\Theta, r)$ and $Par(p^{path}_{\psi,U}(\Theta, r)) = Par(p_{\psi,U}(\Theta, r))$.*

Note that, by Proposition 3.1 and Remark 3.1, R and $me + R$ are strictly increasing uncertainty measures.

Lemma 12.3 *Let $\psi \in \{l, -c, -c_M\}$, U be an uncertainty measure, T be a nondegenerate decision table with n conditional attributes $f_1, \ldots, f_n$, $r = (b_1, \ldots, b_n)$ be a row of T, and Θ be a node of $\Delta(T)$ containing r. Then $Par(p^{path}_{\psi,U}(\Theta, r)) = Par(p_{\psi,U}(\Theta, r))$.*

Proof From Lemma 12.1 it follows that $p_{\psi,U}(\Theta, r) \subseteq p^{path}_{\psi,U}(\Theta, r)$. Let us show that, for any pair $\alpha \in p^{path}_{\psi,U}(\Theta, r)$, there exists a pair $\beta \in p_{\psi,U}(\Theta, r)$ such that $\beta \leq \alpha$. Let $\alpha = (\psi(\Theta, \rho), U(\Theta, \rho))$ where $\rho \in Path(\Theta, r)$. If $\rho \in DR_U(\Theta, r)$ then $\alpha \in p_{\psi,U}(\Theta, r)$, and the considered statement holds.

Let $\rho \notin DR_U(\Theta, r)$ and $\rho = rule(\tau)$ where τ is a directed path from Θ to a node Θ' containing r which edges are labeled with pairs $(f_{i_1}, b_{i_1}), \ldots, (f_{i_m}, b_{i_m})$. Then $\rho = rule(\tau)$ is equal to

$$f_{i_1} = b_{i_1} \wedge \ldots \wedge f_{i_m} = b_{i_m} \rightarrow mcd(\Theta') .$$

Since $\rho \notin DR_U(\Theta, r)$, $m > 0$. Let

$$\Theta^0 = \Theta, \Theta^1 = \Theta(f_{i_1}, b_{i_1}), \ldots, \Theta^m = \Theta(f_{i_1}, b_{i_1}), \ldots, (f_{i_m}, b_{i_m}) = \Theta' .$$

By definition of the graph $\Delta(T)$, $f_{i_j} \in E(\Theta^{j-1})$ for $j = 1, \ldots, m$. Since

$$\rho \notin DR_U(\Theta, r) ,$$

there exists $j \in \{1, \ldots, m-1\}$ such that $U(\Theta^j) \leq U(\Theta^m)$. Let j_0 be the minimum number from $\{1, \ldots, m-1\}$ for which $U(\Theta^{j_0}) \leq U(\Theta^m)$. We denote by ρ' the rule

$$f_{i_1} = b_{i_1} \wedge \ldots \wedge f_{j_0} = b_{j_0} \rightarrow mcd(\Theta^{j_0}) .$$

It is clear that ρ' is a $(U, U(\Theta^{j_0}))$-decision rule for Θ and r. Therefore $\rho' \in DR_U(\Theta, r)$ and $\beta = ((\psi(\Theta, \rho')), U(\Theta, \rho')) \in p_{\psi,U}(\Theta, r)$. We have $U(\Theta, \rho') = U(\Theta^{j_0}) \leq U(\Theta^m) = U(\Theta, \rho)$. If $\psi = l$ then $\psi(\Theta, \rho') = j_0 < m = \psi(\Theta, \rho)$. Let $\psi = -c$. Then

$$\psi(\Theta, \rho') = -N_{mcd(\Theta^{j_0})}(\Theta^{j_0}) \leq -N_{mcd(\Theta^m)}(\Theta^{j_0}) \leq -N_{mcd(\Theta^m)}(\Theta^m) = \psi(\Theta, \rho) .$$

Let $\psi = -c_M$. Then $\psi(\Theta, \rho') = -N^M(\Theta^{j_0}) \leq -N^M(\Theta^m) = \psi(\Theta, \rho)$. Therefore $\beta \leq \alpha$.

Using Lemma 2.2 we obtain $Par(p^{path}_{\psi,U}(\Theta, r)) = Par(p_{\psi,U}(\Theta, r))$. □

So in some cases, when U is a strictly increasing uncertainty measure (in particular, R and $me + R$ are such measures), or $\psi \in \{l, -c, -c_M\}$, we have

$$Par(p_{\psi,U}(T, r)) = Par(p^{path}_{\psi,U}(T, r)) .$$

In these cases we can concentrate on the construction of the set $Par(p^{path}_{\psi,U}(T, r))$.

Let ψ be a strictly increasing cost function for decision rules given by pair of functions ψ^0, F, U be an uncertainty measure, T be a decision table with n conditional attributes $f_1, \ldots, f_n$, and $r = (b_1, \ldots, b_n)$ be a row of T. We now describe an algorithm $\mathscr{A}_{13}$ which constructs the set $Par(p^{path}_{\psi,U}(T, r))$. In fact, this algorithm constructs, for each node Θ of the graph G, the set $B(\Theta, r) = Par(p^{path}_{\psi,U}(\Theta, r))$.

Algorithm $\mathscr{A}_{13}$ (construction of POPs for decision rules, cost versus uncertainty).

Input: Strictly increasing cost function ψ for decision rules given by pair of functions ψ^0, F, an uncertainty measure U, a nonempty decision table T with n conditional attributes $f_1, \ldots, f_n$, a row $r = (b_1, \ldots, b_n)$ of T, and the graph $\Delta(T)$.

Output: The set $Par(p^{path}_{\psi,U}(T, r))$ of Pareto optimal points for the set of pairs $p^{path}_{\psi,U}(T, r) = \{((\psi(T, \rho), U(T, \rho)) : \rho \in Path(T, r)\}$.

1. If all nodes in $\Delta(T)$ containing r are processed, then return the set $B(T, r)$. Otherwise, choose in the graph $\Delta(T)$ a node Θ containing r which is not processed yet and which is either a terminal node of $\Delta(T)$ or a nonterminal node of $\Delta(T)$ such that, for any $f_i \in E(\Theta)$, the node $\Theta(f_i, b_i)$ is already processed, i.e., the set $B(\Theta(f_i, b_i), r)$ is already constructed.
2. If Θ is a terminal node, then set $B(\Theta, r) = \{\{(\psi^0(\Theta), 0)\}\}$. Mark the node Θ as processed and proceed to step 1.
3. If Θ is a nonterminal node then construct $(\psi^0(\Theta), U(\Theta))$, for each $f_i \in E(\Theta)$, construct the set $B(\Theta(f_i, b_i), r)^{FH}$, where $H(x) = x$, and construct the multiset $A(\Theta, r) = \{(\psi^0(\Theta), U(\Theta))\} \cup \bigcup_{f_i \in E(\Theta)} B(\Theta(f_i, b_i), r)^{FH}$ by simple transcription of elements from the sets $B(\Theta(f_i, b_i), r)^{FH}$, $f_i \in E(\Theta)$, and $(\psi^0(\Theta), U(\Theta))$.
4. Apply to $A(\Theta, r)$ the algorithm $\mathscr{A}_1$ which constructs the set $Par(A(\Theta, r))$. Set $B(\Theta, r) = Par(A(\Theta, r))$. Mark the node Θ as processed and proceed to step 1.

Proposition 12.1 *Let ψ be strictly increasing cost function for decision rules given by pair of functions ψ^0, F, U be an uncertainty measure, T be a nonempty decision table with n conditional attributes $f_1, \ldots, f_n$, and $r = (b_1, \ldots, b_n)$ be a row of T. Then, for each node Θ of the graph $\Delta(T)$ containing r, the algorithm $\mathscr{A}_{13}$ constructs the set $B(\Theta, r) = Par(p^{path}_{\psi,U}(\Theta, r))$.*

Proof We prove the considered statement by induction on nodes of G. Let Θ be a terminal node of $\Delta(T)$ containing r. Then $U(\Theta) = 0$, $Path(\Theta, r) = \{\rightarrow mcd(\Theta)\}$, $p^{path}_{\psi,U}(\Theta, r) = Par(p^{path}_{\psi,U}(\Theta, r)) = \{(\psi^0(\Theta), 0)\}$, and $B(\Theta, r) = Par(p^{path}_{\psi,U}(\Theta, r))$.

Let Θ be a nonterminal node of $\Delta(T)$ containing r such that, for any $f_i \in E(\Theta)$, the considered statement holds for the node $\Theta(f_i, b_i)$, i.e.,

$$B(\Theta(f_i, b_i), r) = Par(p_{\psi,U}^{path}(\Theta(f_i, b_i), r)) .$$

One can show that

$$p_{\psi,\varphi}^{path}(\Theta, r) = \{(\psi^0(\Theta), U(\Theta))\} \cup \bigcup_{f_i \in E(\Theta)} p_{\psi,U}^{path}(\Theta(f_i, b_i), r)^{FH}$$

where H is a function from $\mathbb{R}$ to $\mathbb{R}$ such that $H(x) = x$. It is clear that

$$Par(\{(\psi^0(\Theta), U(\Theta))\}) = \{(\psi^0(\Theta), U(\Theta))\}.$$

From Lemma 2.5 it follows that

$$Par(p_{\psi,U}^{path}(\Theta, r)) \subseteq \{(\psi^0(\Theta), U(\Theta))\} \cup \bigcup_{f_i \in E(\Theta)} Par(p_{\psi,U}^{path}(\Theta(f_i, b_i), r)^{FH}) .$$

By Lemma 2.7, $Par(p_{\psi,U}^{path}(\Theta(f_i, b_i), r)^{FH}) = Par(p_{\psi,U}^{path}(\Theta(f_i, b_i), r))^{FH}$ for any $f_i \in E(\Theta)$. Therefore

$$Par(p_{\psi,U}^{path}(\Theta, r)) \subseteq \{(\psi^0(\Theta), U(\Theta))\} \cup \bigcup_{f_i \in E(\Theta)} Par(p_{\psi,U}^{path}(\Theta(f_i, b_i), r))^{FH} \subseteq p_{\psi,U}^{path}(\Theta, r) .$$

Using Lemma 2.4 we obtain

$$Par(p_{\psi,U}^{path}(\Theta, r)) = Par\left(\{(\psi^0(\Theta), U(\Theta))\} \cup \bigcup_{f_i \in E(\Theta)} Par(p_{\psi,U}^{path}(\Theta(f_i, b_i), r))^{FH}\right) .$$

Since $B(\Theta, r) = Par\left(\{(\psi^0(\Theta), U(\Theta))\} \cup \bigcup_{f_i \in E(\Theta)} B(\Theta(f_i, b_i), r)^{FH}\right)$ and

$$B(\Theta(f_i, b_i), r) = Par(p_{\psi,U}^{path}(\Theta(f_i, b_i), r))$$

for any $f_i \in E(\Theta)$, we have $B(\Theta, r) = Par(p_{\psi,U}^{path}(\Theta, r))$. □

We now evaluate the number of elementary operations (computations of F, H, ψ^0, U, and comparisons) made by the algorithm $\mathscr{A}_{13}$. Let us recall that, for a given cost function ψ for decision rules and decision table T,

$$q_{\psi}(T) = |\{\psi(\Theta, \rho) : \Theta \in SEP(T), \rho \in DR(\Theta)\}| \ .$$

In particular, by Lemma 9.1, $q_l(T) \leq n + 1$, $q_{-c}(T) \leq N(T) + 1$, $q_{mc}(T) \leq N(T) + 1$, $q_{-rc}(T) \leq N(T)(N(T) + 1)$, $q_{-c_M}(T) \leq N(T) + 1$, and $q_{rmc}(T) \leq N(T)(N(T) + 1)$.

Proposition 12.2 *Let ψ be a strictly increasing cost function for decision rules given by pair of functions ψ^0, F, H be a function from $\mathbb{R}$ to $\mathbb{R}$ such that $H(x) = x$, U be an uncertainty measure, T be a nonempty decision table with n conditional attributes $f_1, \ldots, f_n$, and $r = (b_1, \ldots, b_n)$ be a row of T. Then, to construct the set $Par(p^{path}_{\psi,U}(\Theta, r))$, the algorithm $\mathcal{A}_{13}$ makes*

$$O(L(\Delta(T))q_{\psi}(T)n \log(q_{\psi}(T)n))$$

elementary operations (computations of F, H, ψ^0, U, and comparisons).

Proof To process a terminal node, the algorithm $\mathcal{A}_{13}$ makes one elementary operation – computes ψ^0. We now evaluate the number of elementary operations under the processing of a nonterminal node Θ.

From Proposition 12.1 it follows that $B(\Theta, r) = Par(p^{path}_{\psi,U}(\Theta, r))$ and

$$B(\Theta(f_i, b_i), r) = Par(p^{path}_{\psi,U}(\Theta(f_i, b_i), r))$$

for any $f_i \in E(\Theta)$. From Lemma 2.3 it follows that $|B(\Theta(f_i, b_i), r)| \leq q_{\psi}(T)$ for any $f_i \in E(\Theta)$. It is clear that $|E(\Theta)| \leq n$,

$$\left|B(\Theta(f_i, b_i), r)^{FH}\right| = |B(\Theta(f_i, b_i), r)|$$

for any $f_i \in E(\Theta)$, and $|A(\Theta, r)| \leq q_{\psi}(T)n$. Therefore, to construct the sets

$$B(\Theta(f_i, b_i), r)^{FH} \ ,$$

$f_i \in E(\Theta)$, from the sets $B(\Theta(f_i, b_i), r)$, $f_i \in E(\Theta)$, the algorithm $\mathcal{A}_{13}$ makes $O(q_{\psi}(T)n)$ computations of F and H. To construct the pair $(\psi^0(\Theta), U(\Theta))$, the algorithm $\mathcal{A}_{13}$ makes two operations – computes ψ^0 and U. To construct the set $Par(A(\Theta, r)) = B(\Theta, r)$ from the set $A(\Theta, r)$, the algorithm $\mathcal{A}_{13}$ makes

$$O(q_{\psi}(T)n \log(q_{\psi}(T)n))$$

comparisons (see Proposition 2.1). Hence, to process a nonterminal node Θ, the algorithm makes

$$O(q_{\psi}(T)n \log(q_{\psi}(T)n))$$

elementary operations.

To construct the set $Par(p^{path}_{\psi,U}(T,r))$, the algorithm $\mathscr{A}_{13}$ makes

$$O(L(\Delta(T))q_\psi(T)n\log(q_\psi(T)n))$$

elementary operations (computations of F, H, ψ^0, U, and comparisons). □

Proposition 12.3 *Let ψ be a strictly increasing cost function for decision rules given by pair of functions ψ^0, F, H be a function from $\mathbb{R}$ to $\mathbb{R}$ such that $H(x) = x$, U be an uncertainty measure, $\psi \in \{l, -c, -rc, -c_M, mc, rmc\}$, $U \in \{me, rme, ent, gini, R, me + R\}$, and $\mathscr{U}$ be a restricted information system. Then the algorithm $\mathscr{A}_{13}$ has polynomial time complexity for decision tables from $\mathscr{T}(\mathscr{U})$ depending on the number of conditional attributes in these tables.*

Proof Since $\psi \in \{l, -c, -rc, -c_M, mc, rmc\}$,

$$\psi^0 \in \{0, -N_{mcd(T)}(T), -N_{mcd(T)}(T)/N(T), -N^M(T), N(T) - N_{mcd(T)}(T), (N(T) - N_{mcd(T)}(T))/N(T)\},$$

and $F, H \in \{x, x+1\}$. From Lemma 9.1 and Proposition 12.2 it follows that, for the algorithm $\mathscr{A}_{13}$, the number of elementary operations (computations of F, H, ψ^0, U, and comparisons) is bounded from above by a polynomial depending on the size of input table T and on the number of separable subtables of T. All operations with numbers are basic ones. The computations of numerical parameters of decision tables used by the algorithm $\mathscr{A}_{13}$ (0, $-N_{mcd(T)}(T)$, $-N_{mcd(T)}(T)/N(T)$, $-N^M(T)$, $N(T) - N_{mcd(T)}(T)$, $(N(T) - N_{mcd(T)}(T))/N(T)$ and $U \in \{me, rme, ent, gini, R, me + R\}$) have polynomial time complexity depending on the size of decision tables.

According to Proposition 3.5, the algorithm $\mathscr{A}_{13}$ has polynomial time complexity for decision tables from $\mathscr{T}(\mathscr{U})$ depending on the number of conditional attributes in these tables. □

12.1.1 Relationships for Decision Rules: Cost Versus Uncertainty

Let U be a strictly increasing uncertainty measure (in particular, R and $me + R$ are such measures) or $\psi \in \{l, -c, -c_M\}$. Then the constructed by the algorithm $\mathscr{A}_{13}$ set $Par(p^{path}_{\psi,U}(T,r))$ is equal to the set $Par(p_{\psi,U}(T,r))$. Using this set we can construct two partial functions $\mathscr{R}^{\psi,U}_{T,r} : \mathbb{R} \to \mathbb{R}$ and $\mathscr{R}^{U,\psi}_{T,r} : \mathbb{R} \to \mathbb{R}$ which describe relationships between cost function ψ and uncertainty measure U on the set $DR_U(T,r)$ of U-decision rules for T and r, and are defined in the following way:

$$\mathscr{R}^{\psi,U}_{T,r}(x) = \min\{U(T,\rho) : \rho \in DR_U(T,r), \psi(T,\rho) \le x\},$$
$$\mathscr{R}^{U,\psi}_{T,r}(x) = \min\{\psi(T,\rho) : \rho \in DR_U(T,r), U(T,\rho) \le x\}.$$

Let $p_{\psi,U}(T,r) = \{(\psi(T,\rho), U(T,\rho)) : \rho \in DR_U(T,r)\}$ and $(a_1, b_1), \ldots, (a_k, b_k)$ be the normal representation of the set $Par(p_{\psi,U}(T,r))$ where $a_1 < \ldots < a_k$ and $b_1 > \ldots > b_k$. By Lemma 2.8 and Remark 2.2, for any $x \in \mathbb{R}$,

$$\mathscr{R}^{\psi,U}_{T,r}(x) = \begin{cases} undefined, & x < a_1 \\ b_1, & a_1 \le x < a_2 \\ \ldots & \ldots \\ b_{k-1}, & a_{k-1} \le x < a_k \\ b_k, & a_k \le x \end{cases},$$

$$\mathscr{R}^{U,\psi}_{T,r}(x) = \begin{cases} undefined, & x < b_k \\ a_k, & b_k \le x < b_{k-1} \\ \ldots & \ldots \\ a_2, & b_2 \le x < b_1 \\ a_1, & b_1 \le x \end{cases}.$$

12.2 Pareto Optimal Points for Systems of Decision Rules: Cost Versus Uncertainty

Let T be a nonempty decision table with n conditional attributes $f_1, \ldots, f_n$ and $N(T)$ rows $r_1, \ldots, r_{N(T)}$, and U be an uncertainty measure.

We denote by $\mathscr{I}_U(T)$ the set $DR_U(T, r_1) \times \ldots \times DR_U(T, r_{N(T)})$ of U-systems of decision rules for T. Let ψ be a strictly increasing cost function for decision rules given by pair of functions ψ^0, F, and f, g be increasing functions from $\mathbb{R}^2$ to $\mathbb{R}$. We consider two parameters of decision rule systems: cost $\psi_f(T, S)$ and uncertainty $U_g(T, S)$ which are defined on pairs T, S where $T \in \mathscr{T}^+$ and $S \in \mathscr{I}_U(T)$. Let $S = (\rho_1, \ldots, \rho_{N(T)})$. Then $\psi_f(T, S) = f(\psi(T, \rho_1), \ldots, \psi(T, \rho_{N(T)}))$ and $U_g(T, S) = g(U(T, \rho_1), \ldots, U(T, \rho_{N(T)}))$ where $f(x_1) = x_1, g(x_1) = x_1$, and, for $k > 2$, $f(x_1, \ldots, x_k) = f(f(x_1, \ldots, x_{k-1}), x_k)$ and $g(x_1, \ldots, x_k) = g(g(x_1, \ldots, x_{k-1}), x_k)$.

We assume that $\psi \in \{l, -c, -c_M\}$ or U be a strictly increasing uncertainty measure (in particular, R and $me + R$ are such measures). According to Corollary 12.1 and Lemma 12.3, in these cases $Par(p^{path}_{\psi,U}(T, r_i)) = Par(p_{\psi,U}(T, r_i))$ for $i = 1, \ldots, N(T)$. We describe now an algorithm which constructs the set of Pareto optimal points for the set of pairs $p^{f,g}_{\psi,U}(T) = \{(\psi_f(T, S), U_g(T, S)) : S \in \mathscr{I}_U(T)\}$.

Algorithm $\mathscr{A}_{14}$ (construction of POPs for decision rule systems, cost versus uncertainty).

Input: Strictly increasing cost function for decision rules ψ given by pair of functions ψ^0, F and an uncertainty measure U such that $\psi \in \{l, -c, -c_M\}$ or U is a strictly increasing uncertainty measure, increasing functions f, g from $\mathbb{R}^2$ to $\mathbb{R}$, a nonempty decision table T with n conditional attributes $f_1, \ldots, f_n$ and $N(T)$ rows $r_1, \ldots, r_{N(T)}$, and the graph $\Delta(T)$.

Output: The set $Par(p^{f,g}_{\psi,U}(T))$ of Pareto optimal points for the set of pairs $p^{f,g}_{\psi,U}(T) = \{(\psi_f(T,S), U_g(T,S)) : S \in \mathscr{I}_U(T)\}$.

1. Using the algorithm $\mathscr{A}_{13}$ construct, for $i = 1, \ldots, N(T)$, the set $Par(P_i)$ where

$$P_i = p^{path}_{\psi,U}(T,r) = \{((\psi(T,\rho), U(T,\rho)) : \rho \in Path(T,r)\} .$$

2. Apply the algorithm $\mathscr{A}_2$ to the functions f, g and the sets $Par(P_1), \ldots, Par(P_{N(T)})$. Set $C(T)$ the output of the algorithm $\mathscr{A}_2$ and return it.

Proposition 12.4 *Let ψ be a strictly increasing cost function for decision rules given by pair of functions ψ^0, F and U be an uncertainty measure such that $\psi \in \{l, -c, -c_M\}$ or U is a strictly increasing uncertainty measure, f, g be increasing functions from $\mathbb{R}^2$ to $\mathbb{R}$, and T be a nonempty decision table with n conditional attributes $f_1, \ldots, f_n$ and $N(T)$ rows $r_1, \ldots, r_{N(T)}$. Then the algorithm $\mathscr{A}_{14}$ constructs the set $C(T) = Par(p^{f,g}_{\psi,U}(T))$.*

Proof For $i = 1, \ldots, N(T)$, denote $P_i = p^{path}_{\psi,U}(T, r_i)$ and $R_i = p_{\psi,U}(T, r_i)$. During the first step, the algorithm $\mathscr{A}_{14}$ constructs (using the algorithm $\mathscr{A}_{13}$) the sets $Par(P_1), \ldots, Par(P_{N(T)})$ (see Proposition 11.1). From Corollary 12.1 and Lemma 12.3 it follows that $Par(P_1) = Par(R_1), \ldots, Par(P_{N(T)}) = Par(R_{N(T)})$. During the second step of the algorithm $\mathscr{A}_{14}$ we apply the algorithm $\mathscr{A}_2$ to the functions f, g and the sets $Par(R_1), \ldots, Par(R_{N(T)})$. The algorithm $\mathscr{A}_2$ constructs the set $C(T) = Par(Q_{N(T)})$ where $Q_1 = R_1$, and, for $i = 2, \ldots, N(T)$, $Q_i = Q_{i-1} \langle fg \rangle R_i$ (see Proposition 2.2). One can show that $Q_{N(T)} = p^{f,g}_{\psi,U}(T))$. Therefore $C(T) = Par(p^{f,g}_{\psi,U}(T))$. □

Let us recall that, for a given cost function ψ and a decision table T, $q_\psi(T) = |\{\psi(\Theta,\rho) : \Theta \in SEP(T), \rho \in DR(\Theta)\}|$. In particular, by Lemma 9.1, $q_l(T) \leq n+1$, $q_{-c}(T) \leq N(T)+1$, $q_{-c_M}(T) \leq N(T)+1$, and $q_{mc}(T) \leq N(T)+1$.

Let us recall also that, for a given cost function ψ for decision rules and a decision table T, $Range_\psi(T) = \{\psi(\Theta,\rho) : \Theta \in SEP(T), \rho \in DR(\Theta)\}$. By Lemma 9.1, $Range_l(T) \subseteq \{0, 1, \ldots, n\}$, $Range_{-c}(T) \subseteq \{0, -1, \ldots, -N(T)\}$, $Range_{-c_M}(T) \subseteq \{0, -1, \ldots, -N(T)\}$, and $Range_{mc}(T) \subseteq \{0, 1, \ldots, N(T)\}$. Let $t_l(T) = n$, $t_{-c}(T) = N(T)$, $t_{-c_M}(T) = N(T)$, and $t_{mc}(T) = N(T)$.

Proposition 12.5 *Let ψ be a strictly increasing cost function for decision rules given by pair of functions ψ^0, F and U be an uncertainty measure such that either $\psi \in \{l, -c, -c_M\}$ or $\psi = mc$ and U is a strictly increasing uncertainty measure, f, g be increasing functions from $\mathbb{R}^2$ to $\mathbb{R}$, $f \in \{x+y, \max(x,y)\}$, H be a function from $\mathbb{R}$ to $\mathbb{R}$ such that $H(x) = x$, and T be a nonempty decision table with n conditional attributes $f_1, \ldots, f_n$ and $N(T)$ rows $r_1, \ldots, r_{N(T)}$. Then, to construct the set $Par(p^{f,g}_{\psi,U}(T))$, the algorithm $\mathscr{A}_{14}$ makes*

$$O(N(T)L(\Delta(T))q_\psi(T)n\log(q_\psi(T)n)) + O(N(T)t_\psi(T)^2\log(t_\psi(T)))$$

elementary operations (computations of F, H, ψ^0, U, f, g *and comparisons) if* $f = \max(x, y)$, *and*

$$O(N(T)L(\Delta(T))q_\psi(T)n\log(q_\psi(T)n)) + O(N(T)^2 t_\psi(T)^2 \log(N(T)t_\psi(T)))$$

elementary operations (computations of F, H, ψ^0, U, f, g *and comparisons) if* $f = x + y$.

Proof For $i = 1, \ldots, N(T)$, denote $P_i = p^{path}_{\psi,U}(T, r)$ and $R_i = p_{\psi,U}(T, r)$. To construct the sets $Par(P_i) = Par(R_i)$, $i = 1, \ldots, N(T)$, the algorithm $\mathscr{A}_{13}$ makes

$$O(N(T)L(\Delta(T))q_\psi(T)n\log(q_\psi(T)n))$$

elementary operations (computations of F, H, ψ^0, U, and comparisons) – see Proposition 12.2.

We now evaluate the number of elementary operations (computations of f, g, and comparisons) made by the algorithm $\mathscr{A}_2$ during the construction of the set $C(T) = Par(p^{f,g}_{\psi,U}(T))$ from the sets $Par(R_i)$, $i = 1, \ldots, N(T)$. We know that $\psi \in \{l, -c, -c_M, mc\}$ and $f \in \{x + y, \max(x, y)\}$.

For $i = 1, \ldots, N(T)$, let $R_i^1 = \{a : (a, b) \in R_i\}$. Since $\psi \in \{l, -c, -c_M, mc\}$, we have $R_i^1 \subseteq \{0, 1, \ldots, t_\psi(T)\}$ for $i = 1, \ldots, N(T)$ or $R_i^1 \subseteq \{0, -1, \ldots, -t_\psi(T)\}$ for $i = 1, \ldots, N(T)$.

Using Proposition 2.3 we obtain the following.

If $f = x + y$ then, to construct the set $C(T)$, the algorithm $\mathscr{A}_2$ makes

$$O(N(T)^2 t_\psi(T)^2 \log(N(T)t_\psi(T)))$$

elementary operations (computations of f, g, and comparisons).

If $f = \max(x, y)$ then, to construct the sets $Par(Q_{N(T)})$, the algorithm $\mathscr{A}_2$ makes

$$O(N(T)t_\psi(T)^2 \log(t_\psi(T)))$$

elementary operations (computations of f, g, and comparisons). □

Proposition 12.6 *Let* ψ *be a strictly increasing cost function for decision rules given by pair of functions* ψ^0, F *and* U *be an uncertainty measure such that either* $\psi \in \{l, -c, -c_M\}$ *and* $U \in \{me, rme, ent, gini, R, me + R\}$, *or* $\psi = mc$ *and* $U \in \{R, me + R\}$, $f, g \in \{\max(x, y), x + y\}$, H *be a function from* $\mathbb{R}$ *to* $\mathbb{R}$ *such that* $H(x) = x$, *and* $\mathscr{U}$ *be a restricted information system. Then the algorithm* $\mathscr{A}_{14}$ *has polynomial time complexity for decision tables from* $\mathscr{T}(\mathscr{U})$ *depending on the number of conditional attributes in these tables.*

Proof We have $\psi^0 \in \{0, -N_{mcd(T)}(T), -N^M(T), N(T) - N_{mcd(T)}(T)\}$, and F, $H \in \{x, x + 1\}$. From Lemma 9.1 and Proposition 12.5 it follows that, for the algorithm $\mathscr{A}_{14}$, the number of elementary operations (computations of F, H, ψ^0, U, f, g, and comparisons) is bounded from above by a polynomial depending on the size of input table T and on the number of separable subtables of T. All operations with numbers are basic ones. The computations of numerical parameters of decision tables used by the algorithm $\mathscr{A}_{14}$ $(0, -N_{mcd(T)}(T), -N^M(T), N(T) - N_{mcd(T)}(T)$, and $U \in \{me, rme, ent, gini, R, me + R\})$ have polynomial time complexity depending on the size of decision tables.

According to Proposition 3.5, the algorithm $\mathscr{A}_{14}$ has polynomial time complexity for decision tables from $\mathscr{T}(\mathscr{U})$ depending on the number of conditional attributes in these tables. □

12.2.1 Relationships for Systems of Decision Rules: Cost Versus Uncertainty

Let ψ be a strictly increasing cost function for decision rules, U be an uncertainty measure, f, g be increasing functions from $\mathbb{R}^2$ to $\mathbb{R}$, and T be a nonempty decision table.

To study relationships between functions ψ_f and U_g which characterize cost and uncertainty of systems of rules from $\mathscr{I}_U(T)$ we can consider two partial functions from $\mathbb{R}$ to $\mathbb{R}$:

$$\mathscr{R}_T^{\psi,f,U,g}(x) = \min\{U_g(T,S) : S \in \mathscr{I}_U(T), \psi_f(T,S) \le x\} ,$$
$$\mathscr{R}_T^{U,g,\psi,f}(x) = \min\{\psi_f(T,S) : S \in \mathscr{I}_U(T), U_g(T,S) \le x\} .$$

Let $p_{\psi,U}^{f,g}(T) = \{(\psi_f(T,S), U_g(T,S)) : S \in \mathscr{I}_U(T)\}$, and $(a_1, b_1), \ldots, (a_k, b_k)$ be the normal representation of the set $Par(p_{\psi,U}^{f,g}(T))$ where $a_1 < \ldots < a_k$ and $b_1 > \ldots > b_k$. By Lemma 2.8 and Remark 2.2, for any $x \in \mathbb{R}$,

$$\mathscr{R}_T^{\psi,f,U,g}(x) = \begin{cases} undefined, & x < a_1 \\ b_1, & a_1 \le x < a_2 \\ \ldots & \ldots \\ b_{k-1}, & a_{k-1} \le x < a_k \\ b_k, & a_k \le x \end{cases} ,$$

$$\mathscr{R}_T^{U,g,\psi,f}(x) = \begin{cases} undefined, & x < b_k \\ a_k, & b_k \le x < b_{k-1} \\ \ldots & \ldots \\ a_2, & b_2 \le x < b_1 \\ a_1, & b_1 \le x \end{cases} .$$

12.3 Illustrative Examples: Cost Versus Uncertainty for Decision Rule Systems

We will now proceed to show examples of trade-offs that exist between cost and uncertainty for systems of decision rules in tables consisting of real-world information. We constructed the set of Pareto optimal points for 16 datasets from the UCI ML Repository comparing the uncertainty measure relative misclassification error with the cost functions length and coverage individually. We considered one rule for each row. The cost of a system was calculated by taking the average cost of all rules while uncertainty was calculated by taking the maximum amongst all rules.

Tables 12.1 and 12.2 present results for comparison of relative misclassification error with coverage and length, respectively. The first three columns of the tables ("Decision table", "Rows", and "Attr") specify the name and dimensions of the decision tables used. The cardinality of the set of Pareto optimal points can be

Table 12.1 Relative misclassification error versus coverage

Decision table	Rows	Attr	# POPs	POP with	
				min rme	max cov
Adult-stretch	16	4	2	(0, 7)	(0.25, 10)
Agaricus-lepiota	8124	22	74	(0, 2135.46)	(0.517971, 4067.25)
Balance-scale	625	4	19	(0, 4.2128)	(0.9216, 269.262)
Breast-cancer	266	9	139	(0, 9.53383)	(0.714286, 157.429)
Cars	1728	6	53	(0, 332.764)	(0.960069, 937.814)
Hayes-roth-data	69	4	11	(0, 6.52174)	(0.73913, 24.1304)
House-votes-84	279	16	121	(0, 73.5233)	(0.670251, 155.674)
Lenses	24	4	5	(0, 7.25)	(0.375, 11.0833)
Lymphography	148	18	122	(0, 21.5405)	(0.587838, 69.6081)
Monks-1-test	432	6	4	(0, 45)	(0.5, 216)
Monks-1-train	124	6	40	(0, 13.4516)	(0.5, 62)
Monks-2-test	432	6	22	(0, 12.3565)	(0.671296, 241.352)
Monks-2-train	169	6	60	(0, 6.3787)	(0.621302, 89.4734)
Monks-3-test	432	6	6	(0, 56)	(0.527778, 216.667)
Monks-3-train	122	6	38	(0, 12.1967)	(0.508197, 61.0164)
Nursery	12960	8	163	(0, 1531.04)	(0.924074, 4114.41)
Shuttle-landing-control	15	6	10	(0, 2.13333)	(0.571429, 7.8)
Soybean-small	47	35	1	(0, 12.5319)	(0, 12.5319)
Spect-test	169	22	32	(0, 58.0473)	(0.864407, 153.757)
Teeth	23	8	1	(0, 1)	(0, 1)
Tic-tac-toe	958	9	34	(0, 66.6806)	(0.653445, 524.113)
Zoo-data	59	16	3	(0, 11.0678)	(0.166667, 11.3729)

Table 12.2 Relative misclassification error versus length

Decision table	Rows	Attr	# POPs	POP with	
				min rme	min len
Adult-stretch	16	4	4	(0, 1.25)	(0.75, 0)
Agaricus-lepiota	8124	22	6	(0, 1.18156)	(0.517971, 0)
Balance-scale	625	4	19	(0, 3.1968)	(0.9216, 0)
Breast-cancer	266	9	69	(0, 2.66541)	(0.714286, 0)
Cars	1728	6	48	(0, 2.43403)	(0.962384, 0)
Hayes-roth-data	69	4	10	(0, 2.14493)	(0.73913, 0)
House-votes-84	279	16	48	(0, 2.53763)	(0.670251, 0)
Lenses	24	4	9	(0, 2)	(0.833333, 0)
Lymphography	148	18	29	(0, 1.99324)	(0.986486, 0)
Monks-1-test	432	6	4	(0, 2.25)	(0.5, 0)
Monks-1-train	124	6	22	(0, 2.26613)	(0.5, 0)
Monks-2-test	432	6	19	(0, 4.52315)	(0.671296, 0)
Monks-2-train	169	6	52	(0, 3.49704)	(0.621302, 0)
Monks-3-test	432	6	6	(0, 1.75)	(0.527778, 0)
Monks-3-train	122	6	24	(0, 2.31148)	(0.508197, 0)
Nursery	12960	8	125	(0, 3.11821)	(0.999846, 0)
Shuttle-landing-control	15	6	8	(0, 1.4)	(0.6, 0)
Soybean-small	47	35	3	(0, 1)	(0.787234, 0)
Spect-test	169	22	29	(0, 1.47929)	(0.952663, 0)
Teeth	23	8	8	(0, 2.26087)	(0.956522, 0)
Tic-tac-toe	958	9	21	(0, 3.0167)	(0.653445, 0)
Zoo-data	59	16	18	(0, 1.55932)	(0.932203, 0)

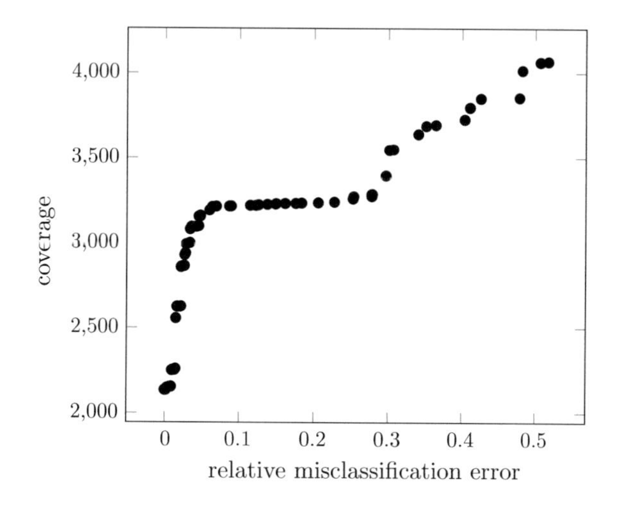

Fig. 12.1 Pareto optimal points for "Agaricus-lepiota": coverage versus relative misclassification error

found in column "# POPs". The tables also present the first and last Pareto optimal points. These points (pairs of values) represent the best cost achievable for systems of rules with minimum uncertainty (found in column "min rme") and the minimum uncertainty achievable for system of rules with optimal cost (found in columns "max cov" and "min len"). In each of these pairs, the first value refers to uncertainty, and the second to cost.

We can also find some examples of the sets of Pareto optimal points for three decision tables ("Agaricus-lepiota", "Spect-test", and "Nursery") with regards to coverage and length in Figs. 12.1, 12.2, 12.3 and 12.4, 12.5, 12.6, respectively. We can learn some interesting things from these diagrams. There are many points in the graphs where there are large changes along one axis with minimal change along the other. These points signify either chances for great improvement in one criterion with

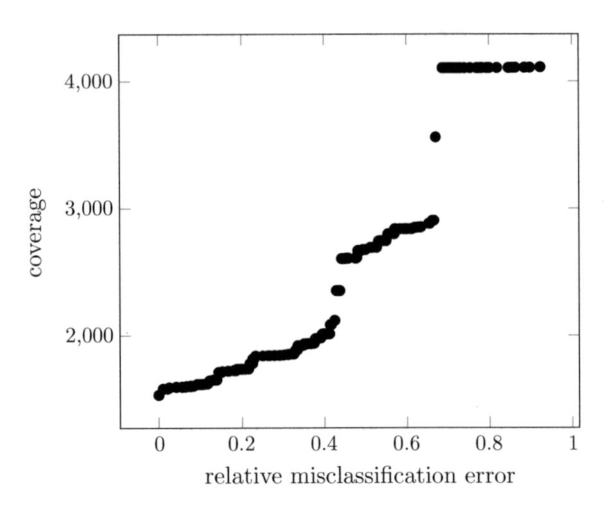

Fig. 12.2 Pareto optimal points for "Nursery": coverage versus relative misclassification error

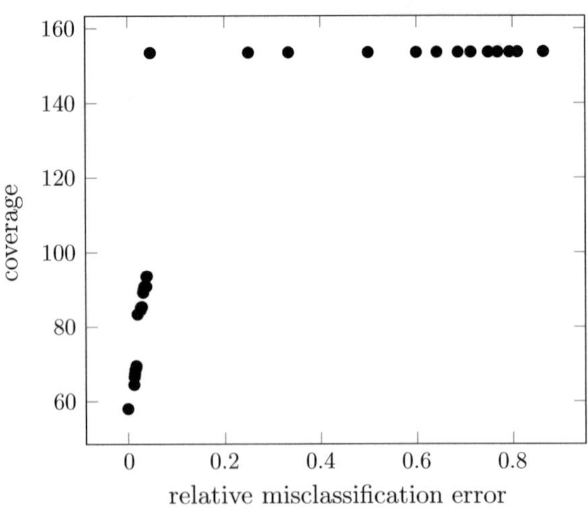

Fig. 12.3 Pareto optimal points for "Spect-test": coverage versus relative misclassification error

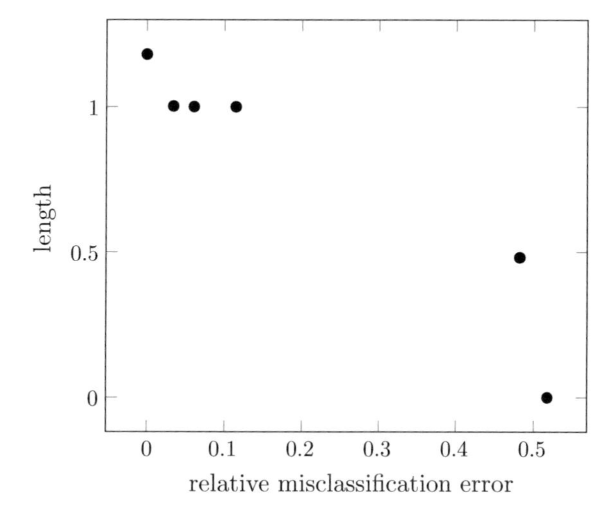

Fig. 12.4 Pareto optimal points for "Agaricus-lepiota": length versus relative misclassification error

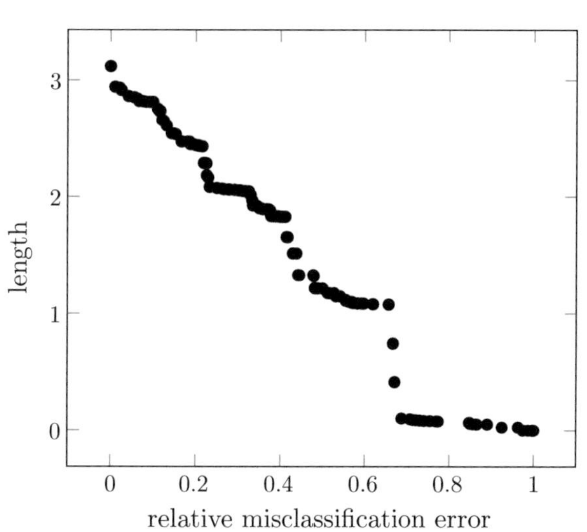

Fig. 12.5 Pareto optimal points for "Nursery": length versus relative misclassification error

minimal loss in the other or great loss in one criterion with minimal improvement in the other. In this manner, such diagrams can be helpful for considering how much uncertainty to tolerate in order to get worthwhile improvements in cost.

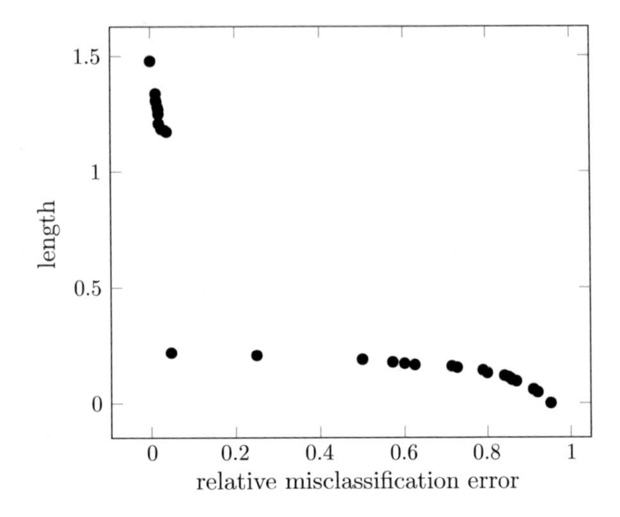

Fig. 12.6 Pareto optimal points for "Spect-test": length versus relative misclassification error

References

1. Bohanec, M., Bratko, I.: Trading accuracy for simplicity in decision trees. Mach. Learn. **15**(3), 223–250 (1994)
2. Fürnkranz, J., Gamberger, D., Lavrac, N.: Foundations of Rule Learning. Cognitive Technologies. Springer, Heidelberg (2012)

Part IV
Element Partition Trees

This part is devoted to the optimization of element partition trees controlling the LU factorization of systems of linear equations. These systems result from the finite element method discretization over two-dimensional meshes with rectangular elements.

We begin by presenting an introduction to the topic followed by the main notions and definitions in Chap. 13. These notions span meshes, element partition trees, and cost functions for element partition trees.

We create and study polynomial time algorithms for the optimization of element partition trees which use only straight lines for the partitioning of rectangular meshes. The considered algorithms are based on the extensions of dynamic programming and allow multi-stage optimization of element partition trees relative to different criteria such as time and memory complexity, and bi-criteria optimization of element partition trees including construction of the set of Pareto optimal points. We describe the multi-stage optimization process and the experimental results in Chap. 14. Polynomial time algorithms for bi-criteria optimization of element partition trees along with possible applications are presented in Chap. 15.

Chapter 13
Element Partition Trees: Main Notions

The *hp* adaptive finite element method is a popular approach used to find approximate solutions for partial differential equations (PDEs). Initially, the problem domain is approximated using a mesh of elements. Then the PDE is transformed into a weak form and discretized using basis functions over the vertices, edges and interior of elements. As a result of the discretization process, we obtain a global system of linear equations. Such system can be decomposed into frontal matrices associated with the mesh elements. Solution of the PDE can be obtained by working with those frontal matrices. An element partition tree prescribes the order that the solver considers while working with the frontal matrices. Using this order we obtain an ordering of the rows of the global matrix representing the system of linear equations.

This chapter starts by formally defining the class of meshes under study. We describe the notion of element partition tree and present an abstract way of defining optimization criteria of element partition trees in terms of cost functions. A definition of a cost function is provided in addition to a few examples of cost functions under study along with some of their properties.

13.1 Introduction

A rectangular mesh can be considered as a finite set of vertical and horizontal line segments, vertices (intersections of line segments) and rectangles bounded by these line segments. An element partition tree is a binary rooted tree that describes the partition of the mesh by line segments. Cost functions for the element partition trees are defined based mainly on the topological information of the mesh. We study the problems of element partition tree optimization for rectangular meshes. The considered problems lie at the intersection of discrete geometry and combinatorial optimization. We begin with a brief discussion of the problem motivation and assumptions which simplify the considered model.

H. AbouEisha et al., *Extensions of Dynamic Programming for Combinatorial Optimization and Data Mining*, Intelligent Systems Reference Library 146,
https://doi.org/10.1007/978-3-319-91839-6_13

The finite element method is a widely used approach to find approximate solutions of partial differential equations (PDEs) specified along with boundary conditions and a solution domain. A mesh with triangular or rectangular elements is created to cover the domain and to approximate the solution over it. Then the weak form of the PDE is discretized using polynomial basis functions spread over the mesh (1 per vertex, p per edge for some nonnegative integer p, and p^2 per element interior) for finite elements of uniform polynomial order p. Let ν be the number of basis functions used. The approximate solution of the discretized PDE can be represented as a linear combination of these ν functions. The coefficients of the linear form can be found as solutions of a system of linear equations $Ax = b$ constructed based on PDE, boundary conditions, and the mesh, where A is a square $\nu \times \nu$ matrix.

Any $\nu \times \nu$ matrix A can be represented in the form $PA = LU$ where P is a permutation matrix which reorders the rows of A, L is a lower triangular matrix in which all elements above the diagonal are zero, and U is an upper triangular matrix in which all elements below the diagonal are zero [17]. The construction process of matrices P, L and U is called LU factorization (with partial pivoting). We can rewrite the equation system $Ax = b$ equivalently as $LUx = Pb$. The solutions of this system can be found in the following way: we solve the equation system $Ly = Pb$ for y (forward substitution) and, after that, we solve the equation system $Ux = y$ for x (backward substitution).

The number of arithmetic floating point operations (additions, subtractions, multiplications, and divisions) for the LU factorization in the worst case is $O(\nu^3)$ if we consider standard algorithms (for example, Doolittle algorithm [10]) and $O(\nu^{2.3728639})$ if we consider the best known (from theoretical point of view) modification of Bunch and Hopcroft algorithm [7] based on Le Gall algorithm for matrix multiplication [14].

This is essentially greater than $O(\nu^2)$ for the forward and backward substitution. The above-mentioned Doolittle algorithm is a simple modification of Gaussian elimination. Let us assume for simplicity that it is applicable to A without the reordering of rows. Then the algorithm constructs the matrix U by a sequential elimination of its rows. Let the first $i-1$, $1 \le i < \nu$, rows are already eliminated: in the first $i-1$ columns, all elements below the diagonal are zero. To eliminate row i we need $(2\nu - 2i + 1)(\nu - i)$ arithmetic floating point operations. To eliminate the first κ rows we need $W(\nu, \kappa) = \sum_{i=1}^{\kappa}(2\nu - 2i + 1)(\nu - i) = 2\nu^2\kappa - 2\nu\kappa^2 - \nu\kappa + \frac{2}{3}\kappa^3 + \frac{1}{2}\kappa^2 - \frac{1}{6}\kappa$ floating point operations, and to construct U we need $W(\nu, \nu)$ operations. All elements of L are extracted from the intermediate results of the algorithm work without additional arithmetic floating point operations. All the entries on the diagonal of L are one. Therefore to keep U and L it is enough the same ν^2 memory as for the matrix A.

The multi-frontal solver is the state-of-the-art algorithm for solving sparse linear systems resulting from finite element method discretization [8, 9]. This algorithm works in both single- and multi-processor environment, either in the shared-memory [11–13] or distributed-memory [3–5] parallel machines.

The matrix A is constructed from element frontal matrices corresponding to particular finite elements of the computational mesh. The order of factorization is given by an element partition tree in which each node corresponds to a submesh of the initial mesh. In particular, the root corresponds to the whole mesh and the terminal nodes (the leaves) correspond to finite elements. In each internal node, the submesh corresponding to this node is divided into two submeshes which correspond to children of the considered node.

In each terminal node, we perform partial factorization of the matrix for a finite element corresponding to this node. Partially factorized matrices for children of a node (in fact, parts of these matrices known as Schur complements [15, 20]) are combined into a matrix which is partially factorized in this node. These operations are repeated recursively until we reach the root of the element partition tree. Partially factorized matrices must be kept for the following forward and backward substitution. These partially factorized matrices can be combined to yield the LU factorization of the original matrix A.

Partial factorization means the following. We have a $\mu \times \mu$ matrix B such that, for the first ρ rows of B, $1 \leq \rho \leq \mu$, variables corresponding to these rows can be expressed (in the considered model) in terms of the variables corresponding to the remaining $\mu - \rho$ rows. These rows are *fully assembled*. In particular, in the matrix corresponding to a finite element at least all variables corresponding to the interior are fully assembled. That is, all interactions of the corresponding basis functions have been fully captured at the present level. Thus, they can be expressed in terms of variables corresponding to edges and vertices of the finite element. The algorithm eliminates the first ρ rows of B as the Doolittle algorithm does, keeps results of partial factorization in memory, and sends to the parent node the submatrix obtained at the intersection of the last $\mu - \rho$ rows and the last $\mu - \rho$ columns (Schur complement).

To simplify the model under consideration, we neglect the complexity of forward and backward substitution, and study only the complexity of the LU factorization. We also neglect the complexity of combining the Schur complements into a matrix which is essentially less than the complexity of partial factorization of this matrix. We assume that the Doolittle algorithm is applicable to the partial factorization of each one of the considered matrices without reordering of rows. We use $W(\mu, \rho)$ as the number of arithmetic floating point operations for partial factorization of a $\mu \times \mu$ matrix with ρ fully assembled rows. We assume that to keep results of partial factorization of a $\mu \times \mu$ matrix we need μ^2 memory, and that we should send $(\mu - \rho)^2$ numbers (corresponding to the Schur complement) to the parent node.

The computational complexity of a multi-frontal solver implementation depends on the element partition tree used. The aim of this part is to optimize element partition trees relative to different cost functions that characterize time or memory complexity. In the general case, when a given mesh can be partitioned in an arbitrary way, the problem of partitioning this mesh to minimize the fill-in is NP-hard [19]. In our case, we only use partitions by straight line segments. We prove that the considered optimization problems have polynomial time complexity.

In this book, we consider a set of rectangular meshes which allow us to describe and study known "benchmark" meshes representing various point and edge singularities. We study a class of element partition trees obtained by recursive partitioning of the mesh along straight line segments. We define two elimination strategies. In the first one, at each elimination step all fully assembled degrees of freedom are eliminated. We call this strategy *as soon as possible* and denote it ASAP. The second strategy eliminates all degrees of freedom from a mesh (submesh) with respect to the boundary. This strategy is called *keep boundary* and we denote it KB. KB assumes that the considered mesh is a submesh of another global rectangular mesh that we seek a good element partition tree for. This submesh usually describes a part of the domain with specific kinds of singularities. The major difference between both cost functions is that the KB cost function assumes that rows associated with the domain boundary are kept, that is they are not eliminated until the root of the elimination tree, so they can be interfaced with the remaining parts of the global mesh. The ASAP cost function assumes that the rows associated with the boundary can be eliminated as soon as possible, since the processed mesh is considered as the global one.

We consider three different cost functions associated with each type of elimination strategy. In particular, we model the memory requirements for each elimination strategy. We also analyze the computational complexity (a proxy of it is execution time) for single- and multi-processor elimination. We introduce the notion of a strictly optimal element partition tree. This is a tree which is optimal for the input mesh and, for each node of the tree, the subtree with root in this node is also optimal for the corresponding submesh. Except for the two cost functions that characterize time for multi-processor computations, the set of strictly optimal elimination trees for the rest of the cost functions coincides with the set of optimal element partition trees. For the two cost functions characterizing the time of multi-processor computations, the set of strictly optimal element partition trees is a subset of the set of optimal element partition trees.

To optimize element partition trees, we use some extensions of dynamic programming which allow us to describe the set of strictly optimal element partition trees by a directed acyclic graph whose nodes are submeshes of the initial mesh. Dynamic programming also allows us to count the number of strictly optimal trees, and to understand either the obtained trees or their prefixes if the strictly optimal trees are too large. We also have the possibility to make multi-stage optimization of element partition trees relative to a sequence of cost functions.

We may be interested in finding a good element partition tree with respect to two criteria. For example, an element partition tree that results in small number of floating point operations by the solver and does not consume much intermediate memory is interesting. The considered criteria are sometimes conflicting so we aim to find element partition trees with acceptable trade-off between the studied cost functions.

We present a bi-criteria optimization algorithm for element partition trees. Such algorithm allows us to construct the set of Pareto optimal points corresponding to values of the two studied cost functions on element partition trees for a given mesh. We refer to this set of points as the Pareto front.

The Pareto front is used to depict the relationship between the two studied criteria. If the Pareto front consists of only one point then totally optimal element partition trees for both cost functions exist. In other words, we have at least one element partition tree that has optimal values of both considered criteria simultaneously.

All the considered algorithms have polynomial time complexity in terms of the input mesh size. We implement these algorithms in Java.

We use the obtained software to study three families of meshes describing different kinds of singularities: point singularities, edge singularities, and point-edge singularities. For these meshes, we investigate the number of optimal element partition trees and the existence of totally optimal element partition trees which are optimal relative to a number of cost functions simultaneously. We use the bi-criteria optimization algorithm to compute the Pareto fronts for these meshes.

The KB mode of the optimization algorithms can be used in large meshes. KB improves the performance of solvers by constructing optimal element partition trees for relatively small but essentially irregular submeshes of the global mesh.

The ASAP optimization algorithm is a research tool which helps us understand the behavior of optimal element partition trees that use only straight line segments as separators. The comparison of the optimal element partition trees with the trees constructed by the known heuristics (MUMPS [6] solver with METIS [16] library) allowed us to propose a new heuristic [1, 18] which outperforms existing state-of-the-art alternatives.

The first attempt to analyze mathematically the considered dynamic programming algorithms was done in [2]. The investigated model was less accurate, we did not distinguish ASAP and KB modes. We did not consider the notion of strictly optimal element partition trees, and worked only with a cost function which characterizes the time of single-processor computation.

This chapter discusses the main notions connected with meshes, element partition trees, and cost functions.

13.2 Meshes and Element Partition Trees

We describe the class of finite element meshes studied, and define the notions of an element partition tree and a cost function for element partition trees.

13.2.1 Meshes

The class of *meshes* investigated is constructed as follows. We start with a rectangle. We call the sides of this initial rectangle *boundary sides* and its vertices are called *corners*. Further vertical and horizontal straight line segments (we will refer to these segments including sides of the initial rectangle as *lines*) may be added as follows. We select two points that lie on different existing parallel lines and which can be connected by a vertical or horizontal line and connect them by adding a line. The process may be repeated until the desired mesh structure is obtained. After this construction process is finished, we have a mesh that is mainly a rectangle with a set of vertical and horizontal lines that lies inside it.

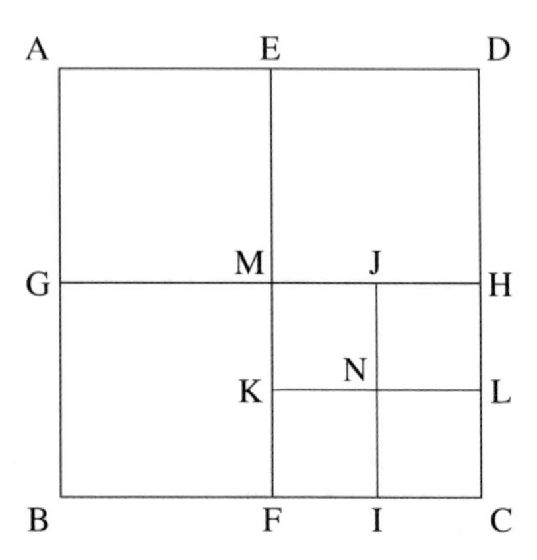

Fig. 13.1 Example of mesh

The following example illustrates the description of an instance of this class of meshes presented in Fig. 13.1. Initially, the rectangle boundary *ABCD* is specified. The order of points is assumed to follow counter-clockwise order, and the selection of the starting point is arbitrary. First, line *EF* is added and its end points belong to the horizontal sides of the mesh's boundary (*AD* and *BC*). Next, *GM* is drawn between *AB* and *EF* while *MH* is drawn between *EF* and *CD*. Finally, lines *KL* and *JI* are added.

As a result of this construction process, we obtain a set of points that are either vertices of the initial rectangle, endpoints of the lines added or points that result from the intersection of the lines added. We call this set of points *mesh points*. Any segment of a straight horizontal or vertical line in the mesh that connects two different mesh points is called a *mesh line*. A *boundary line* is a mesh line which belongs to a boundary side. A *maximal* mesh line is a mesh line which does not belong to any other mesh line. For example, *KE*, *FK*, *CH* and *GH* are mesh lines, *CH* is a boundary line, and *AB*, *GH* and *JI* are maximal mesh lines (see Fig. 13.1).

We define *dividing lines* that are used to partition a given mesh. We denote the set of vertical lines that extend between the borders of a mesh M by $DL_V(M)$, horizontal lines that extend between the borders of M by $DL_H(M)$ and the union of both sets by $DL(M)$. We do not consider vertical border sides of the mesh M among $DL_V(M)$ and similarly for its horizontal border sides. The mesh M can be partitioned using a dividing line l that belongs to the set $DL(M)$. This partitioning step results in two submeshes: $M(l, 0)$ that represents the submesh which lies below the horizontal (left of the vertical) line l and $M(l, 1)$ denotes the submesh which is above the horizontal (right of the vertical) line l. In Fig. 13.1, $DL_V(ABCD) = \{FE\}$, $DL_H(ABCD) = \{GH\}$ and $DL(ABCD) = \{GH, FE\}$. The mesh *ABCD* can be partitioned using the dividing line *FE* resulting in the two submeshes: $ABFE = ABCD(FE, 0)$ and $EFCD = ABCD(FE, 1)$.

We describe an arbitrary *submesh* N of M by a sequence of binary partitioning steps. Formally, a submesh N of a mesh M is an expression of the kind

$$N = M(l_1, \delta_1) \dots (l_n, \delta_n)$$

where $\delta_1, \dots, \delta_n \in \{0, 1\}$, $l_1 \in DL(M)$ and $l_i \in DL(M(l_1, \delta_1) \dots (l_{i-1}, \delta_{i-1}))$ for $i = 2, \dots, n$.

The resulting submesh is described as follows. First, a dividing line l_1 is used to partition the mesh M then the submesh $M(l_1, \delta_1)$ is acquired. The line l_2 is a dividing line of this submesh ($l_2 \in DL(M(l_1, \delta_1))$) which is used to partition $M(l_1, \delta_1)$ again until the desired submesh is obtained, etc. For example, $MHDE = ABCD(EF, 1)(MH, 1)$. We denote by $\text{SUB}(M)$ the set of submeshes of the mesh M including the mesh M. In our example,

$$\begin{aligned} \text{SUB}(ABCD) = \{&BFMG, BIJG, BCHG, ABFE, ABCD, AGME, AGHD, \\ &FINK, FIJM, FCLK, FCHM, FCDE, ICLN, ICHJ, \\ &KNJM, KLHM, KLDE, NLHJ, MHDE\} \, . \end{aligned}$$

The submesh N can be considered as a mesh that is represented as a rectangle, whose sides are called *border sides*, that may contain a set of vertical and horizontal lines inside it. For the submesh N, we can define the notion of dividing lines in a similar way as for M, so we can use the notation $DL_V(N)$, $DL_H(N)$ and $DL(N)$. As for the initial mesh M, it is possible that a dividing line is constructed indirectly in a series of steps. For example, $DL_V(KLHM) = \{NJ\}$ and $DL_H(FIJM) = \{KN\}$ (see Fig. 13.1).

A submesh N of the mesh M is a *unitary submesh* if and only if it does not have any dividing lines, i.e., $DL(N) = \emptyset$. We assume that each unitary submesh N of M has a unique *identifier* $\varphi(N)$. Unitary submeshes correspond to finite elements.

13.2.2 Mesh and Line Parameters

Let M be a mesh. For each mesh line l of M, we describe a set $P(M, l)$ of *distinct points of the line* l. If l is a boundary line, then the set of points includes the endpoints of l and any point of l that results from any other mesh line touching l. If l is not a boundary line, then the set of points includes all the points of l that result from another mesh line cutting l in addition to the endpoints of l. End points of other lines that start or finish on l are not included. That is, a touching line does not generate a point for l. For example, $P(ABCD, AD) = \{A, E, D\}$, $P(ABCD, EF) = \{F, M, E\}$. Each pair of consecutive points in $P(M, l)$ forms an *edge*. The number of edges on a line l is denoted by $E(M, l)$ and $E(M, l) = |P(M, l)| - 1$. For example, $E(ABCD, EF) = 2$ for the mesh in Fig. 13.1.

Let N be a submesh of M and $l \in DL(N)$. The line l represents a *common border side* of two submeshes: $N(l, 0)$ and $N(l, 1)$. We define now a number of parameters of the submesh N and dividing line l for N:

- $EP(M, N, l)$ represents the number of endpoints of l that lie in a boundary side.
- $B(M, N)$ is the number of edges on the border sides of N. These edges result from mesh lines that cut the border sides of N or touch them if they are on boundary sides, i.e., border sides of the initial mesh M.
- $BE(M, N)$ is the number of edges that lie in the border sides of N which are on the boundary sides, where $0 \leq BE(M, N) \leq B(M, N)$.
- $BV(M, N)$ represents the number of vertices of N that are endpoints of boundary sides, i.e., corners, where $0 \leq BV(M, N) \leq 4$.

For example in Fig. 13.1, some values of the parameters above are:

$$EP(ABCD, FIJM, KN) = 0, \qquad B(ABCD, FIJM) = 5,$$
$$EP(ABCD, FCHM, KL) = 1, \qquad BE(ABCD, BFMG) = 2,$$
$$EP(ABCD, ABCD, EF) = 2, \qquad BV(ABCD, BFMG) = 1.$$

We denote by $s(M)$ the number of maximal mesh lines in M. The number $s(M)$ is the *size* of the mesh M. For example, $s(ABCD) = 8$ for the mesh in Fig. 13.1.

Lemma 13.1 *For any mesh M, $|SUB(M)| \leq s(M)^4$.*

Proof Each submesh of M is defined by its four border sides, i.e., by four straight lines. The number of possible straight lines is at most the size $s(M)$ of the mesh M. Therefore $|\mathrm{SUB}(M)| \leq s(M)^4$. □

13.2.3 Element Partition Trees

An element partition tree is a labeled finite directed tree with a root. We define the notion of an *element partition tree* for a submesh N of the mesh M by induction. Let N be a unitary mesh. Then there exists only one element partition tree for N that contains exactly one node which is labeled with $\varphi(N)$. We denote this element partition tree by $etree(\varphi(N))$. Let N be a nonunitary mesh. Then any element partition tree for N can be represented in the form $etree(l, \Gamma_0, \Gamma_1)$ where $l \in DL(N)$, Γ_δ is an element partition tree for the submesh $N(l, \delta)$, $\delta \in \{0, 1\}$, and $etree(l, \Gamma_0, \Gamma_1)$ is a tree in which the root is labeled with l, and two edges start from the root which are labeled with 0 and 1 and enter the roots of element partition trees Γ_0 and Γ_1, respectively. We denote the set of all element partition trees for the submesh N of the mesh M by $ET(M, N)$.

Let Γ be an element partition tree for the submesh N of the mesh M. Any terminal node (leaf) of this tree is labeled as a unitary submesh of N. Any internal node is labeled with a line. Each internal node has exactly two edges that start from it and are labeled with 0 and 1, respectively. Figure 13.2 shows an element partition tree for the mesh presented in Fig. 13.1.

We now associate to each node v of the element partition tree Γ a submesh $N_\Gamma(v)$ of the submesh N. If v is the root of Γ then $N_\Gamma(v) = N$. If v is not the root and the path

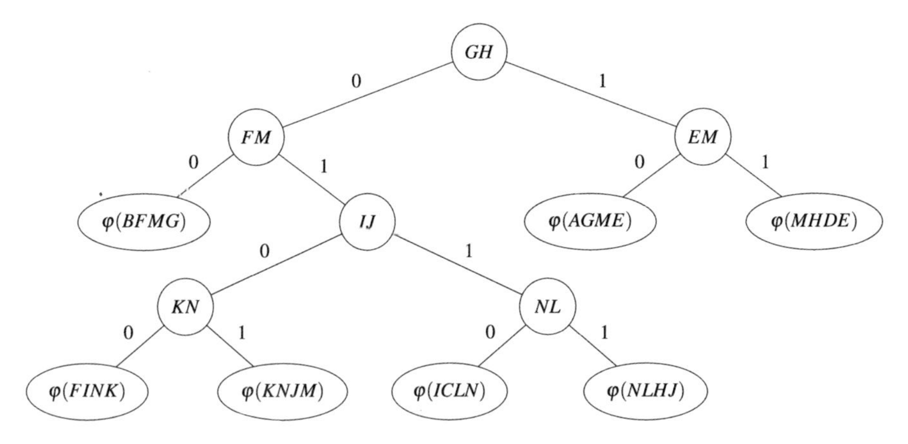

Fig. 13.2 Element partition tree for the mesh in Fig. 13.1

from the root to v consists of nodes labeled with lines $l_1, \ldots, l_m$ and edges labeled with the numbers $\delta_1, \ldots, \delta_m$, respectively, then $N_\Gamma(v) = N(l_1, \delta_1) \ldots (l_m, \delta_m)$.

For each internal node v of Γ, this node is labeled with a line from $DL(N_\Gamma(v))$. For each terminal node v, the submesh $N_\Gamma(v)$ is a unitary mesh and v is labeled with the identifier $\varphi(N_\Gamma(v))$ of $N_\Gamma(v)$.

For a node v of Γ, we denote by $\Gamma(v)$ the subtree of Γ with the root in v. Thus, $\Gamma(v)$ is an element partition tree for the submesh $N_\Gamma(v)$.

Let M be a mesh and N be a nonunitary submesh of M. For each $l \in DL(N)$, let $ET(M, N, l) = \{etree(l, \Gamma_0, \Gamma_1) : \Gamma_\delta \in ET(M, N(l, \delta)), \delta = 0, 1\}$. The next statement follows immediately from the definition of element partition tree.

Proposition 13.1 *Let M be a mesh and N be a submesh of M. Then $ET(M, N) = \{etree(\varphi(N))\}$ if N is unitary, and $ET(M, N) = \bigcup_{l \in DL(N)} ET(M, N, l)$ if N is nonunitary.*

13.2.4 Cost Functions for Element Partition Trees

In this section, we define the notion of a *cost function for element partition trees*.

Each cost function ψ has values from the set $\mathbb{R}$ of real numbers and is defined on triples (M, N, Γ) where M is a mesh, N is a submesh of M, and Γ is an element partition tree for N. The function ψ is specified by three functions ψ^0, F and w. Such cost functions are defined inductively as follows.

If N is a unitary submesh of M and $\Gamma = etree(\varphi(N))$, then $\psi(M, N, \Gamma) = \psi^0(M, N)$ where ψ^0 is a function which is defined on pairs (M, N), where M is a mesh and N is a unitary submesh of M, and has values from the set $\omega = \{0, 1, 2, \ldots\}$ of nonnegative integers.

Let N be a nonunitary mesh and $\Gamma = etree(l, \Gamma_0, \Gamma_1)$ where $l \in DL(N)$, $\Gamma_\delta \in ET(M, N(l, \delta))$, $\delta \in \{0, 1\}$. Then

$$\psi(M, N, \Gamma) = F(\psi(M, N(l, 0), \Gamma_0), \psi(M, N(l, 1), \Gamma_1)) + w(M, N, l)$$

where F is a function which is defined on ω^2 and has values in ω while w is a function that maps a triplet(M, N, l) to ω where M is a mesh, N is a submesh of M and $l \in DL(N)$.

Using this inductive definition, we can compute the cost of given element partition tree beginning from terminal nodes and finishing at the root.

Let $\leq$ be a partial order on the set ω^2: $(x_1, x_2) \leq (y_1, y_2)$ if $x_1 \leq y_1$ and $x_2 \leq y_2$. A function $f : \omega^2 \to \omega$ is called *increasing* if $f(x) \leq f(y)$ for any $x, y \in \omega^2$ such that $x \leq y$. The function f is called *strictly increasing* if $f(x) < f(y)$ for any $x, y \in \omega^2$ such that $x \leq y$ and $x \neq y$. For example, $\max(x_1, x_2)$ and $x_1 + x_2$ are increasing functions, and $x_1 + x_2$ is a strictly increasing function. We consider the cost function ψ to be *increasing* if the function F is an increasing function. Similarly, ψ is considered a *strictly increasing* cost function if F is a strictly increasing one. Each strictly increasing cost function is an increasing cost function.

We study different increasing cost functions that model various aspects of complexities. Each of these cost functions is associated with one of two modes of solvers: *ASAP* and *KB*. The mode determines the value of some of the parameters (r, n, t, and m) used by the function.

All the discussed cost functions depend on a parameter p. Some of these cost functions also depend on a parameter q. The parameter p (the polynomial order) is a nonnegative integer which identifies the number of basis functions spread over vertices, edges, and interiors of unitary submeshes (finite elements): one per vertex, p per edge, and p^2 per interior. The parameter q is a nonnegative integer that characterizes the time of one memory transfer operation relative to the cost of performing an arithmetic operation. Both parameters p and q does not depend on the mode of the solver but on the considered architecture and problem size.

We use $W(\mu, \rho) = \sum_{i=1}^{\rho}(2\mu - 2i + 1)(\mu - i) = 2\mu^2\rho - 2\mu\rho^2 - \mu\rho + \frac{2}{3}\rho^3 + \frac{1}{2}\rho^2 - \frac{1}{6}\rho$ as the number of arithmetic floating point operations for partial factorization of a $\mu \times \mu$ matrix with ρ fully assembled rows. It is clear that $W(\mu, \rho) \geq 0$. We assume that to keep results of the partial factorization of a $\mu \times \mu$ matrix we need μ^2 memory. We also assume that we should send $(\mu - \rho)^2$ numbers (corresponding to the Schur complement) to the parent node. Note that $\rho \leq \mu$ and $W(\mu, \rho) \leq 4 \times \mu^3$.

Let N be a unitary submesh of M, $\Gamma = etree(\varphi(N))$, and A' be an $n \times n$ matrix with r fully assembled rows corresponding to N. Then, during the processing of the submesh N, the algorithm makes $W(n, r)$ arithmetic floating point operations for partial factorization of A'. We send $(n - r)^2$ numbers corresponding to the Schur complement to the parent node if $N \neq M$ and keep the n^2 numbers which result from factorization. We define a parameter $\sigma \in \{0, 1\}$ to indicate whether current submesh has a parent or not, i.e., σ is set to 1 if and only if $N \neq M$.

Let N be a nonunitary submesh, $\Gamma = etree(l, \Gamma_0, \Gamma_1)$ where $l \in DL(N)$ and $\Gamma_\delta \in ET(M, N(l, \delta))$, $\delta \in \{0, 1\}$, and A'' be an $m \times m$ matrix with t fully assembled rows corresponding to N. Then, during the processing of the submesh N, the algorithm makes $W(m, t)$ arithmetic floating point operations during partial factorization of A'', it should send $(m - t)^2$ numbers corresponding to the Schur complement to the parent node if $N \neq M$ and store m^2 numbers to keep the factorization.

We present the definitions of several cost functions describing different complexity measures.

- Sequential time $time^{1,p,q}_{mode}$ is a cost function describing the complexity of a sequential solver following a given element partition tree. For this cost function, $\psi^0(M, N) = \sigma q(n - r)^2 + W(n, r)$, $F(x, y) = x + y$, and $w(M, N, l) = \sigma q(m - t)^2 + W(m, t)$. The considered cost function is strictly increasing.
- Parallel time $time^{\infty,p,q}_{mode}$ is a cost function modeling the complexity of a parallel solver following a given element partition tree. This solver solves the children of a given node in parallel and merges the solution sequentially. For this cost function, $\psi^0(M, N) = \sigma q(n - r)^2 + W(n, r)$, $F(x, y) = \max(x_1, x_2)$, and $w(M, N, l) = \sigma q(m - t)^2 + W(m, t)$. The considered cost function is increasing.
- Memory $memory^p_{mode}$ is a cost function that measures the space complexity of the solver. For this cost function, $\psi^0(M, N) = n^2$, $F(x, y) = x_1 + x_2$, and $w(M, N, l) = m^2$. The considered cost function is strictly increasing.

The parameters depending on the mode (r, n, t, and m) are defined as follows:
If *mode* is *ASAP*, we have:

$$r = p^2 + BE(M, N) \times p + BV(M, N) , \tag{13.1}$$

$$n = p^2 + 4p + 4 , \tag{13.2}$$

$$t = E(M, l) \times (p + 1) + EP(M, N, l) - 1 , \tag{13.3}$$

$$m = \begin{cases} (B(M, N) - BE(M, N) + E(M, l)) & \\ \times (p + 1) + EP(M, N, l) + \gamma , & 1 \leq BE(M, N) < B(M, N) , \\ (B(M, N) - BE(M, N) + E(M, l)) & \\ \times (p + 1) + EP(M, N, l) - 1 , & \text{otherwise} , \end{cases} \tag{13.4}$$

where $\gamma \in \{0, 1\}$ and $\gamma = 1$ if and only if N has exactly two boundary sides which are parallel.

If *mode* is *KB*, we have:

$$r = p^2 , \tag{13.5}$$

$$n = p^2 + 4p + 4 , \tag{13.6}$$

$$t = E(M, l) \times (p + 1) - 1 , \tag{13.7}$$

$$m = (B(M, N) + E(M, l))(p + 1) - 1 . \tag{13.8}$$

The parameters $B(M, N)$, $BE(M, N)$, $BV(M, N)$, $E(M, l)$, $EP(M, N, l)$ and γ can be computed in polynomial time with respect to $s(M)$. As a result, all of the aforementioned cost functions can be evaluated in polynomial time with respect to the mesh size $s(M)$. Since F is either $\max(x, y)$ or $x + y$, the time complexity of F computation is $O(1)$. As a result, we have the following statement.

Proposition 13.2 *For each cost function*

$$\psi \in \{time^{1,p,q}_{mode}, time^{\infty,p,q}_{mode}, memory^{p}_{mode} : mode \in \{KB, ASAP\}\}$$

defined by the functions ψ^0, F *and* w, ψ^0, F *and* w *can be computed in polynomial time depending on the size of the mesh* $s(M)$.

Let $\psi \in \{time^{1,p,q}_{mode}, time^{\infty,p,q}_{mode}, memory^{p}_{mode} : mode \in \{KB, ASAP\}\}$ and ψ is defined by the functions ψ^0, F and w.

It is clear that independent of the mode of the solver, $0 \leq r \leq n$ and $0 \leq t \leq m$. In addition, we recall from the definitions of the mesh and line parameters that $0 \leq BE(M, N) \leq B(M, N)$, $0 \leq EP(M, N, l) \leq 2$, $B(M, N) \leq 4(s(M) + 1)$ and $E(M, l) \leq s(M) + 1$. We use this information to show that there exists an upper bound on the value of ψ which is polynomial on the size of mesh $s(M)$ and the parameters p and q.

Let N be a submesh of a mesh M and Γ be an element partition tree for N. If N is a unitary submesh then Γ consists of one node, and the cost of Γ is nonnegative. Otherwise, we notice that $w(M, N, l)$ is a nonnegative value for any nonunitary submesh N of M and $l \in DL(N)$. As a result, the cost of an element partition tree Γ' for mesh M with maximum cost forms an upper bound on the cost of any element partition tree for any submesh N of M.

We denote

$$ub(\psi, M) = \max\{\psi(M, M, \Gamma) : \Gamma \in ET(M, M)\} ,$$

$$v'(\psi, M) = \max\{\psi^0(M, N) : N \text{ is a unitary submesh of } M\}$$

and

$$w'(\psi, M) = \max\{w(M, N, l) : N \text{ is a nonunitary submesh of } M, l \in DL(N)\} .$$

Consider $L_t(M)$ to be the number of unitary submeshes of M and hence the number of leaves for any element partition tree for M.

Lemma 13.2 *Let* $\psi \in \{time^{1,p,q}_{mode}, time^{\infty,p,q}_{mode}, memory^{p}_{mode} : mode \in \{KB, ASAP\}\}$ *and* M *be a mesh. Then* $ub(\psi, M) \leq s(M)^2(v'(\psi, M) + w'(\psi, M))$.

Proof Any element partition tree for the mesh M has all unitary submeshes of M as leaves. As a result, any element partition tree for M has exactly $L_t(M)$ leaves.

Since the element partition tree is a binary tree whose each node either has zero or two children, the number of internal nodes of any element partition tree for M is $L_t(M) - 1$.

The cost of any element partition tree for M is at most the sum of the costs of its leaves in addition to the sum of the values $w(M, N, l)$ computed at each node N. Therefore

$$\begin{aligned} ub(\psi, M) &\leq L_t(M)v'(\psi, M) + (L_t(M) - 1)w'(\psi, M) \\ &\leq L_t(M)(v'(\psi, M) + w'(\psi, M)) \ . \end{aligned}$$

One can show that there exists at most $s(M)^2$ unitary submeshes of the mesh M. Thus

$$ub(\psi, M) \leq s(M)^2(v'(\psi, M) + w'(\psi, M)) \ . \qquad \square$$

Lemma 13.3 *Let M be a mesh and $\psi \in \{time^{1,p,q}_{mode}, time^{\infty,p,q}_{mode} : mode \in \{KB, ASAP\}\}$. Then $ub(\psi, M) = O(s(M)^4 qp^4 + s(M)^5 p^6)$.*

Proof We know that $v'(\psi, M) = \psi^0(M, N)$ for some unitary submesh N of M. Therefore

$$v'(\psi, M) = \sigma q(n - r)^2 + W(n, r) \ .$$

Since $\sigma \in \{0, 1\}$ and $W(n, r) \leq 4n^3$, we have

$$v'(\psi, M) \leq qn^2 + 4n^3 \ .$$

From (13.2) and (13.6) we deduce that $n = p^2 + 4p + 4$, and hence

$$v'(\psi, M) \leq q(p^2 + 4p + 4)^2 + 4(p^2 + 4p + 4)^3 = q(p + 2)^4 + 4(p + 2)^6 \ .$$

We know that $w'(\psi, M) = w(M, N, l)$ for some nonunitary submesh N of M and $l \in DL(N)$. Therefore

$$w'(\psi, M) = \sigma q(m - t)^2 + W(m, t) \ .$$

By (13.4) and (13.8), $m \leq (B(M, N) + E(M, l))(p + 1) + 3$. In addition, since $0 \leq B(M, N) \leq 4s(M)$ and $E(M, l) \leq s(M)$,

$$\begin{aligned} m &\leq 5s(M)(p + 1) + 3 \\ &\leq 5s(M)(p + 2) \ . \end{aligned}$$

On the other hand,

$$\begin{aligned} w'(\psi, M) &= \sigma q(m-t)^2 + W(m,t) \\ &\le qm^2 + 4m^3 \\ &\le 25qs(M)^2(p+2)^2 + 500s(M)^3(p+2)^3 . \end{aligned}$$

From Lemma 13.2 it follows that

$$\begin{aligned} ub(\psi, M) &\le s(M)^2(v'(\psi, M) + w'(\psi, M)) \\ &\le s(M)^2(q(p+2)^4 + 4(p+2)^6 + 25q(p+2)^2 s(M)^2 \\ &\quad + 500s(M)^3(p+2)^3) \\ &\le qs(M)^2(p+2)^4 + 4s(M)^2(p+2)^6 + 25qs(M)^4(p+2)^2 \\ &\quad + 500s(M)^5(p+2)^3 . \end{aligned}$$

Therefore $ub(\psi, M) = O(s(M)^4 qp^4 + s(M)^5 p^6)$. □

Lemma 13.4 *Let M be a mesh and $\psi \in \{memory^p_{mode} : mode \in \{KB, ASAP\}\}$. Then $ub(\psi, M) = O(s(M)^4 p^4)$.*

Proof As in the proof of Lemma 13.3, we can show that $v'(\psi, M) = n^2$ where $n = (p+2)^2$, and $w'(\psi, M) = m^2$ where $m \le 5s(M)(p+2)$. Applying Lemma 13.2, we obtain that

$$\begin{aligned} ub(\psi, M) &\le s(M)^2(v'(\psi, M) + w'(\psi, M)) \\ &\le s(M)^2((p+2)^4 + 25s(M)^2(p+2)^2) \\ &= s(M)^2(p+2)^4 + 25s(M)^4(p+2)^2 . \end{aligned}$$

Therefore $ub(\psi, M) = O(s(M)^4 p^4)$. □

References

1. AbouEisha, H., Gurgul, P., Paszynska, A., Paszynski, M., Kuznik, K., Moshkov, M.: An automatic way of finding robust elimination trees for a multi-frontal sparse solver for radical 2D hierarchical meshes. In: Wyrzykowski, R., Dongarra, J., Karczewski, K., Wasniewski, J. (eds.) Parallel Processing and Applied Mathematics – 10th International Conference, PPAM 2013, Warsaw, Poland, September 8–11, 2013, Revised Selected Papers, Part II, Lecture Notes in Computer Science, vol. 8385, pp. 531–540. Springer (2013)
2. AbouEisha, H., Moshkov, M., Calo, V.M., Paszynski, M., Goik, D., Jopek, K.: Dynamic programming algorithm for generation of optimal elimination trees for multi-frontal direct solver over h-refined grids. In: Abramson, D., Lees, M., Krzhizhanovskaya, V.V., Dongarra, J., Sloot, P.M.A. (eds.) International Conference on Computational Science, ICCS 2014, Cairns, Queensland, Australia, June 10–12, 2014, Procedia Computer Science, vol. 29, pp. 947–959. Elsevier (2014)

3. Amestoy, P., Duff, I., L'Excellent, J.Y.: Multifrontal parallel distributed symmetric and unsymmetric solvers. Comput. Methods Appl. Mech. Eng. **184**(2–4), 501–520 (2000)
4. Amestoy, P., Duff, I.S., L'Excellent, J., Koster, J.: A fully asynchronous multifrontal solver using distributed dynamic scheduling. SIAM J. Matrix Anal. Appl. **23**(1), 15–41 (2001)
5. Amestoy, P., Guermouche, A., L'Excellent, J., Pralet, S.: Hybrid scheduling for the parallel solution of linear systems. Parallel Comput. **32**(2), 136–156 (2006)
6. Amestoy, P.R., Duff, I.S., L'Excellent, J.Y.: MUMPS MUltifrontal Massively Parallel Solver Version 2.0. Research Report RT/APO/98/3 (1998). https://hal.inria.fr/hal-00856859
7. Bunch, J., Hopcroft, J.: Triangular factorization and inversion by fast matrix multiplication. Math. Comput. **28**(125), 231–236 (1974)
8. Duff, I.S., Erisman, A.M., Reid, J.K.: Direct Methods for Sparse Matrices. Oxford University Press, New York (1986)
9. Duff, I.S., Reid, J.K.: The multifrontal solution of indefinite sparse symmetric linear systems. ACM Trans. Math. Softw. **9**(3), 302–325 (1983)
10. Dwyer, P.S.: The Doolittle technique. Ann. Math. Stat. **12**, 449–458 (1941)
11. Fialko, S.: A block sparse shared-memory multifrontal finite element solver for problems of structural mechanics. Comput. Assist. Mech. Eng. Sci. **16**(2), 117–131 (2009)
12. Fialko, S.: The block subtracture multifrontal method for solution of large finite element equation sets. Tech. Trans. **1-NP**(8) (2009)
13. Fialko, S.: PARFES: A method for solving finite element linear equations on multi-core computers. Adv. Eng. Softw. **41**(12), 1256–1265 (2010)
14. Gall, F.L.: Powers of tensors and fast matrix multiplication. In: Nabeshima, K., Nagasaka, K., Winkler, F., Szántó, Á. (eds.) International Symposium on Symbolic and Algebraic Computation, ISSAC '14, Kobe, Japan, July 23–25, 2014, pp. 296–303. ACM (2014)
15. Gurgul, P.: A linear complexity direct solver for h-adaptive grids with point singularities. In: Abramson, D., Lees, M., Krzhizhanovskaya, V.V., Dongarra, J., Sloot, P.M.A. (eds.) International Conference on Computational Science, ICCS 2014, Cairns, Queensland, Australia, June 10–12, 2014, Procedia Computer Science, vol. 29, pp. 1090–1099. Elsevier (2014)
16. Karypis, G., Kumar, V.: METIS-Serial Graph Partitioning And Fill-reducing Matrix Ordering. http://glaros.dtc.umn.edu/gkhome/metis/metis/overview (2012)
17. Okunev, P., Johnson, C.R.: Necessary and sufficient conditions for existence of the LU factorization of an arbitrary matrix. https://arxiv.org/pdf/math/0506382v1.pdf (1997)
18. Paszynska, A., Paszynski, M., Jopek, K., Wozniak, M., Goik, D., Gurgul, P., AbouEisha, H., Moshkov, M., Calo, V.M., Lenharth, A., Nguyen, D., Pingali, K.: Quasi-optimal elimination trees for 2D grids with singularities. Sci. Program. **2015**, 303,024:1–303,024:18 (2015)
19. Yannakakis, M.: Computing the minimum fill-in is NP-complete. SIAM J. Algebr. Discret. Methods **2**(1), 77–79 (1981)
20. Zhang, F. (ed.): The Schur Complement and Its Applications, Numerical Methods and Algorithms, vol. 4. Springer, New York (2005)

Chapter 14
Multi-stage Optimization of Element Partition Trees

An element partition tree describes an ordering of rows in the matrix representing a system of linear equations. This ordering is essential for solvers to find a solution for our system. The quality of an element partition tree can be evaluated using different criteria. For example, one may be interested in finding an ordering that allows the solver to solve the system of equations using the fewest number of arithmetic operations. Another criterion may focus on the memory usage of the solver and others can focus on the data transfer and the energy consumption [2].

This chapter presents tools and applications of multi-stage optimization of element partition trees. It starts by describing how the set of element partition trees can be represented compactly in the form of a directed acyclic graph (DAG). It then suggests algorithms to count the number of element partition trees represented by a DAG and to optimize the corresponding set of element partition trees with respect to some criterion. The latter algorithm can be used for multi-stage optimization of element partition trees relative to a sequence of cost functions and for the study of totally optimal element partition trees that are optimal with respect to multiple criteria simultaneously. Finally, we present results of experiments on three different classes of adaptive meshes.

Note that in the paper [1] the procedure of element partition tree optimization was extended to a more general class of cost functions.

14.1 Directed Acyclic Graph $\Delta(M)$

Let M be a mesh. We describe an algorithm $\mathscr{A}_{15}$ for the construction of a directed acyclic graph $\Delta(M)$ which is used to describe and optimize the element partition trees for M. The set of nodes of this graph coincides with the set $\mathrm{SUB}(M)$ of submeshes of the mesh M. During each iteration (with the exception of the last one) the algorithm processes one node. It starts with the graph that consists of one node M which is not processed yet and finishes when all nodes of the constructed graph are processed.

H. AbouEisha et al., *Extensions of Dynamic Programming for Combinatorial Optimization and Data Mining*, Intelligent Systems Reference Library 146,
https://doi.org/10.1007/978-3-319-91839-6_14

Algorithm $\mathscr{A}_{15}$ (construction of DAG $\Delta(M)$).

Input: Mesh M.
Output: Directed acyclic graph $\Delta(M)$.

1. Construct the graph that consists of one node M which is not marked as processed.
2. If all nodes of the graph are processed then the work of the algorithm is finished. Return the resulting graph as $\Delta(M)$. Otherwise, choose a node (submesh) N that has not been processed yet.
3. If N is a unitary mesh then mark N as processed and proceed to the step 2.
4. If N is nonunitary then, for each $l \in DL(N)$, draw two edges from the node N (this pair of edges is called *l-pair*) and label these edges with pairs $(l, 0)$ and $(l, 1)$. These edges enter nodes $N(l, 0)$ and $N(l, 1)$, respectively. If some of the nodes $N(l, 0)$, $N(l, 1)$ are not present in the graph then add these nodes to the graph. Mark the node N as processed and proceed to the step 2.

Proposition 14.1 *The algorithm $\mathscr{A}_{15}$ has polynomial time complexity with respect to the size of the input mesh.*

Proof The number of iterations of the algorithm $\mathscr{A}_{15}$ (each iteration includes step 2 and step 3 or 4) is at most $|\mathrm{SUB}(M)| + 1$. By Lemma 13.1, $|\mathrm{SUB}(M)| \leq s(M)^4$. One can show that the time complexity of each iteration is polynomial with respect to the size of M. Therefore the algorithm $\mathscr{A}_{15}$ has polynomial time complexity with respect to the size of the input mesh. □

A node of a directed graph is called *terminal* if there are no edges starting in this node, and *internal* otherwise. A node N of the graph $\Delta(M)$ is terminal if and only if N is a unitary mesh. A *proper subgraph* of the graph $\Delta(M)$ is a graph G obtained from $\Delta(M)$ by removal of some l-pairs of edges such that each internal node of $\Delta(M)$ keeps at least one l-pair of edges starting from this node. By definition, $\Delta(M)$ is a proper subgraph of $\Delta(M)$. Proper subgraphs of the graph $\Delta(M)$ arise as results of the optimization procedure applied to $\Delta(M)$ or to its proper subgraphs.

Let G be a proper subgraph of the graph $\Delta(M)$. For each internal node N of the graph G, we denote by $DL_G(N)$ the set of lines l from $DL(N)$ such that an l-pair of edges starts from N in G. For each node N of the graph G, we define the set $Etree(G, N)$ of elimination trees in the following way. If N is a terminal node of G (in this case N is unitary), then $Etree(G, N) = \{etree(\varphi(N))\}$. Let N be an internal node of G (in this case N is nonunitary) and $l \in DL_G(N)$. We denote $Etree(G, N, l) = \{etree(l, \Gamma_0, \Gamma_1) : \Gamma_\delta \in Etree(G, N(l, \delta)), \delta = 0, 1\}$. Then $Etree(G, N) = \bigcup_{l \in DL_G(N)} Etree(G, N, l)$.

Proposition 14.2 *Let M be a mesh. Then the equality*

$$Etree(\Delta(M), N) = ET(M, N)$$

holds for any node N of the graph $\Delta(M)$.

Proof We prove this statement by induction on the nodes of $\Delta(M)$. Let N be a terminal node of $\Delta(M)$. Then $Etree(\Delta(M), N) = \{etree(\varphi(N))\} = ET(M, N)$. On the other hand, consider N to be an internal node of $\Delta(M)$, and let us assume that $Etree(\Delta(M), N(l, \delta)) = ET(M, N(l, \delta))$ for any $l \in DL(N)$ and $\delta \in \{0, 1\}$. Then, for any $l \in DL(N)$, we have $Etree(\Delta(M), N, l) = ET(M, N, l)$, where $DL_{\Delta(M)}(N) = DL(N)$. Using Proposition 13.1, we obtain $Etree(\Delta(M), N) = ET(M, N)$. □

So the set of element partition trees for a mesh M is equal to $Etree(\Delta(M), M)$. We show later that the set of strictly optimal element partition trees for a mesh M relative to an increasing cost function ψ can be represented in the form $Etree(G, M)$ where G is a proper subgraph of the graph $\Delta(M)$. In the next section, we describe an algorithm which counts the cardinality of the set $Etree(G, M)$ for a proper subgraph G of the graph $\Delta(M)$.

14.2 Cardinality of the Set $Etree(G, M)$

Let M be a mesh, and G be a proper subgraph of the graph $\Delta(M)$. We describe an algorithm which counts, for each node N of the graph G, the cardinality $C(N)$ of the set $Etree(G, N)$, and returns the number $C(M) = |Etree(G, M)|$.

Algorithm $\mathscr{A}_{16}$ (counting the number of element partition trees).

Input: A proper subgraph G of the graph $\Delta(M)$ for some mesh M.
Output: The number $|Etree(G, M)|$.

1. If all nodes of the graph G are processed then return the number $C(M)$ and finish the work of the algorithm. Otherwise, choose a node N which is not processed yet and which is either a terminal node of G or an internal node of G such that all nodes $N(l, \delta)$ are processed for each $l \in DL_G(N)$ and $\delta \in \{0, 1\}$.
2. If N is a terminal node (unitary mesh), set $C(N) = 1$, mark N as processed and proceed to the step 1.
3. If N is an internal node (nonunitary mesh) then set

$$C(N) = \sum_{l \in DL_G(N)} C(N(l, 0)) \times C(N(l, 1)) ,$$

mark the node N as processed and proceed to the step 1.

Proposition 14.3 *Let M be a mesh, and G be a proper subgraph of the graph $\Delta(M)$. Then the algorithm $\mathscr{A}_{16}$ returns the cardinality of the set $Etree(G, M)$ and makes at most $2s(M)^5$ operations of multiplication and addition.*

Proof We prove by induction on the nodes of the graph G that $C(N) = |Etree(G, N)|$ for each node N of G. Let N be a terminal node of G. Then

$Etree(G, N) = \{etree(\varphi(N))\}$ and $|Etree(G, N)| = 1$. Therefore the considered statement holds for N. Let now N be an internal node of G such that the considered statement holds for its children. We know that $Etree(G, N) = \bigcup_{l \in DL_G(N)} Etree(G, N, l)$ where, for $l \in DL_G(N)$,

$$Etree(G, N, l) = \{etree(l, \Gamma_0, \Gamma_1) : \Gamma_\delta \in Etree(G, N(l, \delta)), \delta = 0, 1\} .$$

Then, for any $l \in DL_G(N)$,

$$|Etree(G, N, l)| = |Etree(G, N(l, 0))| \times |Etree(G, N(l, 1))| ,$$

and $|Etree(G, N)| = \sum_{l \in DL_G(N)} |Etree(G, N, l)|$. By the inductive hypothesis,

$$|Etree(G, N(l, \delta))| = C(N(l, \delta))$$

for each $l \in DL_G(N)$ and $\delta \in \{0, 1\}$. Therefore $C(N) = |Etree(G, N)|$. Hence the considered statement holds. From here it follows that $C(M) = |Etree(G, M)|$, i.e., the algorithm $\mathscr{A}_{16}$ returns the cardinality of the set $Etree(G, M)$.

We evaluate the number of arithmetic operations made by the algorithm $\mathscr{A}_{16}$. By Lemma 13.1, $|\mathrm{SUB}(M)| \leq s(M)^4$. Therefore the number of internal nodes in G is at most $s(M)^4$. In each internal node N of G, the algorithm $\mathscr{A}_{16}$ makes $|DL_G(N)|$ operations of multiplication and $|DL_G(N)| - 1$ operations of addition. One can show that $|DL_G(N)| \leq s(M)$. Therefore, the algorithm $\mathscr{A}_{16}$ makes at most $2s(M)^5$ arithmetic operations. □

14.3 Optimization Procedure

Let M be a mesh, G be a proper subgraph of the graph $\Delta(M)$, and ψ be an increasing cost function for element partition trees defined by functions ψ^0, F and w.

Let N be a node of G and $\Gamma \in Etree(G, N)$. One can show that, for any node v of Γ, the element partition tree $\Gamma(v)$ belongs to the set $Etree(G, N_\Gamma(v))$.

An element partition tree Γ from $Etree(G, N)$ is called an *optimal element partition tree for N relative to ψ and G* if

$$\psi(M, N, \Gamma) = \min\{\psi(M, N, \Gamma') : \Gamma' \in Etree(G, N)\} .$$

An element partition tree Γ from $Etree(G, N)$ is called a *strictly optimal element partition tree for N relative to ψ and G* if, for any node v of Γ, the element partition tree $\Gamma(v)$ is an optimal element partition tree for $N_\Gamma(v)$ relative to ψ and G.

We denote by $Etree_{\psi}^{opt}(G, N)$ the set of optimal element partition trees for N relative to ψ and G. We denote by $Etree_{\psi}^{s-opt}(G, N)$ the set of strictly optimal element partition trees for N relative to ψ and G.

Let $\Gamma \in Etree_{\psi}^{opt}(G, N)$ and $\Gamma = etree(l, \Gamma_0, \Gamma_1)$. Then $\Gamma \in Etree_{\psi}^{s-opt}(G, N)$ if and only if $\Gamma_{\delta} \in Etree_{\psi}^{s-opt}(G, N(l, \delta))$ for $\delta = 0, 1$.

Proposition 14.4 *Let ψ be a strictly increasing cost function for element partition trees, M be a mesh, and G be a proper subgraph of the graph $\Delta(M)$. Then, for any node N of the graph G, $Etree_{\psi}^{opt}(G, N) = Etree_{\psi}^{s-opt}(G, N)$.*

Proof It is clear that $Etree_{\psi}^{s-opt}(G, N) \subseteq Etree_{\psi}^{opt}(G, N)$. Let $\Gamma \in Etree_{\psi}^{opt}(G, N)$ and assume that $\Gamma \notin Etree_{\psi}^{s-opt}(G, N)$. Then there is a node v of Γ such that $\Gamma(v) \notin Etree_{\psi}^{opt}(G, N_{\Gamma}(v))$. Let $\Gamma_0 \in Etree_{\psi}^{opt}(G, N_{\Gamma}(v))$ and Γ' be the element partition tree obtained from Γ by replacement of $\Gamma(v)$ with Γ_0. One can show that $\Gamma' \in Etree(G, N)$. Since ψ is strictly increasing and

$$\psi(M, N_{\Gamma}(v), \Gamma_0) < \psi(M, N_{\Gamma}(v), \Gamma(v)) ,$$

we have $\psi(M, N, \Gamma') < \psi(M, N, \Gamma)$. Therefore $\Gamma \notin Etree_{\psi}^{opt}(G, N)$, but this is impossible. Thus $Etree_{\psi}^{opt}(G, N) \subseteq Etree_{\psi}^{s-opt}(G, N)$. □

We present an algorithm $\mathscr{A}_{17}$ which is a *procedure of element partition tree optimization relative to a cost function ψ*. This algorithm attaches to each node N of G the number $c(N) = \min\{\psi(M, N, \Gamma) : \Gamma \in Etree(G, N)\}$ and, probably, removes some l-pairs of edges starting from internal nodes of G. As a result, we obtain a proper subgraph G^{ψ} of the graph G. By construction, G^{ψ} is also a proper subgraph of the graph $\Delta(M)$.

Algorithm $\mathscr{A}_{17}$ (procedure of element partition tree optimization).

Input: A proper subgraph G of the graph $\Delta(M)$ for some mesh M, and an increasing cost function ψ for element partition trees defined by functions ψ^0, F and w.

Output: Proper subgraph G^{ψ} of the graph G.

1. If all nodes of the graph G are processed then return the obtained graph as G^{ψ} and finish. Otherwise, choose a node N which is not processed yet and which is either a terminal node of G or an internal node of G for which all its children have been processed.
2. If N is a terminal node (unitary mesh) then set $c(N) = \psi^0(M, N)$, mark N as processed and proceed to the step 1.
3. If N is an internal node (nonunitary mesh) then, for each $l \in DL_G(N)$, compute the value
$$c(N, l) = F(c(N(l, 0)), c(N(l, 1))) + w(M, N, l)$$

and set $c(N) = \min\{c(N,l) : l \in DL_G(N)\}$. Remove all l-pairs of edges starting from N for which $c(N) < c(N,l)$. Mark the node N as processed and proceed to the step 1.

Proposition 14.5 *Let the algorithm $\mathscr{A}_{17}$ use a cost function ψ specified by functions ψ^0, F and w which have polynomial time complexity with respect to the size of the input mesh. Then the algorithm $\mathscr{A}_{17}$ has polynomial time complexity with respect to the size of the input mesh.*

Proof By Lemma 13.1, $|\text{SUB}(M)| \leq s(M)^4$. Therefore the number of nodes in G is at most $s(M)^4$. In each terminal node of the graph G, the algorithm $\mathscr{A}_{17}$ computes the value of ψ^0. In each internal node N of G the algorithm $\mathscr{A}_{17}$ computes $|DL_G(N)|$ times the value of F and w and makes $|DL_G(N)| - 1$ comparisons. One can show that $|DL_G(N)| \leq s(M)$. Therefore, the algorithm $\mathscr{A}_{17}$ makes at most $s(M)^5$ comparisons and at most $s(M)^5$ computations of the functions ψ^0, F and w. If ψ^0, F, and w have polynomial time complexity with respect to $s(M)$ then the algorithm $\mathscr{A}_{17}$ has polynomial time complexity with respect to $s(M)$. □

For any node N of the graph G and for any $l \in DL_G(N)$ we denote $\psi_G(N) = \min\{\psi(M,N,\Gamma) : \Gamma \in Etree(G,N)\}$ and

$$\psi_G(N,l) = \min\{\psi(M,N,\Gamma) : \Gamma \in Etree(G,N,l)\} .$$

Lemma 14.1 *Let ψ be an increasing cost function for element partition trees specified by functions ψ^0, F and w, M be a mesh, and G be a proper subgraph of the graph $\Delta(M)$. Then, for any node N of the graph G and for any $l \in DL_G(N)$, the algorithm $\mathscr{A}_{17}$ computes values $c(N) = \psi_G(N)$ and $c(N,l) = \psi_G(N,l)$.*

Proof We prove the considered statement by induction on the nodes of the graph G. Let N be a terminal node of G. Then $Etree(G,N) = \{etree(\varphi(N))\}$ and $\psi_G(N) = \psi^0(M,N)$. Therefore $c(N) = \psi_G(N)$. Let N be an internal node of G for which the considered statement holds for each node $N(l,\delta)$ such that $l \in DL_G(N)$ and $\delta \in \{0,1\}$. We know that

$$Etree(G,N) = \bigcup_{l \in DL_G(N)} Etree(G,N,l)$$

and, for each $l \in DL_G(N)$,

$$Etree(G,N,l) = \{etree(l,\Gamma_0,\Gamma_1) : \Gamma_\delta \in Etree(G,N(l,\delta)), \delta = 0,1\} .$$

Since ψ is an increasing cost function,

$$\psi_G(N,l) = F(\psi_G(N(l,0)), \psi_G(N(l,1))) + w(M,N,l) .$$

By the induction hypothesis, $\psi_G(N(l,\delta)) = c(N(l,\delta))$ for each $l \in DL_G(N)$ and $\delta \in \{0,1\}$. Therefore $c(N,l) = \psi_G(N,l)$ for each $l \in DL_G(N)$, and $c(N) = \psi_G(N)$. $\square$

Theorem 14.1 *Let ψ be an increasing cost function for element partition trees specified by functions ψ^0, F and w, M be a mesh, and G be a proper subgraph of the graph $\Delta(M)$. Then, for any node N of the graph G^ψ, the following equality holds: $Etree(G^\psi, N) = Etree_\psi^{s-opt}(G,N)$.*

Proof We prove the considered statement by induction on nodes of G^ψ. We use Lemma 14.1 which shows that, for each node N of the graph G and for each $l \in DL_G(N), c(N) = \psi_G(N)$ and $c(N,l) = \psi_G(N,l)$. Let N be a terminal node of G^ψ. Then $Etree(G^\psi, N) = \{etree(\varphi(N))\}$ with $Etree(G^\psi, N) = Etree_\psi^{s-opt}(G,N)$.

Let N be an internal node of G^ψ. Then

$$Etree(G^\psi, N) = \bigcup_{l \in DL_{G^\psi}(N)} Etree(G^\psi, N, l)$$

and, for each $l \in DL_{G^\psi}(N)$,

$$Etree(G^\psi, N, l) = \{etree(l, \Gamma_0, \Gamma_1) : \Gamma_\delta \in Etree(G^\psi, N(l,\delta)), \delta = 0, 1\} .$$

Let us assume that, for any $l \in DL_{G^\psi}(N)$ and $\delta \in \{0,1\}$,

$$Etree(G^\psi, N(l,\delta)) = Etree_\psi^{s-opt}(G, N(l,\delta)) .$$

We know that

$$DL_{G^\psi}(N) = \{l : l \in DL_G(N), \psi_G(N,l) = \psi_G(N)\} .$$

Let $l \in DL_{G^\psi}(N)$, and $\Gamma \in Etree(G^\psi, N, l)$. Then $\Gamma = etree(l, \Gamma_0, \Gamma_1)$, where

$$\Gamma_\delta \in Etree(G^\psi, N(l,\delta))$$

for $\delta = 0, 1$. According to the inductive hypothesis,

$$Etree(G^\psi, N(l,\delta)) = Etree_\psi^{s-opt}(G, N(l,\delta))$$

and

$$\Gamma_\delta \in Etree_\psi^{s-opt}(G, N(l,\delta))$$

for $\delta = 0, 1$. In particular, $\psi(M, N(l,\delta), \Gamma_\delta) = \psi_G(N(l,\delta))$ for $\delta = 0, 1$. Since $\psi_G(N,l) = \psi_G(N)$ we have $F(\psi_G(N(l,0)), \psi_G(N(l,1))) + w(M,N,l,) = \psi_G(N)$. Therefore $\Gamma \in Etree_\psi^{opt}(G,N)$, $\Gamma \in Etree_\psi^{s-opt}(G,N)$, and

$$Etree(G^{\psi}, N) \subseteq Etree_{\psi}^{s-opt}(G, N) .$$

Let $\Gamma \in Etree_{\psi}^{s-opt}(G, N)$. Since N is an internal node, Γ can be represented in the form $etree(l, \Gamma_0, \Gamma_1)$ where $l \in DL_G(N)$, and $\Gamma_\delta \in Etree_{\psi}^{s-opt}(G, N(l, \delta))$ for $\delta = 0, 1$. Since $\Gamma \in Etree_{\psi}^{s-opt}(G, N)$, $\psi_G(N, l) = \psi_G(N)$ and $l \in DL_{G^{\psi}}(N)$. According to the inductive hypothesis,

$$Etree(G^{\psi}, N(l, \delta)) = Etree_{\psi}^{s-opt}(G, N(l, \delta))$$

for $\delta = 0, 1$. Therefore $\Gamma \in Etree(G^{\psi}, N, l) \subseteq Etree(G^{\psi}, N)$. As a result, we have

$$Etree_{\psi}^{s-opt}(G, N) \subseteq Etree(G^{\psi}, N) .$$

□

Corollary 14.1 *Let ψ be a strictly increasing cost function for element partition trees, M be a mesh, and G be a proper subgraph of the graph $\Delta(M)$. Then, for any node N of G^{ψ}, $Etree(G^{\psi}, N) = Etree_{\psi}^{opt}(G, N)$.*

This corollary follows immediately from Proposition 14.4 and Theorem 14.1.

14.4 Multi-stage Optimization

We use a multi-stage optimization for the element partition trees for M relative to a sequence of strictly increasing cost functions $\psi_1, \ldots, \psi_l$. We begin from the graph $G = \Delta(M)$ and apply to it the procedure of optimization (the algorithm $\mathscr{A}_{17}$) relative to the cost function ψ_1. As a result, we obtain a proper subgraph G^{ψ_1} of the graph $G = \Delta(M)$. By Proposition 14.2, the set $Etree(G, M)$ is equal to the set $ET(M, M)$ of all element partition trees for M. Using Corollary 14.1, we obtain that the set $Etree(G^{\psi_1}, M)$ coincides with the set $Etree_{\psi_1}^{opt}(G, M)$ of all element partition trees from $Etree(G, M)$ which have minimum cost relative to ψ_1 among all trees from $Etree(G, M)$. Next we apply to G^{ψ_1} the procedure of optimization relative to the cost function ψ_2. As a result, we obtain a proper subgraph G^{ψ_1,ψ_2} of the graph G^{ψ_1} (and of the graph $G = \Delta(M)$). By Corollary 14.1, the set $Etree(G^{\psi_1,\psi_2}, M)$ coincides with the set $Etree_{\psi_2}^{opt}(G^{\psi_1}, M)$ of all element partition trees from $Etree(G^{\psi_1}, M)$ which have minimum cost relative to ψ_2 among all trees from $Etree(G^{\psi_1}, M)$, etc.

If one of the cost functions ψ_i is not strictly increasing but increasing only, then the set $Etree(G^{\psi_1,\ldots,\psi_i}, M)$ coincides with the set $Etree_{\psi_i}^{s-opt}(G^{\psi_1,\ldots,\psi_{i-1}}, M)$ which is a subset of the set of all element partition trees from $Etree(G^{\psi_1,\ldots,\psi_{i-1}}, M)$ that have minimum cost relative to ψ_i among all trees from $Etree(G^{\psi_1,\ldots,\psi_{i-1}}, M)$.

14.5 Totally Optimal Trees

Let M be a mesh. For a cost function ψ, we denote $\psi(M) = \min\{\psi(M, M, \Gamma) : \Gamma \in ET(M, M)\}$, i.e., $\psi(M)$ is the minimum cost of an element partition tree for M relative to the cost function ψ. An element partition tree Γ for M is called *totally optimal relative to cost functions* ψ_1 *and* ψ_2 if $\psi_1(M, M, \Gamma) = \psi_1(M)$ and $\psi_2(M, M, \Gamma) = \psi_2(M)$, i.e., Γ is optimal relative to ψ_1 and ψ_2 simultaneously.

Let us assume ψ_1 is strictly increasing and ψ_2 is increasing or strictly increasing. We now describe how to recognize the existence of an element partition tree for M which is totally optimal relative to cost functions ψ_1 and ψ_2.

Let $G = \Delta(M)$. First, we apply to G the procedure of optimization relative to ψ_1. As a result, we obtain the graph G^{ψ_1} and the number $\psi_1(M)$ attached to the node M of G^{ψ_1}. Next, we apply to G the procedure of optimization relative to ψ_2. As a result, we obtain the graph G^{ψ_2} and the number $\psi_2(M)$ attached to the node M of G^{ψ_2}. After that, we apply to G^{ψ_1} the procedure of optimization relative to ψ_2. As a result, we obtain the graph G^{ψ_1,ψ_2} and the number $\min\{\psi_2(M, M, \Gamma) : \Gamma \in Etree(G^{\psi_1}, M)\}$ attached to the node M of G^{ψ_1,ψ_2}. One can show that a totally optimal relative to ψ_1 and ψ_2 element partition tree for M exists if and only if $\min\{\psi_2(M, M, \Gamma) : \Gamma \in Etree(G^{\psi_1}, M)\} = \psi_2(M)$.

In some cases, totally optimal trees do not exist and what we can seek is to find optimal trees under some constraints. For example, most of the time we are looking for a tree which is optimal with respect to time complexity, but when the problem becomes large enough, the solver may fail due to the memory constraints. At this point, we can use multi-stage optimization to find fastest trees among those that respect the memory constraints. Thus, we can determine trees that are fastest within trees with the least memory needs. These trees may not be totally optimal trees, however they may answer important practical questions.

14.6 Experimental Study of Three Types of Meshes

In this section, we consider results of experimental study of optimal and totally optimal element partition trees for three types of meshes.

14.6.1 Experimental Settings

For each solver *mode* (ASAP and KB) we use five specific cases of the cost functions described in Sect. 13.2.4. Two cost functions characterize the number of arithmetic floating point operations for solvers working in single-processor environment ($time^{1,p,0}_{mode}$) and multi-processor environment ($time^{\infty,p,0}_{mode}$). It is known that memory transfer operations can be more expensive than arithmetic operations. So we consider two other cost functions counting memory transfer operations in addition to arithmetic operations in single-processor environment ($time_{mode}^{1,p,107}$) and multi-processor environment ($time_{mode}^{\infty,p,107}$). We assume, based on some experimental work with GPU computations, that a memory transfer operation is as expensive as 107 arithmetic operations. The last cost function ($memory_{mode}^{p}$) measures the memory consumption of solver following a given element partition tree. All cost functions under consideration are strictly increasing except $time_{mode}^{\infty,p,0}$ and $time_{mode}^{\infty,p,107}$ that are only increasing.

A finite element mesh may be refined in some regions where the relative error of approximation is higher as done in the adaptive finite element method [3, 4]. We consider a refinement step as the step where all elements nearest to the singularity position are divided into four finite elements. This refinement step improves the quality of approximation however it increases the complexity of computing the approximation as the mesh size increases.

We consider three types of meshes that differ according to the singularity type. All of them start with a mesh that contains only one element, and a number of refinement steps k is specified and performed. The first class of meshes (*Point-singular mesh*) has the bottom right corner as the position of the singularity. The second class (*Edge-singular mesh*) has the bottom edge as the singularity position. While the last class (*Angle-singular mesh*) has its bottom and right edges as positions of singularity in addition to their intersection point. Figures 14.1, 14.2, and 14.3 present schematic examples of each of the previous classes of meshes. Here P_k, E_k and A_k refer to a *Point-singular mesh, Edge-singular mesh,* and *Angle-singular mesh* with k refinement steps, respectively.

Another parameter that affects the complexity of the element partition tree optimization problem is the polynomial order p. We specify a given optimization problem by the mesh type, the number of refinement steps k, and the polynomial order p which is one of the parameters of the considered cost functions. For each type of mesh, we work with $1 \le k \le 10$ refinements and $1 \le p \le 5$ polynomial orders, i.e., with 50 optimization problems.

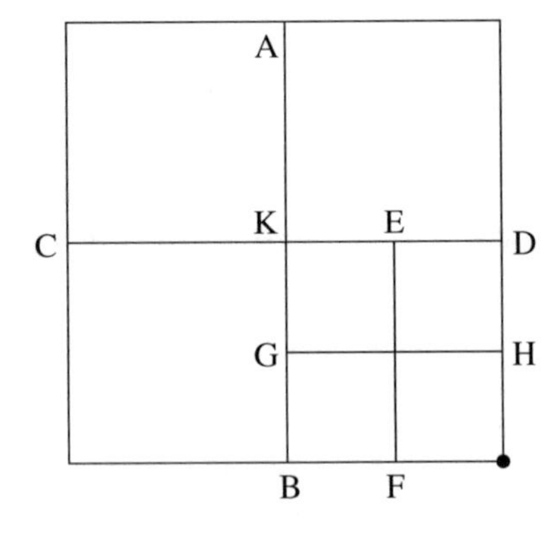

Fig. 14.1 P_2 mesh

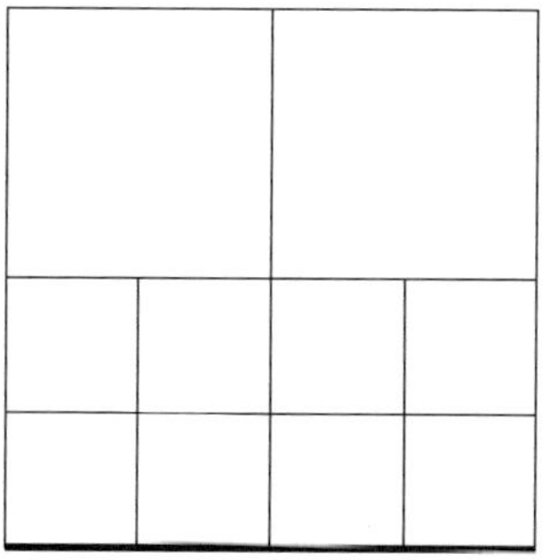

Fig. 14.2 E_2 mesh

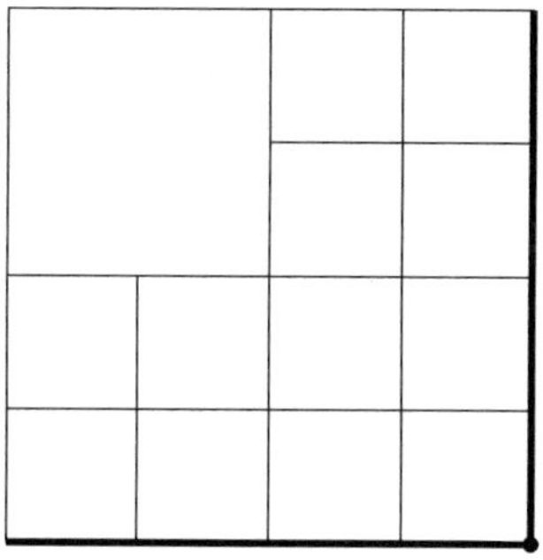

Fig. 14.3 A_2 mesh

14.6.2 Optimal Element Partition Trees

This section shows the results of optimizing the element partition trees with respect to a single cost function. We find the number of optimal element partition trees for all meshes and cost functions under consideration with the exception of the cost functions $time_{mode}^{\infty,p,0}$ and $time_{mode}^{\infty,p,107}$. The aforementioned cost functions are increasing so our algorithm can describe only a subset of the set of optimal element partition trees for those cost functions, more precisely, the set of strictly optimal element partition trees.

As the number of results is too large to include here, we present only the results associated with $time_{mode}^{1,p,107}$ cost functions for the studied mesh classes. The column "# all trees" in Tables 14.1, 14.3, and 14.5 contains the number of all element partition trees for the meshes P_k, E_k, and A_k, respectively. This number does not depend on the mode ASAP or KB. The number of trees increases too fast with the number of refinement steps. For example, the number of element partition trees for A_{10} is more than 10^{1603}. This prevents a naive search algorithm from solving the problem.

Tables 14.1 and 14.2 summarize the number of optimal elimination trees for P_k meshes with different polynomial orders p for KB and ASAP solvers, respectively. The P_k meshes are simpler than the other investigated classes so we are able to understand the behavior of any such optimal trees.

Table 14.1 Number of optimal element partition trees for mesh P_k and cost function $time_{KB}^{1,p,107}$

		p				
k	# all trees	1	2	3	4	5
1	2	2	2	2	2	2
2	6	4	4	4	4	4
3	18	4	4	4	4	4
4	54	4	4	4	4	4
5	162	4	4	4	4	4
6	486	4	4	4	4	4
7	1458	4	4	4	4	4
8	4374	4	4	4	4	4
9	13122	4	4	4	4	4
10	39366	4	4	4	4	4

Table 14.2 Number of optimal element partition trees for mesh P_k and cost function $time_{ASAP}^{1,p,107}$

	p				
k	1	2	3	4	5
1	2	2	2	2	2
2	4	4	4	4	4
3	8	8	8	8	8
4	16	16	16	16	16
5	32	32	32	32	32
6	64	64	64	64	64
7	128	128	128	128	128
8	256	256	256	256	256
9	512	512	512	512	512
10	1024	1024	1024	1024	1024

The optimal tree for $time_{ASAP}^{1,p,107}$ can be described in a recursive manner. For the P_1 mesh, there are only two steps possible and their order does not matter. For a general P_k mesh, the optimal strategy uses a pair of steps based on vertical and horizontal middle lines in any order (e.g, AB and CD in Fig. 14.1). It then obtains three unitary submeshes that are not further processed and a P_{k-1} submesh which is partitioned using the same strategy. Since going from a P_k mesh to a P_{k-1} submesh, the tree takes two steps in a non-deterministic order then the number of optimal trees for P_k is 2^k as computed by our algorithm.

An optimal tree for $time_{KB}^{1,p,107}$ and a P_k mesh is harder to describe. For a P_1 mesh, we have the same optimal trees as before. For a general P_k mesh, the optimal strategy chooses at the beginning one of the horizontal and vertical middle lines and then it continues deterministically until it obtains a P_1 mesh. The last P_1 mesh obtained is

Table 14.3 Number of optimal element partition trees for mesh E_k and cost function $time_{KB}^{1,p,107}$

		p				
k	# all trees	1	2	3	4	5
1	2	2	2	2	2	2
2	84	1	1	1	1	1
3	144859	1	1	1	1	1
4	2.3×10^{11}	1	1	1	1	1
5	9.1×10^{23}	1	1	1	1	1
6	1.5×10^{49}	1	1	1	1	1
7	3.2×10^{99}	1	1	1	1	1
8	1.6×10^{200}	1	1	1	1	1
9	4.7×10^{401}	1	1	1	1	1
10	3.2×10^{804}	1	1	1	1	1

Table 14.4 Number of optimal element partition trees for mesh E_k and cost function $time_{ASAP}^{1,p,107}$

	p				
k	1	2	3	4	5
1	2	2	2	2	2
2	1	1	1	1	1
3	1	1	1	1	1
4	1	1	1	1	1
5	1	1	1	1	1
6	1	1	1	1	1
7	1	1	1	1	1
8	1	1	1	1	1
9	1	1	1	1	1
10	1	1	1	1	1

dealt with as described above. The number of optimal trees is four because initially there are two possible first steps and two possible pairs of last steps (e.g., (EF, GH) and (GH, EF) in Fig. 14.1).

Tables 14.3 and 14.4 depict the number of optimal element partition trees for E_k meshes. There is exactly one optimal element partition tree for $k > 1$ however those element partition trees may change while varying p. Moreover, the optimal element partition tree for an E_k mesh and $time_{ASAP}^{1,p,107}$ is different from that for an E_k mesh and $time_{KB}^{1,p,107}$.

For A_k meshes, the situation is more complicated. Tables 14.5 and 14.6 present the number of optimal trees for an A_k mesh and cost functions $time_{KB}^{1,p,107}$ and $time_{ASAP}^{1,p,107}$, respectively. We are not able to find a clear pattern describing the optimal strategy for such meshes.

Table 14.5 Number of optimal element partition trees for mesh A_k and cost function $time_{KB}^{1,p,107}$

		p				
k	# all trees	1	2	3	4	5
1	2	2	2	2	2	2
2	580	4	4	4	4	4
3	2.6×10^8	16	16	16	12	12
4	2.9×10^{20}	48	48	48	48	48
5	1×10^{45}	24	24	24	24	24
6	6.6×10^{94}	12	12	12	12	12
7	9.2×10^{194}	12	12	12	12	12
8	6.5×10^{395}	12	12	12	12	12
9	1.2×10^{798}	12	12	12	12	12
10	1.7×10^{1603}	12	12	12	12	12

Table 14.6 Number of optimal element partition trees for mesh A_k and cost function $time_{ASAP}^{1,p,107}$

	p				
k	1	2	3	4	5
1	2	2	2	2	2
2	6	6	6	6	6
3	8	8	8	8	8
4	8	8	8	8	8
5	48	48	48	48	48
6	32	32	32	32	32
7	64	64	64	64	64
8	128	128	128	128	128
9	256	256	256	256	256
10	512	512	2048	2048	2048

14.6.3 Totally Optimal Element Partition Trees

This section contains experimental results for totally optimal element partition trees. The summary of the results is found in Tables 14.7, 14.9, and 14.11. For each type of mesh and each of the considered eight pairs of cost functions these tables contain a number of optimization problems (pairs k and p) for which a totally optimal tree relative to the considered pair of cost functions exists. As we mentioned before, the whole number of optimization problems (pairs k and p) is equal to 50.

For a P_k mesh, we find in many cases totally optimal element partition trees for all of the considered problems (see Table 14.7). We focus on the pair of cost functions $(time_{KB}^{1,p,107}, time_{KB}^{\infty,p,0})$ where we have totally optimal element partition trees for 42 from 50 cases (see Table 14.8 where 1 means existence and 0 means non-existence

Table 14.7 Existence of totally optimal element partition trees for P_k meshes

	Mode	
Pair of cost functions	ASAP	KB
$memory^{p}_{mode}, time^{1,p,0}_{mode}$	50	10
$memory^{p}_{mode}, time^{\infty,p,0}_{mode}$	10	42
$memory^{p}_{mode}, time^{1,p,107}_{mode}$	50	50
$memory^{p}_{mode}, time^{\infty,p,107}_{mode}$	10	50
$time^{1,p,0}_{mode}, time^{\infty,p,0}_{mode}$	10	11
$time^{1,p,0}_{mode}, time^{\infty,p,107}_{mode}$	10	10
$time^{1,p,107}_{mode}, time^{\infty,p,0}_{mode}$	10	**42**
$time^{1,p,107}_{mode}, time^{\infty,p,107}_{mode}$	10	50

Table 14.8 Existence of totally optimal trees for P_k relative to $time^{1,p,107}_{KB}, time^{\infty,p,0}_{KB}$

	p				
k	1	2	3	4	5
1	1	1	1	1	1
2	1	1	1	1	1
3	0	1	1	1	1
4	0	1	1	1	1
5	0	1	1	1	1
6	0	1	1	1	1
7	0	1	1	1	1
8	0	1	1	1	1
9	0	1	1	1	1
10	0	1	1	1	1

of a totally optimal tree). There exist totally optimal trees for all the optimization problems other than those where $p = 1$ and $k > 2$.

For an E_k mesh, we find totally optimal element partition trees for all 50 problems only for $memory^{p}_{KB}$ and $time^{1,p,107}_{KB}$ (see Table 14.9). We highlight a case where totally optimal trees exist for 23 of the 50 problems in Table 14.10. This is noticed with the pair of cost functions ($time^{1,p,107}_{KB}, time^{\infty,p,107}_{KB}$).

Finally, for A_k meshes, there does not exist a pair of cost functions that have totally optimal elimination trees for all 50 problems (see Table 14.11). However, the pair of cost functions ($memory^{p}_{KB}, time^{1,p,107}_{KB}$) have totally optimal trees for 38 of the investigated 50 problems. We present which problems have totally optimal element partition trees for the pair ($time^{1,p,107}_{KB}, time^{\infty,p,0}_{KB}$) in Table 14.12.

Table 14.9 Existence of totally optimal element partition trees for E_k meshes

	Mode	
Pair of cost functions	ASAP	KB
$memory^{p}_{mode}, time^{1,p,0}_{mode}$	10	10
$memory^{p}_{mode}, time^{\infty,p,0}_{mode}$	9	10
$memory^{p}_{mode}, time^{1,p,107}_{mode}$	43	50
$memory^{p}_{mode}, time^{\infty,p,107}_{mode}$	10	23
$time^{1,p,0}_{mode}, time^{\infty,p,0}_{mode}$	9	31
$time^{1,p,0}_{mode}, time^{\infty,p,107}_{mode}$	10	5
$time^{1,p,107}_{mode}, time^{\infty,p,0}_{mode}$	9	10
$time^{1,p,107}_{mode}, time^{\infty,p,107}_{mode}$	10	**23**

Table 14.10 Existence of totally optimal trees for E_k relative to $time^{1,p,107}_{KB}, time^{\infty,p,107}_{KB}$

	p				
k	1	2	3	4	5
1	1	1	1	1	1
2	0	0	0	0	0
3	1	1	1	1	1
4	1	1	1	1	1
5	0	0	0	0	0
6	0	0	0	0	0
7	1	1	1	1	1
8	0	0	0	0	0
9	0	0	0	0	0
10	1	1	1	0	0

Table 14.11 Existence of totally optimal element partition trees for A_k meshes

	Mode	
Pair of cost functions	ASAP	KB
$memory^{p}_{mode}, time^{1,p,0}_{mode}$	10	10
$memory^{p}_{mode}, time^{\infty,p,0}_{mode}$	10	15
$memory^{p}_{mode}, time^{1,p,107}_{mode}$	36	38
$memory^{p}_{mode}, time^{\infty,p,107}_{mode}$	10	15
$time^{1,p,0}_{mode}, time^{\infty,p,0}_{mode}$	10	15
$time^{1,p,0}_{mode}, time^{\infty,p,107}_{mode}$	10	10
$time^{1,p,107}_{mode}, time^{\infty,p,0}_{mode}$	10	**17**
$time^{1,p,107}_{mode}, time^{\infty,p,107}_{mode}$	10	15

Table 14.12 Existence of totally optimal trees for A_k relative to $time_{KB}^{1,p,107}$, $time_{KB}^{\infty,p,0}$

	p				
k	1	2	3	4	5
1	1	1	1	1	1
2	1	1	1	1	1
3	0	0	0	1	1
4	1	1	1	1	1
5	0	0	0	0	0
6	0	0	0	0	0
7	0	0	0	0	0
8	0	0	0	0	0
9	0	0	0	0	0
10	0	0	0	0	0

References

1. AbouEisha, H., Calo, V.M., Jopek, K., Moshkov, M., Paszynska, A., Paszynski, M., Skotniczny, M.: Element partition trees for h-refined meshes to optimize direct solver performance. Part I: dynamic programming. Appl. Math. Comput. Sci. **27**(2), 351–365 (2017)
2. AbouEisha, H., Moshkov, M., Jopek, K., Gepner, P., Kitowski, J., Paszynski, M.: Towards green multi-frontal solver for adaptive finite element method. In: Koziel, S., Leifsson, L., Lees, M., Krzhizhanovskaya, V.V., Dongarra, J., Sloot, P.M.A. (eds.) International Conference on Computational Science, ICCS 2015, Computational Science at the Gates of Nature, Reykjavík, Iceland, June 1–3, 2015, Procedia Computer Science, vol. 51, pp. 984–993. Elsevier (2015)
3. Demkowicz, L.: Computing with hp-Adaptive Finite Elements, Vol. 1: One and Two Dimensional Elliptic and Maxwell Problems. Chapman & Hall/CRC Applied Mathematics and Nonlinear Science Series. Chapman & Hall/CRC, Boca Raton (2006)
4. Demkowicz, L., Kurtz, J., Pardo, D., Paszynski, M., Rachowicz, W., Zdunek, A.: Computing with hp-Adaptive Finite Elements, Vol. 2: Frontiers: Three Dimensional Elliptic and Maxwell Problems with Applications. Chapman & Hall/CRC Applied Mathematics and Nonlinear Science Series. Chapman & Hall/CRC, Boca Raton (2007)

Chapter 15
Bi-criteria Optimization of Element Partition Trees

Sometimes it is important to take into consideration two or more criteria when evaluating the quality of an element partition tree. Moreover, the relationship between different criteria is interesting for different problems. The minimization of the number of arithmetic operations is shown to be different than minimizing the fill-in in [2]. The relationships between improving the runtime of compiler performance and the energy consumption were studied in [3].

In the previous chapter, we discussed totally optimal element partition trees. Those are trees that are optimal with respect to multiple criteria simultaneously. For example, we might have element partition trees that are optimal from the point of view of both time and memory complexity. However, it is not a common case to have such trees. Thus, it becomes important to find Pareto optimal element partition trees with respect to different criteria.

This chapter presents an algorithm for finding Pareto optimal points corresponding to Pareto optimal element partition trees, and some applications of this algorithm. An additional application of improving the runtime of iterative solver is discussed in [1].

15.1 Bi-criteria Optimization

In this section, we describe an algorithm that computes the set of Pareto optimal points for bi-criteria optimization problems of element partition trees. In addition, we show a natural way to create a graph describing the relationship between the considered cost functions based on the set of Pareto optimal points found.

Let F, H be increasing functions from $\mathbb{R}^2$ to $\mathbb{R}$, and A, B be nonempty finite subsets of the set $\mathbb{R}^2$. Note that

$$A \langle FH \rangle B = \{(F(a,c), H(b,d)) : (a,b) \in A, (c,d) \in B\} .$$

H. AbouEisha et al., *Extensions of Dynamic Programming for Combinatorial Optimization and Data Mining*, Intelligent Systems Reference Library 146,
https://doi.org/10.1007/978-3-319-91839-6_15

Let ψ and λ be increasing cost functions for element partition trees specified by triples of functions ψ^0, F, w and λ^0, H, u, respectively. Consider M to be a mesh and $G = \Delta(M)$. For each node N of the graph G, we define the set

$$s_{\psi,\lambda}(G, N) = \{(\psi(M, N, \Gamma), \lambda(M, N, \Gamma)) : \Gamma \in Etree(G, N)\} .$$

Note that, by Proposition 14.2, the set $Etree(G, N)$ is equal to $ET(M, N)$. We denote by $Par(s_{\psi,\lambda}(G, N))$ the set of Pareto optimal points for $s_{\psi,\lambda}(G, N)$. We present an algorithm $\mathscr{A}_{18}$ that builds the set $Par(s_{\psi,\lambda}(G, M))$. More precisely, it constructs the set $B(N) = Par(s_{\psi,\lambda}(G, N))$ for each node N of G.

Algorithm $\mathscr{A}_{18}$ (construction of POPs for element partition trees).

Input: Increasing cost functions for element partition trees ψ and λ given by triples of functions ψ^0, F, w and λ^0, H, u, respectively, a mesh M, and the graph $G = \Delta(M)$.

Output: The set $Par(s_{\psi,\lambda}(G, M))$ of Pareto optimal points for the set of pairs $s_{\psi,\lambda}(G, M) = \{(\psi(M, M, \Gamma), \lambda(M, M, \Gamma)) : \Gamma \in Etree(G, M)\}$.

1. If all nodes in G are processed, then return the set $B(M)$. Otherwise, choose a node N in the graph G which is not processed yet and which is either a terminal node (unitary submesh of M) or an internal node (nonunitary submesh of M) such that, for any $l \in DL(N)$ and any $\delta \in \{0, 1\}$, the node $N(l, \delta)$ is already processed, i.e. the set $B(N(l, \delta))$ is already constructed.
2. If N is a terminal node, then set $B(N) = \{(\psi^0(M, N), \lambda^0(M, N))\}$. Mark the node N as processed and proceed to the step 1.
3. If N is an internal node then, for each $l \in DL(N)$, compute

$$D(N, l) = B(N(l, 0)) \langle FH \rangle B(N(l, 1)) ,$$

apply algorithm $\mathscr{A}_1$ to $D(N, l)$ to obtain $C(N, l) = Par(D(N, l))$, and compute

$$\begin{aligned} B(N, l) &= C(N, l) \langle ++ \rangle \{(w(M, N, l), u(M, N, l))\} \\ &= \{(a + w(M, N, l), b + u(M, N, l)) : (a, b) \in C(N, l)\} . \end{aligned}$$

4. Construct the set $A(N) = \bigcup_{l \in DL(N)} B(N, l)$ and apply the algorithm $\mathscr{A}_1$ to $A(N)$ producing the set $B(N) = Par(A(N))$. Mark the node N as processed and return to the step 1.

Proposition 15.1 *Let ψ and λ be increasing cost functions for element partition trees defined by triples of functions ψ^0, F, w and λ^0, H, u, respectively, M be a mesh, and $G = \Delta(M)$. Then, for each node N of the graph G, the algorithm $\mathscr{A}_{18}$ constructs the set $B(N) = Par(s_{\psi,\lambda}(G, N))$.*

Proof We prove the considered statement by induction on nodes of G. Let N be a terminal node of G. Then $Etree(G, N) = \{etree(\varphi(N))\}$,

$$s_{\psi,\lambda}(G, N) = Par(s_{\psi,\lambda}(G, N)) = \{(\psi^0(M, N), \lambda^0(M, N))\} ,$$

and $B(N) = Par(s_{\psi,\lambda}(G, N))$.

Let N be an internal node of G such that, for any $l \in DL(N)$ and any $\delta \in \{0, 1\}$, the considered statement holds for the node $N(l, \delta)$, i.e.,

$$B(N(l, \delta)) = Par(s_{\psi,\lambda}(G, N(l, \delta))) .$$

Let $l \in DL(N)$, $P(l, \delta) = s_{\psi,\lambda}(G, N(l, \delta))$ for $\delta = 0, 1$, and

$$P(l) = (P(l, 0) \langle FH \rangle P(l, 1)) \langle ++ \rangle \{(w(M, N, l), u(M, N, l))\} .$$

One can show that

$$Par(P(l)) = Par(P(l, 0) \langle FH \rangle P(l, 1)) \langle ++ \rangle \{(w(M, N, l), u(M, N, l))\} .$$

By Lemma 2.6, $Par(P(l, 0) \langle FH \rangle P(l, 1)) \subseteq Par(P(l, 0)) \langle FH \rangle Par(P(l, 1))$. It is clear that $Par(P(l, 0)) \langle FH \rangle Par(P(l, 1)) \subseteq P(l, 0) \langle FH \rangle P(l, 1)$. Using Lemma 2.4, we obtain

$$Par(P(l, 0) \langle FH \rangle P(l, 1)) = Par(Par(P(l, 0)) \langle FH \rangle Par(P(l, 1))) .$$

According to the inductive hypothesis, $B(N(l, \delta)) = Par(P(l, \delta))$ for $\delta = 0, 1$. Therefore

$$Par(P(l)) = Par(B(N(l, 0)) \langle FH \rangle B(N(l, 1))) \langle ++ \rangle \{(w(M, N, l), u(M, N, l))\}$$

and $Par(P(l)) = B(N, l)$.

It is clear that $s_{\psi,\lambda}(G, N) = \bigcup_{l \in DL(N)} P(l)$. By Lemma 2.5,

$$Par(s_{\psi,\lambda}(G, N)) = Par\left(\bigcup_{l \in DL(N)} P(l)\right) \subseteq \bigcup_{l \in DL(N)} Par(P(l)) .$$

Since $\bigcup_{l \in DL(N)} Par(P(l)) \subseteq s_{\psi,\lambda}(G, N)$, we deduce by Lemma 2.4 that

$$Par(s_{\psi,\lambda}(G, N)) = Par\left(\bigcup_{l \in DL(N)} Par(P(l))\right) = Par(A(N)) = B(N) .$$

□

In the proof of the following statement we use notation from the description of the algorithm $\mathscr{A}_{18}$.

Proposition 15.2 *Let ψ and λ be increasing cost functions for element partition trees defined by triples of functions ψ^0, F, w and λ^0, H, u, respectively, and functions ψ^0,*

F, w, λ^0, H, u can be computed in polynomial time depending on the size $s(M)$ of an input mesh M. Let the function $ub(\psi, M)$ be bounded from above by a polynomial on $s(M)$. Then the time complexity of the algorithm $\mathscr{A}_{18}$ is polynomial depending on $s(M)$.

Proof We consider computations of the functions ψ^0, F, w, λ^0, H, u, comparisons and additions as *elementary operations*. Let M be a mesh, $G = \Delta(M)$, and $s_{\psi,\lambda}(G, N)^{(1)} = \{a : (a, b) \in s_{\psi,\lambda}(G, N)\}$ for any node N of G. It is clear that $s_{\psi,\lambda}(G, N)^{(1)} \subseteq \{0, 1, \ldots, ub(\psi, M)\}$. From Proposition 15.1 it follows that $B(N) = Par(s_{\psi,\lambda}(G, N))$. Using Lemma 2.3 we obtain that $|B(N)| \leq ub(\psi, M) + 1$.

To process a terminal node N of G (to construct the set $B(N)$), the algorithm $\mathscr{A}_{18}$ computes values $\psi^0(M, N)$ and $\lambda^0(M, N)$, i.e., makes two elementary operations.

Let N be an internal node of G such that, for any $l \in DL(N)$ and any $\delta \in \{0, 1\}$, the set $B(N(l, \delta))$ is already constructed. We know that $|B(N(l, \delta))| \leq ub(\psi, M) + 1$ for each $l \in DL(N)$ and $\delta \in \{0, 1\}$. Let $l \in DL(N)$. The algorithm $\mathscr{A}_{18}$ performs $O(ub(\psi, M)^2)$ computations of F and H to construct $D(N, l)$. The cardinality of the set $D(N, l)$ is at most $(ub(\psi, M) + 1)^2$. To find the set $C(N, l)$, the algorithm $\mathscr{A}_{18}$ makes $O(ub(\psi, M)^2 \log ub(\psi, M))$ comparisons (see Proposition 2.1). To construct the set $B(N, l)$, the algorithm $\mathscr{A}_{18}$ computes the values $w(M, N, l)$ and $u(M, N, l)$, and makes $O(ub(\psi, M)^2)$ additions. It is clear that $|DL(N)| \leq s(M)$. Therefore, to construct the set $A(N)$, the algorithm $\mathscr{A}_{18}$ makes $O(s(M)ub(\psi, M)^2 \log ub(\psi, M))$ elementary operations. It is clear that $|A(N)| \leq s(M)(ub(\psi, M) + 1)^2$. To construct the set $B(N)$ from the set $A(N)$, the algorithm $\mathscr{A}_{18}$ makes $O(s(M)ub(\psi, M)^2(\log s(M) + \log ub(\psi, M)))$ comparisons (see Proposition 2.1). Therefore, to construct the set $B(N)$ from the sets $B(N(l, \delta))$, the algorithm $\mathscr{A}_{18}$ makes

$$O(s(M)ub(\psi, M)^2(\log s(M) + \log ub(\psi, M)))$$

elementary operations.

The graph G contains $|\mathrm{SUB}(M)|$ nodes. By Lemma 13.1, $|\mathrm{SUB}(M)| \leq s(M)^4$. Therefore, to construct the set $B(M) = Par(s_{\psi,\lambda}(G, M))$, the algorithm $\mathscr{A}_{18}$ makes $O(s(M)^5 ub(\psi, M)^2(\log s(M) + \log ub(\psi, M)))$ elementary operations. Since each elementary operation can be done in polynomial time depending on $s(M)$, and $ub(\psi, M)$ is bounded from above by a polynomial on $s(M)$, the time complexity of the algorithm $\mathscr{A}_{18}$ is polynomial depending on $s(M)$. □

Corollary 15.1 *Let parameters p and q be fixed, and*

$$\psi, \lambda \in \{time_{mode}^{1,p,q}, time_{mode}^{\infty,p,q}, memory_{mode}^{p} : mode \in \{KB, ASAP\}\}\ .$$

Then the algorithm $\mathscr{A}_{18}$ has polynomial time complexity depending on the size $s(M)$ of the input mesh M.

Proof Let ψ and λ be defined by triples of functions ψ^0, F, w and λ^0, H, u, respectively. From Proposition 13.2 it follows that the functions ψ^0, F, w, λ^0, H, and u

can be computed in polynomial time depending on the size $s(M)$ of the input mesh M. In addition, Lemmas 13.3 and 13.4 show that the function $ub(\psi, M)$ is bounded from above by a polynomial depending on $s(M)$. Using Proposition 15.2 we obtain that the algorithm $\mathscr{A}_{18}$ has polynomial time complexity depending on the size $s(M)$ of the input mesh M. □

Let ψ and λ be increasing cost functions for element partition trees, M be a mesh, and $G = \Delta(M)$.

To study relationships between cost functions ψ and λ on the set of element partition trees $Etree(G, M)$ we consider partial functions $\mathscr{E}_M^{\psi,\lambda} : \mathbb{R} \to \mathbb{R}$ and $\mathscr{E}_{G,T}^{\lambda,\psi} : \mathbb{R} \to \mathbb{R}$ defined as follows:

$$\mathscr{E}_M^{\psi,\lambda}(x) = \min\{\lambda(M, M, \Gamma) : \Gamma \in Etree(G, M), \psi(M, M, \Gamma) \leq x\} ,$$
$$\mathscr{E}_M^{\lambda,\psi}(x) = \min\{\psi(M, M, \Gamma) : \Gamma \in Etree(G, M), \lambda(M, M, \Gamma) \leq x\} .$$

Let $(a_1, b_1), \ldots, (a_k, b_k)$ be the normal representation of the set $Par(s_{\psi,\lambda}(G, M))$ where $a_1 < \ldots < a_k$ and $b_1 > \ldots > b_k$. By Lemma 2.8 and Remark 2.2, for any $x \in \mathbb{R}$,

$$\mathscr{E}_M^{\psi,\lambda}(x) = \begin{cases} undefined, & x < a_1 \\ b_1, & a_1 \leq x < a_2 \\ \ldots & \ldots \\ b_{k-1}, & a_{k-1} \leq x < a_k \\ b_k, & a_k \leq x \end{cases} ,$$

$$\mathscr{E}_M^{\lambda,\psi}(x) = \begin{cases} undefined, & x < b_k \\ a_k, & b_k \leq x < b_{k-1} \\ \ldots & \ldots \\ a_2, & b_2 \leq x < b_1 \\ a_1, & b_1 \leq x \end{cases} .$$

15.2 Experimental Results

In this section, we describe some experiments that show the usage of our presented tools. We consider a pair of criteria reflecting the number of floating point operations done in parallel and the memory consumption respectively. We compute the Pareto front (the set of Pareto optimal points) of these two criteria in both KB and $ASAP$ mode for the three classes of meshes studied in previous chapter: P_k, E_k and A_k. We present in Tables 15.1, 15.2, 15.3, 15.4, 15.5, and 15.6 the number of points in each Pareto front. We show in Figs. 15.1, 15.2, and 15.3 some samples of the Pareto front computed as well as relationship described in Sect. 15.1.

The number of points in Pareto front is equal to one if and only if there exists a totally optimal element partition tree relative to the considered cost functions. The

Table 15.1 Size of Pareto front: $time^{\infty,p,107}_{ASAP}$ versus $memory^{p}_{ASAP}$ for P_k

k	p				
	1	2	3	4	5
1	1	1	1	1	1
2	1	1	1	1	1
3	2	2	2	2	2
4	2	2	2	2	2
5	3	3	3	3	3
6	4	4	4	4	4
7	5	5	5	5	5
8	6	6	6	6	6
9	7	7	7	7	7
10	8	8	8	8	8

Table 15.2 Size of Pareto front: $time^{\infty,p,107}_{ASAP}$ versus $memory^{p}_{ASAP}$ for E_k

k	p				
	1	2	3	4	5
1	1	1	1	1	1
2	1	1	1	1	1
3	4	4	4	4	4
4	10	11	11	12	12
5	13	14	13	15	15
6	28	24	27	28	26
7	24	23	24	27	28
8	21	23	28	29	29
9	12	17	17	21	26
10	20	15	11	12	13

Table 15.3 Size of Pareto front: $time^{\infty,p,107}_{ASAP}$ versus $memory^{p}_{ASAP}$ for A_k

k	p				
	1	2	3	4	5
1	1	1	1	1	1
2	1	1	1	1	1
3	4	4	4	4	4
4	5	4	4	4	4
5	21	17	18	18	18
6	31	28	25	23	23
7	38	27	26	23	22
8	9	8	9	14	16
9	19	16	11	19	21
10	30	17	10	15	19

Table 15.4 Size of Pareto front: $time_{KB}^{\infty,p,107}$ versus $memory_{KB}^{p}$ for P_k

k	p				
	1	2	3	4	5
1	1	1	1	1	1
2	1	1	1	1	1
3	1	1	1	1	1
4	1	1	1	1	1
5	1	1	1	1	1
6	1	1	1	1	1
7	1	1	1	1	1
8	1	1	1	1	1
9	1	1	1	1	1
10	1	1	1	1	1

Table 15.5 Size of Pareto front: $time_{KB}^{\infty,p,107}$ versus $memory_{KB}^{p}$ for E_k

k	p				
	1	2	3	4	5
1	1	1	1	1	1
2	3	3	3	3	3
3	1	1	1	1	1
4	1	1	1	1	1
5	3	3	3	3	3
6	2	2	2	2	2
7	1	1	1	1	1
8	3	3	3	3	3
9	2	2	2	2	3
10	1	1	1	4	4

Table 15.6 Size of Pareto front: $time_{KB}^{\infty,p,107}$ versus $memory_{KB}^{p}$ for A_k

k	p				
	1	2	3	4	5
1	1	1	1	1	1
2	1	1	1	1	1
3	2	2	2	2	2
4	2	2	2	2	2
5	4	4	4	4	4
6	2	2	2	2	2
7	3	3	3	3	3
8	2	2	2	2	2
9	1	1	1	1	1
10	3	3	3	3	3

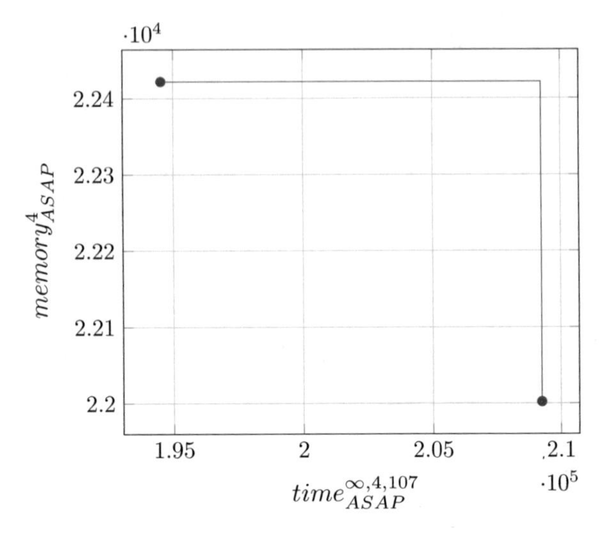

Fig. 15.1 Pareto front: $time^{\infty,4,107}_{ASAP}$ versus $memory^4_{ASAP}$ for P_4

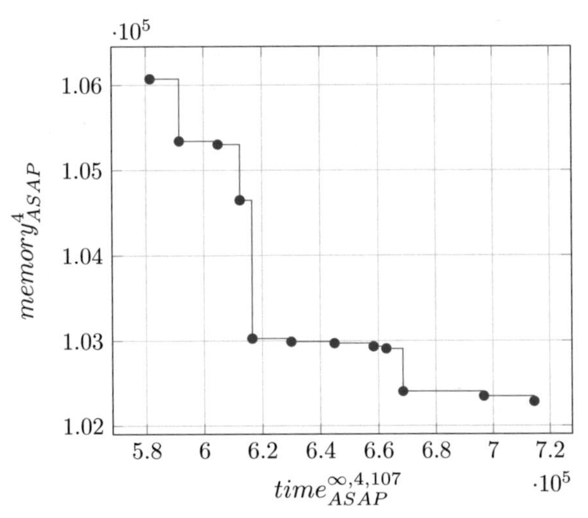

Fig. 15.2 Pareto front: $time^{\infty,4,107}_{ASAP}$ versus $memory^4_{ASAP}$ for E_4

obtained results for totally optimal element partition trees are consistent with the results obtained by multi-stage optimization (see Tables 14.7, 14.9, and 14.11).

The size of Pareto front for E_k and A_k is irregular (see Tables 15.2, 15.3, 15.5, and 15.6).

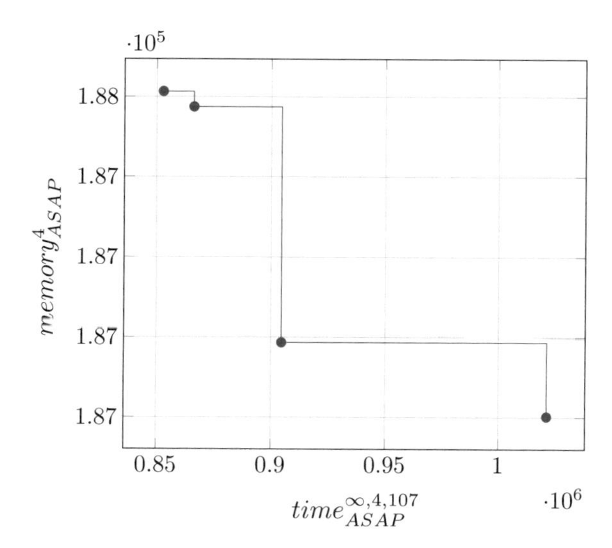

Fig. 15.3 Pareto front: $time^{\infty,4,107}_{ASAP}$ versus $memory^4_{ASAP}$ for A_4

References

1. AbouEisha, H., Jopek, K., Medygrał, B., Nosek, S., Moshkov, M., Paszynska, A., Paszynski, M., Pingali, K.: Hybrid direct and iterative solver with library of multi-criteria optimal orderings for h adaptive finite element method computations. In: Connolly, M. (ed.) International Conference on Computational Science 2016, ICCS 2016, June 6–8, 2016, San Diego, California, USA. Procedia Computer Science, vol. 80, pp. 865–874. Elsevier (2016)
2. Luce, R., Ng, E.G.: On the minimum FLOPs problem in the sparse Cholesky factorization. SIAM J. Matrix Anal. Appl. **35**(1), 1–21 (2014)
3. Valluri, M., John, L.K.: Is compiling for performance - compiling for power? In: Lee, G., Yew, P.C. (eds.) Interaction between Compilers and Computer Architectures, pp. 101–115. Springer, New York (2001)

Part V
Combinatorial Optimization Problems

This part presents an extension of dynamic programming for combinatorial optimization problems. We considered four different classic optimization problems (which have simple dynamic programming solutions):

1. Matrix chain multiplication,
2. Binary search trees,
3. Global sequence alignment, and
4. Shortest paths.

In the following, we discuss, in detail, the multi-stage optimization for all these four problems.

Chapter 16
Matrix Chain Multiplication

Matrix chain multiplication is a classic optimization problem in computer science. For a given sequence $A_1, A_2, \ldots, A_n$ of matrices, we need to compute the product of these n matrices using minimum number of scalar multiplications on a single processor (sequential) machine. Other optimization scenarios are also possible (some of corresponding optimization criteria are discussed here). This classic problem was introduced by Godbole in [2]. Hu and Shing [3] present an $O(n \log n)$ algorithm for this problem in contrast to $O(n^3)$ algorithm proposed by Godbole in [2].

We should find the product $A_1 \times A_2 \times \cdots \times A_n$. It is well known that matrix multiplication is associative, i.e., $A \times (B \times C) = (A \times B) \times C$. The cost of multiplying a chain of n matrices depends on the order of multiplications. Each possible ordering of multiplication of n matrices corresponds to a different parenthesization. It is well known that the total number of ways to fully parenthesize n matrices is equal to the nth Catalan number bounded from below by $\Omega(4^n/n^{1.5})$. We consider different cost functions for parenthesizations. Our aim is to find a parenthesization with minimum cost. Furthermore, we present multi-stage optimization of matrix chain multiplication relative to a sequence of cost functions. The results considered in this chapter were published in [1].

16.1 Cost Functions

Let $A_1, A_2, \ldots, A_n$ be matrices with dimensions $m_0 \times m_1, m_1 \times m_2, \ldots, m_{n-1} \times m_n$, respectively. We consider the problem of matrix chain multiplication of these matrices and denote it as $S(1, n)$. Furthermore, we consider subproblems of the initial problem. For $1 \leq i \leq j \leq n$, we denote by $S(i, j)$ the problem of multiplication of matrices $A_i \times A_{i+1} \times \cdots \times A_j$.

We describe inductively the set $P(i, j)$ of parenthesizations for $S(i, j)$. We have $P(i, i) = \{A_i\}$ for all $i = 1, \ldots, n$. For $i < j$ we define:

H. AbouEisha et al., *Extensions of Dynamic Programming for Combinatorial Optimization and Data Mining*, Intelligent Systems Reference Library 146,
https://doi.org/10.1007/978-3-319-91839-6_16

$$P(i,j)=\bigcup_{k=i}^{j-1}\{(p_1\times p_2):p_1\in P(i,k),\, p_2\in P(k+1,j)\}\,.$$

We implement a parenthesization using either one processor π or n processors $\pi_1, \pi_2, \ldots, \pi_n$. In the latter case, initially, the processor π_i contains the matrix A_i, $i = 1, \ldots, n$. Suppose we should compute $(p_1 \times p_2)$, where p_1 is computed by the processor π_{i_1} and p_2 is computed by the processor π_{i_2}. At this point, there are two possibilities, i.e., either the processor π_{i_1} sends the matrix corresponding to p_1 to the processor π_{i_2} or vice versa, where the receiving processor computes the product of matrices corresponding to p_1 and p_2.

We describe inductively the notion of cost function ψ which associates to each parenthesization p a nonnegative integer $\psi(p)$, which can be interpreted as the cost of the parenthesization. Let $F(x_1, x_2, y_1, y_2, y_3)$ be a function from ω^5 into ω, i.e., $F : \omega^5 \to \omega$, where $\omega = \{0, 1, 2, 3, \ldots\}$.

Now, for $i = 1, \ldots, n$, we have $\psi(A_i) = 0$. Let $1 \le i \le k < j \le n$, $p_1 \in P(i,k)$, $p_2 \in P(k+1, j)$ and $p = (p_1 \times p_2)$, then

$$\psi(p) = F(\psi(p_1), \psi(p_2), m_{i-1}, m_k, m_j)\,.$$

Note that p_1 describes a matrix of dimension $m_{i-1} \times m_k$ and p_2 describes a matrix of dimension $m_k \times m_j$.

The function F (and corresponding cost function ψ) is called *increasing* if, for $x_1' \le x_1$ and $x_2' \le x_2$, the following inequality holds:

$$F(x_1', x_2', y_1, y_2, y_3) \le F(x_1, x_2, y_1, y_2, y_3)\,.$$

The function F (and corresponding cost function ψ) is called *strictly increasing* if it is increasing and, for $x_1' \le x_1$ and $x_2' \le x_2$ such that $x_1' < x_1$ or $x_2' < x_2$, we have

$$F(x_1', x_2', y_1, y_2, y_3) < F(x_1, x_2, y_1, y_2, y_3)\,.$$

Let us consider examples of functions F and corresponding cost functions ψ.

1. Functions F_1 and $\psi^{(1)}$ where

$$\begin{aligned}\psi^{(1)}(p) &= F_1(\psi^{(1)}(p_1), \psi^{(1)}(p_2), m_{i-1}, m_k, m_j)\\ &= \psi^{(1)}(p_1) + \psi^{(1)}(p_2) + m_{i-1}m_km_j\,.\end{aligned}$$

Parenthesizations p_1 and p_2 represent matrices of dimensions $m_{i-1} \times m_k$ and $m_k \times m_j$, respectively. For multiplication of these matrices we need to make $m_{i-1}m_km_j$ scalar multiplications. So $\psi^{(1)}(p)$ is the total number of scalar multiplications required to compute the product $A_i \times \cdots \times A_j$ according to parenthesization $p = (p_1 \times p_2)$. We should add that $\psi^{(1)}(p)$ can be considered as time complexity of computation of p (when we count only scalar multiplications) using one processor. It is clear that F_1 and $\psi^{(1)}$ are strictly increasing.

2. Functions F_2 and $\psi^{(2)}$ where

$$\begin{aligned}\psi^{(2)}(p) &= F_2(\psi^{(2)}(p_1), \psi^{(2)}(p_2), m_{i-1}, m_k, m_j) \\ &= \max\{\psi^{(2)}(p_1), \psi^{(2)}(p_2)\} + m_{i-1}m_k m_j \ .\end{aligned}$$

This cost function describes time complexity of computation of p (when we count only scalar multiplications) using n processors. The functions F_2 and $\psi^{(2)}$ are increasing.

3. Functions F_3 and $\psi^{(3)}$ where

$$\begin{aligned}\psi^{(3)}(p) &= F_3(\psi^{(3)}(p_1), \psi^{(3)}(p_2), m_{i-1}, m_k, m_j) \\ &= \psi^{(3)}(p_1) + \psi^{(3)}(p_2) + \min\{m_{i-1}m_k, m_k m_j\} \ .\end{aligned}$$

This cost function describes the total cost of sending matrices between the processors when we compute p by n processors. We have the following situation: either processor π_{i_1} can send the $m_{i-1} \times m_k$ matrix to π_{i_2}, or π_{i_2} can send the $m_k \times m_j$ matrix to π_{i_1}. The number of elements in the first matrix is equal to $m_{i-1}m_k$ and the number of elements in the second matrix is equal to $m_k m_j$. To minimize the number of elements that should be sent we must choose minimum between $m_{i-1}m_k$ and $m_k m_j$. It is clear that F_3 and $\psi^{(3)}$ are strictly increasing.

4. Functions F_4 and $\psi^{(4)}$ where

$$\begin{aligned}\psi^{(4)}(p) &= F_4(\psi^{(4)}(p_1), \psi^{(4)}(p_2), m_{i-1}, m_k, m_j) \\ &= \max\{\psi^{(4)}(p_1), \psi^{(4)}(p_2)\} + \min\{m_{i-1}m_k, m_k m_j\} \ .\end{aligned}$$

This cost function describes the cost (time) of sending matrices between processors in the worst-case where we use n processors. The functions F_4 and $\psi^{(4)}$ are increasing.

16.2 Procedure of Optimization

Now we describe a *directed acyclic graph* (DAG) G_0 which allows us to represent all parenthesizations $P(i, j)$ for each subproblem $S(i, j)$, $1 \leq i \leq j \leq n$. The set of vertices of this graph coincides with the set $\{S(i, j) : 1 \leq i \leq j \leq n\}$. If $i = j$ then $S(i, j)$ has no outgoing edges. For $i < j$, $S(i, j)$ has exactly $2(j - i)$ outgoing edges. For $k = i, \ldots, j - 1$, exactly two edges start from $S(i, j)$ and finish in $S(i, k)$ and $S(k + 1, j)$, respectively. These edges are labeled with the index k, we call these edges a *rigid pair* of edges with index k.

Let G be a subgraph of G_0 which is obtained from G_0 by removal of some rigid pairs of edges such that, for each vertex $S(i, j)$ with $i < j$, at least one rigid pair outgoing from $S(i, j)$ remains intact.

Now, for each vertex $S(i, j)$, we define, by induction, the set $P_G(i, j)$ of parenthesizations corresponding to $S(i, j)$ in G. For every i, we have $P_G(i, i) = \{A_i\}$. For $i < j$, let $K_G(i, j)$ be the set of indexes of remaining rigid pairs outgoing from $S(i, j)$ in G, then $P_G(i, j) = \bigcup_{k \in K_G(i,j)} \{(p_1 \times p_2) : p_1 \in P_G(i, k), p_2 \in P_G(k + 1, j)\}$.

Let ψ be a cost function, we consider the procedure of optimization of parenthesizations corresponding to vertices of G relative to ψ. We know that $\psi(A_j) = 0$ for $i = 1, \ldots, n$. For $1 \le i \le k < j \le n$, $p_1 \in P(i, k)$, $p_2 \in P(k + 1, j)$ and $p = (p_1 \times p_2)$, we have $\psi(p) = F(\psi(p_1), \psi(p_2), m_{i-1}, m_k, m_j)$.

For each vertex $S(i, j)$ of the graph G, we mark this vertex by the number $\psi_{i,j}$ which can be interpreted as minimum value of ψ on $P_G(i, j)$. For $i = 1, \ldots, n$, we have $\psi_{i,i} = 0$. If, for $i < j$ and for each $k \in K_G(i, j)$, we know values $\psi_{i,k}$ and $\psi_{k+1,j}$, we can compute the value $\psi_{i,j}$: $\psi_{i,j} = \min_{k \in K_G(i,j)} F(\psi_{i,k}, \psi_{k+1,j}, m_{i-1}, m_k, m_j)$. Let $\Psi(i, k, j)$ be defined as follows: $\Psi(i, k, j) = F(\psi_{i,k}, \psi_{k+1,j}, m_{i-1}, m_k, m_j)$. Now we should remove from G each rigid pair with index k such that $\Psi(i, k, j) > \psi_{i,j}$.

We denote by G^ψ the resulting subgraph obtained from G. It is clear that, for each vertex $S(i, j)$, $i < j$, at least one rigid pair outgoing from $S(i, j)$ was left intact.

The following theorem summarizes the procedure of optimization for increasing cost function.

Theorem 16.1 *Let ψ be an increasing cost function. Then, for any i and j, $1 \le i \le j \le n$, $\psi_{i,j}$ is the minimum cost of a parenthesization from $P_G(i, j)$ and each parenthesization from $P_{G^\psi}(i, j)$ has the cost equal to $\psi_{i,j}$.*

Proof The proof is by induction on $j - i$. If $i = j$ then $P_G(i, j) = \{A_i\}$, $P_{G^\psi}(i, j) = \{A_i\}$, $\psi(A_i) = 0$ and $\psi_{i,j} = 0$ by definition. So if $j - i = 0$ the considered statement holds. Let, for some $t > 0$ and for each pair (i, j) such that $j - i \le t$, this statement hold.

Let us consider a pair (i, j) such that $j - i = t + 1$. Then

$$\psi_{i,j} = \min_{k \in K_G(i,j)} \Psi(i, k, j) ,$$

where $\Psi(i, k, j) = F(\psi_{i,k}, \psi_{k+1,j}, m_{i-1}, m_k, m_j)$. Since $i \le k < j$, $k - i \le t$ and $j - k - 1 \le t$. By the inductive hypothesis, the considered statement holds for the pairs (i, k) and $(k + 1, j)$ for each $k \in K_G(i, j)$. Furthermore, for $k \in K_G(i, j)$, let

$$P_G^k(i, j) = \{(p_1 \times p_2) : p_1 \in P_G(i, k), p_2 \in P_G(k + 1, j)\} .$$

It is clear that

$$P_G(i, j) = \bigcup_{k \in K_G(i,j)} P_G^k(i, j) . \tag{16.1}$$

Since F is increasing, $\Psi(i,k,j)$ is the minimum cost of an element from $P_G^k(i,j)$. From this and (16.1) it follows that $\psi_{i,j}$ is the minimum cost of an element from $P_G(i,j)$. In G^ψ the only rigid pairs outgoing from $S(i,j)$ are those for which $\Psi(i,k,j) = \psi_{i,j}$ where k is the index of the considered pair.

It is clear that the cost of each element from the set

$$P_{G^\psi}^k(i,j) = \{(p_1 \times p_2) : p_1 \in P_{G^\psi}(i,k),\, p_2 \in P_{G^\psi}(k+1,j)\}$$

is equal to $F(\psi_{i,k}, \psi_{k+1,j}, m_{i-1}, m_k, m_j) = \Psi(i,k,j)$ since the cost of each element from $P_{G^\psi}(i,k)$ is equal to $\psi_{i,k}$ and the cost of each element from $P_{G^\psi}(k+1,j)$ is equal to $\psi_{k+1,j}$. Since

$$P_{G^\psi}(i,j) = \bigcup_{\substack{k \in K_G(i,j) \\ \Psi(i,k,j)=\psi_{i,j}}} P_{G^\psi}^k(i,j)\ ,$$

we have that each parenthesization from $P_{G^\psi}(i,j)$ has the cost equal to $\psi_{i,j}$. □

In case of a strictly increasing cost function we have stronger result.

Theorem 16.2 *Let ψ be a strictly increasing cost function. Then, for any i and j, $1 \le i \le j \le n$, $\psi_{i,j}$ is the minimum cost of a parenthesization from $P_G(i,j)$ and $P_{G^\psi}(i,j)$ coincides with the set of all elements $p \in P_G(i,j)$ for which $\psi(p) = \psi_{i,j}$.*

Proof From Theorem 16.1 it follows that $\psi_{i,j}$ is the minimum cost of parenthesization in $P_G(i,j)$ and every parenthesization in $P_{G^\psi}(i,j)$ has the cost $\psi_{i,j}$. To prove this theorem it is enough to show that an arbitrary parenthesization p in $P_G(i,j)$ with optimal cost $\psi_{i,j}$ also belongs to $P_{G^\psi}(i,j)$.

We prove this statement by induction on $j-i$. If $j=i$ then $P_G(i,j) = \{A_i\}$ and $P_{G^\psi}(i,j) = \{A_i\}$. So, for $j-i=0$, the considered statement holds. Let this statement hold for some $t>0$ and for each pair (i,j) such that $j-i \le t$. Let us consider a pair (i,j) such that $j-i = t+1$.

Let $p \in P_G(i,j)$ and $\psi(p) = \psi_{i,j}$. Since $p \in P_G(i,j)$ then there is k, $i \le k < j$, such that $p = (p_1 \times p_2)$ where $p_1 \in P_G(i,k)$ and $p_2 \in P_G(k+1,j)$. Also, we know that $\psi(p) = F(\psi(p_1), \psi(p_2), m_{i-1}, m_k, m_j)$ and $\psi(p) = \psi_{i,j}$. Let us assume that $\psi(p_1) > \psi_{i,k}$ or $\psi(p_2) > \psi_{k+1,j}$. Since ψ and F are strictly increasing, we have

$$\psi(p) > F(\psi_{i,k}, \psi_{k+1,j}, m_{i-1}, m_k, m_j) = \Psi(i,k,j) \ge \psi_{i,j}\ ,$$

however, this is impossible. So we have $\Psi(i,k,j) = \psi_{i,j}$, $\psi(p_1) = \psi_{i,k}$ and $\psi(p_2) = \psi_{k+1,j}$. Since $\Psi(i,k,j) = \psi_{i,j}$, we have that the graph G^ψ has a rigid pair with index k outgoing from the vertex $S(i,j)$.

From the inductive hypothesis it follows that $p_1 \in P_{G^\psi}(i,k)$ and $p_2 \in P_{G^\psi}(k+1,j)$. Therefore $p \in P_{G^\psi}(i,j)$. □

16.3 Multi-stage Optimization

We can apply several cost functions *sequentially* one after another to optimize matrix multiplication for several optimization criteria.

Initially, we get a DAG $G = G_0$. This DAG G can be optimized according to any cost function ψ (increasing or strictly increasing). We can use the resulting DAG G^{ψ} as input to further optimize for any other cost function say ψ' resulting in a DAG that is sequentially optimized according to two different cost functions ψ and ψ', i.e., $G^{\psi,\psi'} = (G^{\psi})^{\psi'}$. We can continue optimizing on the resulting graph for further cost functions. We can guarantee the consideration of all optimal parenthesizations only in the case when all cost functions used in the multi-stage optimization are strictly increasing.

It is important to note here that in the process of optimization we retain the vertices and only remove the edges. However, in this process there may be some vertices which become unreachable from the starting vertex (that represents the initial basic problem).

In the next section we discuss an example of matrix chain multiplication. We optimize it for several cost functions one after another, and do not consider vertices that are unreachable from the initial vertex $S(1, n)$.

16.4 Example

We consider an example of matrix chain multiplication. Let A_1, A_2, A_3, A_4, and A_5 be five matrices with dimensions 2×2, 2×5, 5×5, 5×2, and 2×2, respectively, i.e., $m_0 = 2$, $m_1 = 2$, $m_2 = 5$, $m_3 = 5$, $m_4 = 2$, and $m_5 = 2$.

We use the cost function $\psi^{(1)}$ and obtain the following DAG (as shown in Fig. 16.1). This DAG is a subgraph of $G^{\psi^{(1)}}$ (note that $G = G_0$) as it does not include vertices $S(i, j)$ (representing subproblems) which are unreachable from the vertex $S(1, 5)$ (representing the main problem).

The resulting directed acyclic graph $G^{\psi^{(1)}}$ describes the following two different parenthesizations, which are optimal from the point of view of $\psi^{(1)}$:

1. $(A_1 \times ((A_2 \times (A_3 \times A_4)) \times A_5))$,
2. $((A_1 \times (A_2 \times (A_3 \times A_4))) \times A_5)$.

We can apply another cost function say $\psi^{(3)}$ (which is also a strictly increasing function) to our DAG $G^{\psi^{(1)}}$ and we obtain $G^{\psi^{(1)},\psi^{(3)}}$. Interestingly, we get exactly the same graph as there are no edges to remove from $G^{\psi^{(1)}}$.

We apply yet another cost function $\psi^{(4)}$ ($\psi^{(4)}$ is an increasing function and not a strictly increasing function). However, we again get the same DAG as depicted in Fig. 16.1.

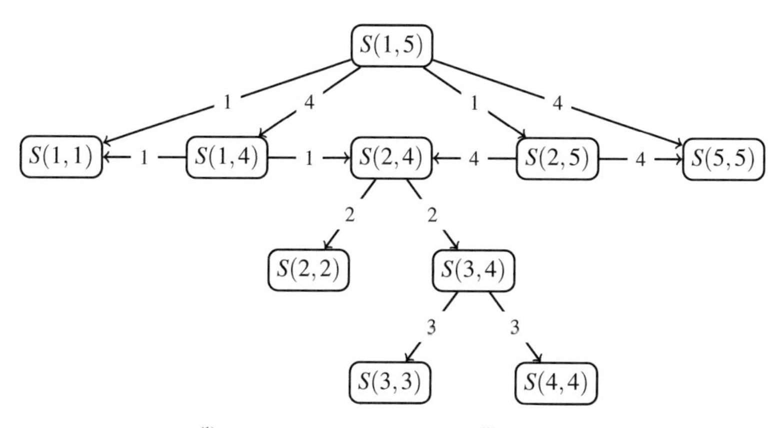

Fig. 16.1 Subgraph of $G^{\psi^{(1)}}$ containing only nodes of $G^{\psi^{(1)}}$ which are reachable from $S(1,5)$

16.5 Computational Complexity of Optimization

We assume that, for each cost function ψ, there exists a constant c_ψ such that to compute the value of ψ we need to make at most c_ψ arithmetic operations. The comparison operation is also considered as an arithmetic operation. In particular, we can choose $c_{\psi^{(1)}} = c_{\psi^{(2)}} = 4$ and $c_{\psi^{(3)}} = c_{\psi^{(4)}} = 5$.

Initially we get a DAG G_0. After optimization relative to a sequence of cost functions we get a DAG G, which is obtained from G_0 by removal of some rigid pairs. Let us now consider the process of optimization of G relative to a cost function ψ.

The number of subproblems $S(i,j)$ (vertices in G) is $O(n^2)$. It is enough to make $O(n)$ arithmetic operations to compute the value $\psi_{i,j}$ and to remove all rigid pairs for which $\Psi(i,k,j) > \psi_{i,j}$, if the values $\psi_{i,k}$ and $\psi_{k+1,j}$ are already known for $k \in K_G(i,j)$. Therefore, the total number of arithmetic operations to obtain G^ψ from G is $O(n^3)$.

References

1. Chikalov, I., Hussain, S., Moshkov, M.: Sequential optimization of matrix chain multiplication relative to different cost functions. In: Cerná, I., Gyimóthy, T., Hromkovic, J., Jeffery, K.G., Královic, R., Vukolic, M., Wolf S. (eds.) SOFSEM 2011: Theory and Practice of Computer Science – 37th Conference on Current Trends in Theory and Practice of Computer Science, Nový Smokovec, Slovakia, January 22–28, 2011, Lecture Notes in Computer Science, vol. 6543, pp. 157–165. Springer, Berlin (2011)
2. Godbole, S.: On efficient computation of matrix chain products. IEEE Trans. Comput. **22**(9), 864–866 (1973)
3. Hu, T.C., Shing, M.T.: Computation of matrix chain products. Part I. SIAM J. Comput. **11**(2), 362–373 (1982)

Chapter 17
Binary Search Trees

The *binary search tree* is an important data structure (see [4] or [2] for more discussion about binary search trees), which can be used for a variety of applications including a computerized translation of words from one language to another [2] or a lookup table for words with known frequencies of occurrences (see [3]), etc.

A typical computerized word translator application uses keys as well as dummy keys. The keys mean the words in a language to be translated and dummy keys represent the words for which the system cannot translate. Here, the usual problem is, for the given probabilities of keys and dummy keys, to find a binary search tree with minimum average depth.

We further generalize this problem: consider weights of keys (which correspond to complexity of key "computation") and try to construct a binary search tree with minimum average weighted depth. We also study binary search trees with minimum weighted depth to minimize the worst-case time complexity of the trees.

We represent all possible binary search trees by a directed acyclic graph G and apply to this graph a procedure of optimization relative to average weighted depth ψ. As a result, we obtain a subgraph G^{ψ} of G which describes the whole set $\mathscr{P}$ of binary search trees with minimum average weighted depth. We further apply to this graph the procedure of optimization relative to weighted depth φ. As a result, we obtain a subgraph $(G^{\psi})^{\varphi}$ of G^{ψ} which describes a subset of the set of all binary search trees from $\mathscr{P}$ with minimum weighted depth. The second step of optimization can improve the worst-case time complexity of the constructed trees. The results considered in this chapter were published in [1].

17.1 Preliminaries

Let we have a sequence of n distinct keys $k_1, k_2, \ldots, k_n$ (we assume that $k_1 < k_2 < \cdots < k_n$) with probabilities $p_1, p_2, \ldots, p_n$ and $n+1$ dummy keys $d_0, d_1, \ldots, d_n$ with probabilities $q_0, q_1, \ldots, q_n$ such that, for $i = 1, 2, \ldots, n-1$, the dummy key

H. AbouEisha et al., *Extensions of Dynamic Programming for Combinatorial Optimization and Data Mining*, Intelligent Systems Reference Library 146,
https://doi.org/10.1007/978-3-319-91839-6_17

Fig. 17.1 Binary search tree from $T(i, i)$

d_i

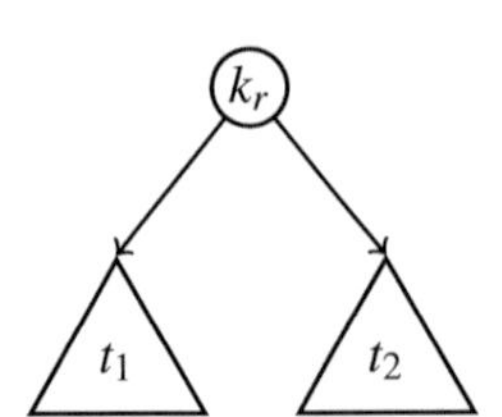

Fig. 17.2 Binary search tree $\text{TREE}(k_r, t_1, t_2)$

d_i corresponds to all values between k_i and k_{i+1} with d_0 representing all values less than k_1 and similarly d_n representing all values greater than k_n. The sum of probabilities is one, i.e., $\sum_{i=1}^{n} p_i + \sum_{i=0}^{n} q_i = 1$.

A *binary search tree* consists of keys as the internal nodes and dummy keys as the leaves. A search for some x is considered *successful* if x is one of the keys k_i and *unsuccessful* otherwise. Each key k_i, $1 \leq i \leq n$, has an associated weight w_i that is a positive number which can be interpreted as the complexity of comparison of k_i with x.

We consider the problem of optimizing binary search trees and denote it by $S(0, n)$ for $k_1, \ldots, k_n$ and $d_0, d_1, \ldots, d_n$. Furthermore, we consider the subproblems of initial problem $S(i, j)$ for $0 \leq i \leq j \leq n$. For $i = j$, the subproblem $S(i, i)$ is the optimization problem for the binary search trees for only one dummy key d_i.

For $i < j$, the subproblem $S(i, j)$ is the optimization problem for binary search trees for the keys $k_{i+1}, \ldots, k_j$ and dummy keys $d_i, d_{i+1}, \ldots, d_j$.

We define inductively the set $T(i, j)$ of all binary search trees for $S(i, j)$. For $i = j$, the set $T(i, i)$ contains exactly one binary search tree as depicted in Fig. 17.1.

For $i < j$, we have

$$T(i, j) = \bigcup_{r=i+1}^{j} \{\text{TREE}(k_r, t_1, t_2) : t_1 \in T(i, r-1), t_2 \in T(r, j)\} ,$$

where $\text{TREE}(k_r, t_1, t_2)$ is depicted in Fig. 17.2.

17.2 Cost Functions

We consider the cost functions which are given in the following way: values of considered *cost function* ψ (nonnegative function), are defined by induction on pair $(S(i, j), t)$, where $S(i, j)$ is a subproblem, $0 \leq i \leq j \leq n$, and $t \in T(i, j)$. Let $i = j$ and t be a tree represented as in Fig. 17.1 then $\psi(S(i, i), t) = 0$. Let $i < j$ and t be a tree depicted as in Fig. 17.2 where $r \in \{i + 1, \ldots, j\}$, $t_1 \in T(i, r - 1)$, and $t_2 \in T(r, j)$. Then $\psi(S(i, j), t) = F(\pi(i, j), w_r, \psi(S(i, r - 1), t_1), \psi(S(r, j), t_2))$, where $\pi(i, j)$ is the sum of probabilities of keys and dummy keys from $S(i, j)$, i.e.,

$$\pi(i, j) = \sum_{l=i+1}^{j} p_l + \sum_{l=i}^{j} q_l ,$$

w_r is the weight of the key k_r, and $F(\pi, w, \psi_1, \psi_2)$ is an operator which transforms the considered tuple of nonnegative numbers into a nonnegative number.

The considered cost function is called *increasing* if, for arbitrary nonnegative numbers a, b, c_1, c_2, d_1 and d_2, from inequalities $c_1 \leq d_1$ and $c_2 \leq d_2$ the inequality $F(a, b, c_1, c_2) \leq F(a, b, d_1, d_2)$ follows.

The considered cost function is called *strictly increasing* if it is increasing and, for arbitrary nonnegative numbers a, b, c_1, c_2, d_1 and d_2, from inequalities $c_1 \leq d_1$, $c_2 \leq d_2$, and inequality $c_i < d_i$ (which is true for some $i \in \{1, 2\}$), the inequality $F(a, b, c_1, c_2) < F(a, b, d_1, d_2)$ follows.

Now we take a closer view of two cost functions.

Weighted depth. We attach a weight to each path from the root to each node of the tree t which is equal to the sum of weights of the keys in this path. Then $\psi(S(i, j), t)$ is the maximum weight of a path from the root to a node of t. For this cost function, $F(\pi, w, \psi_1, \psi_2) = w + \max\{\psi_1, \psi_2\}$. The considered cost function is increasing.

Average weighted depth. For an arbitrary key or a dummy key h from $S(i, j)$, we denote $\Pr(h)$ its probability and by $w(h)$ we denote the weight of the path from the root of t to h. Then $\psi(S(i, j), t) = \sum_h w(h) \cdot \Pr(h)$, where we take the sum on all keys and dummy keys h from $S(i, j)$. For this cost function, $F(\pi, w, \psi_1, \psi_2) = w \cdot \pi + \psi_1 + \psi_2$. The considered cost function is strictly increasing.

17.2.1 Representing All Possible Binary Search Trees

Now we describe a *directed acyclic graph* (DAG) G_0 which allows us to represent all binary search trees from $T(i, j)$ for each subproblem $S(i, j)$, $0 \leq i \leq j \leq n$. The set of vertices of this graph coincides with the set $\{S(i, j) : 0 \leq i \leq j \leq n\}$. If $i = j$ then $S(i, i)$ has no outgoing edges. For $i < j$, $S(i, j)$ has exactly $2(j - i)$ outgoing edges. For $r = i + 1, \ldots, j$, exactly two edges start from $S(i, j)$ and finish in $S(i, r - 1)$ and $S(r, j)$, respectively. These edges are labeled with the index r, we call these edges a *rigid pair* of edges with index r.

Let G be a subgraph of G_0 which is obtained from G_0 by removal of some rigid pairs of edges such that, for each vertex $S(i, j)$ with $i < j$, at least one rigid pair outgoing from $S(i, j)$ remains intact.

Now, for each vertex $S(i, j)$, we define by induction the set $T_G(i, j)$ of binary search trees corresponding to $S(i, j)$ in G. For $i = j$, we have $T_G(i, i)$ containing only the trivial binary search tree as depicted in Fig. 17.1. For $i < j$, let $R_G(i, j)$ be the set of indexes of remaining rigid pairs outgoing from $S(i, j)$, then

$$T_G(i,j) = \bigcup_{r \in R_G(i,j)} \{\text{TREE}(k_r, t_1, t_2) : t_1 \in T_G(i, r-1), t_2 \in T_G(r, j)\} ,$$

where $\text{TREE}(k_r, t_1, t_2)$ is as depicted in Fig. 17.2.

One can show that $T_{G_0}(i, j) = T(i, j)$ for $0 \le i \le j \le n$.

17.3 Procedure of Optimization

Let ψ be a cost function, we now consider the procedure of optimization of binary search trees corresponding to vertices in G relative to ψ.

For each vertex $S(i, j)$ of the graph G, we mark this vertex by the number $\psi^*_{(i,j)}$, which can be interpreted as the minimum value of ψ on $T_G(i, j)$. We can compute the value $\psi^*_{(i,j)}$ for $i < j$ as $\psi^*_{(i,j)} = \min_{r \in R_G(i,j)} F(\pi(i, j), w_r, \psi^*_{(i,r-1)}, \psi^*_{(r,j)})$, where $\psi^*_{(i,i)} = 0$.

Let $\Psi_{(i,r,j)}$ be defined as following: $\Psi_{(i,r,j)} = F(\pi(i, j), w_r, \psi^*_{(i,r-1)}, \psi^*_{(r,j)})$. Now we remove from G each rigid pair with the index r starting in $S(i, j)$ such that

$$\Psi_{(i,r,j)} > \psi^*_{(i,j)} .$$

Let G^ψ be the resulting subgraph obtained from G. It is clear that, for each vertex $S(i, j), i < j$, at least one rigid pair outgoing from $S(i, j)$ is left intact.

We denote $T^*_{G,\psi}(i, j)$ the set of all trees from $T_G(i, j)$ that have minimum cost relative to ψ.

Following theorem summarizes the procedure of optimization for increasing cost function.

Theorem 17.1 *Let ψ be an increasing cost function. Then, for every i and j, $0 \le i \le j \le n$, $\psi^*_{(i,j)}$ is the minimum cost of binary search trees in $T_G(i, j)$ and $T_{G^\psi}(i, j) \subseteq T^*_{G,\psi}(i, j)$.*

Proof Proof is by induction on $j - i$. If $i = j$ then $T_G(i, j), T^*_{G,\psi}(i, j)$, and $T_{G^\psi}(i, j)$ contain only one tree as depicted in Fig. 17.1. Furthermore, the cost of this tree is zero as well as $\psi^*_{(i,j)} = 0$ by definition. So if $j - i = 0$ the considered statement holds. Let, for some $m \ge 0$ and for each pair (i, j) such that $j - i \le m$, this statement hold.

Let us consider a pair (i, j) such that $j - i = m + 1$. Then

$$\psi^*_{(i,j)} = \min_{r \in R_G(i,j)} \Psi_{(i,r,j)} ,$$

since $i < r \le j$, $r - 1 - i \le m$ and $j - r \le m$. By the inductive hypothesis, the considered statement holds for the pairs $(i, r - 1)$ and (r, j) for each $r \in R_G(i, j)$. Furthermore, for $r \in R_G(i, j)$, let

$$T^r_G(i, j) = \left\{\text{TREE}(k_r, t_1, t_2) : t_1 \in T_G(i, r-1), t_2 \in T_G(r, j)\right\} .$$

It is clear that

$$T_G(i, j) = \bigcup_{r \in R_G(i,j)} T_G^r(i, j) . \tag{17.1}$$

Since the considered cost function is increasing, we obtain that $\Psi_{(i,r,j)}$ is the minimum cost of a tree from $T_G^r(i, j)$. From here and (17.1) it follows that $\psi^*_{(i,j)}$ is the minimum cost of a tree in $T_G(i, j)$. In G^ψ the only rigid pairs outgoing from $S(i, j)$ are those for which $\Psi_{(i,r,j)} = \psi^*_{(i,j)}$, where r is the index of the considered pair. Let $R_{G^\psi}(i, j)$ be the set of indexes r of rigid pairs outgoing from $S(i, j)$ in G^ψ.

It is clear that the cost of each element in the set

$$T^r_{G^\psi}(i, j) = \{\text{TREE}(k_r, t_1, t_2) : t_1 \in T_{G^\psi}(i, r-1), t_2 \in T_{G^\psi}(r, j)\}$$

is equal to $\Psi_{(i,r,j)} = F(\pi(i, j), w_r, \psi^*_{(i,r-1)}, \psi^*_{(r,j)})$, since the cost of each tree in $T_{G^\psi}(i, r-1)$ is equal to $\psi^*_{(i,r-1)}$ and the cost of each tree in $T_{G^\psi}(r, j)$ is equal to $\psi^*_{(r,j)}$. Furthermore, $T_{G^\psi}(i, j) = \bigcup_{r \in R_{G^\psi}(i,j)} T^r_{G^\psi}(i, j)$, we have that each binary search tree in $T_{G^\psi}(i, j)$ has the cost equal to $\psi^*_{(i,j)}$. □

In case of a strictly increasing cost function we have stronger result.

Theorem 17.2 *Let ψ be a strictly increasing cost function. Then, for every i and j, $0 \le i \le j \le n$, $\psi^*_{(i,j)}$ is the minimum cost of binary search trees in $T_G(i, j)$ and $T_{G^\psi}(i, j) = T^*_{G,\psi}(i, j)$.*

Proof Since ψ is a strictly increasing cost function, ψ is also an increasing cost function. By Theorem 17.1, $\psi^*_{(i,j)}$ is the minimum cost of binary search trees in $T_G(i, j)$ and $T_{G^\psi}(i, j) \subseteq T^*_{G,\psi}(i, j)$. We only need to prove that $T^*_{G,\psi}(i, j) \subseteq T_{G^\psi}(i, j)$. We prove this statement by induction on $j - i$. If $j - i = 0$ then both sets $T^*_{G,\psi}(i, j)$ and $T_{G^\psi}(i, j)$ contain the only tree as depicted in Fig. 17.1. So $T^*_{G,\psi}(i, j) = T_{G^\psi}(i, j)$ and the considered statement holds for $j - i = 0$. Let, for some $m \ge 0$ and for each pair (i, j) such that $j - i \le m$, this statement hold.

Let us consider a pair (i, j) such that $j - i = m + 1$. Then (we use here the notation from the proof of Theorem 17.1) $T_G(i, j) = \bigcup_{r \in R_G(i,j)} T_G^r(i, j)$. Let $t \in T^*_{G,\psi}(i, j)$, then there exists $r \in R_G(i, j)$ such that $t \in T_G^r(i, j)$ and t can be represented in the form $\text{TREE}(k_r, t_1, t_2)$ (see Fig. 17.2) where $t_1 \in T_G(i, r-1)$ and $t_2 \in T_G(r, j)$. Therefore the minimum cost $\Psi_{(i,r,j)}$ of trees from $T_G^r(i, j)$ is equal to $\psi^*_{(i,j)}$. Hence $r \in R_{G_\psi}(i, j)$.

Let us assume that the cost of t_1 is greater than $\psi^*_{(i,r-1)}$ or the cost of t_2 is greater than $\psi^*_{(r,j)}$. Since ψ is strictly increasing function we have that the cost of t is greater than $\psi^*_{(i,j)}$, which is impossible. Thus $t_1 \in T^*_{G,\psi}(i, r-1)$ and $t_2 \in T^*_{G,\psi}(r, j)$. It is clear that $j - r \le m$ and $r - 1 - i \le m$. Therefore, by induction hypothesis, $T_{G^\psi}(i, r-1) \supseteq T^*_{G,\psi}(i, r-1)$ and $T_{G^\psi}(r, j) \supseteq T^*_{G,\psi}(r, j)$. From here it follows that $t_1 \in T_{G^\psi}(i, r-1)$, $t_2 \in T_{G^\psi}(r, j)$ and $t \in T^r_{G^\psi}(i, j)$. Since $r \in R_{G^\psi}(i, j)$ we have $t \in T_{G^\psi}(i, j)$ and $T^*_{G,\psi}(i, j) \subseteq T_{G,\psi}(i, j)$. □

17.4 Example

We consider an example of a simple optimization problem with two keys k_1, k_2 and three dummy keys d_0, d_1 and d_2 with probabilities as $p_1 = 0.4$, $p_2 = 0.3$ and $q_0 = q_1 = q_2 = 0.1$. We also associate weights with keys: $w_1 = 6$ and $w_2 = 4$. We get the DAG $G = G_0$ (see Fig. 17.3).

We apply the procedure of optimization to the graph G and use the average weighted depth ψ as the cost function. As a result, we obtain the subgraph G^ψ of the graph G (see Fig. 17.4) The binary search tree as depicted in Fig. 17.5 corresponds to the vertex $S(0, 2)$ of the graph G^ψ. This is the only optimal binary search tree relative to ψ for the considered problem.

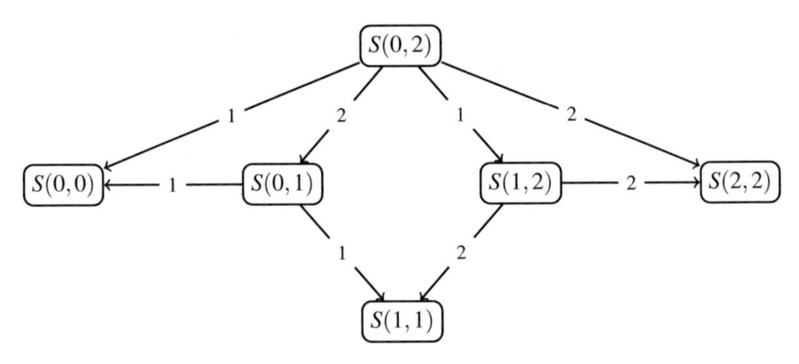

Fig. 17.3 Initial DAG $G = G_0$

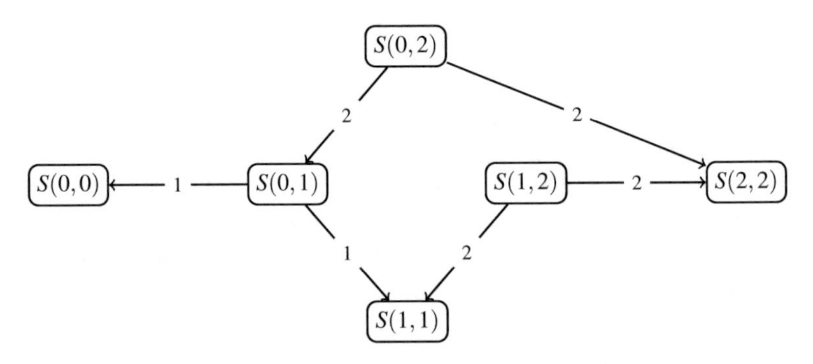

Fig. 17.4 Graph G^ψ

Fig. 17.5 Binary search tree corresponding to the vertex $S(0, 2)$ in G^ψ (see Fig. 17.4)

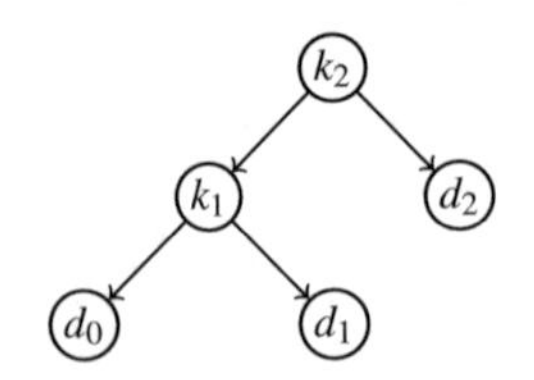

17.5 Multi-stage Optimization

The optimization procedure presented in this section allows multi-stage optimization relative to different cost functions. Let us consider one of the possible scenario.

Initially, we get a DAG $G = G_0$. We apply to the the graph G the procedure of optimization relative to the average weighted depth ψ. As a result, we obtain the subgraph G^ψ of the graph G. By Theorem 17.2, the set of binary search trees corresponding to the vertex $S(0, n)$ of G^ψ coincides with the set of all binary search trees which have minimum cost relative to ψ. We denote this set by $\mathscr{P}$.

For each vertex $S(i, j), i < j$, in G^ψ at least one rigid pair outgoing from $S(i, j)$ is left intact. So we can now apply to the graph G^ψ the procedure of optimization relative to the weighted depth φ. As a result, we obtain the subgraph $(G^\psi)^\varphi$. By Theorem 17.1, the set of binary search trees corresponding to the vertex $S(0, n)$ of $(G^\psi)^\varphi$ is a subset of the set of all trees from $\mathscr{P}$ which have minimum cost relative to φ.

17.6 Computational Complexity

Initially, we get a DAG G_0. After optimization relative to a sequence of cost functions we get *subgraph* G of G_0, which is obtained from G_0 by removal of some edges (rigid pairs). It is clear that the number of subproblems $S(i, j)$ for a problem with n keys is $O(n^2)$. We study two types of cost functions, i.e., weighted depth and average weighted depth.

Let us consider the procedure of optimization of G relative to one of these cost functions ψ. This procedure (as outlined in this section) requires to compute $\psi^*_{(i,j)}$ for each subproblem $S(i, j)$. It is enough to make $O(n)$ arithmetic operations to compute the value $\psi^*_{(i,j)}$ and to remove all rigid pairs for which $\Psi_{(i,r,j)} > \psi^*_{(i,j)}$, if the values $\pi(i, j)$, $\psi^*_{(i,r-1)}$, and $\psi^*_{(r,j)}$ are already known for each $r \in R_G(i, j)$. It is clear that $\pi(i, i) = q_i$ and $\pi(i, j) = \pi(i, i) + p_{i+1} + \pi(i + 1, j)$, if $j > i$. So to compute all values $\pi(i, j)$, $0 \le i \le j \le n$, it is enough to make $O(n^2)$ operations of additions. Therefore the total number of arithmetic operations required to obtain G^ψ from G is $O(n^3)$.

References

1. Alnafie, M., Chikalov, I., Hussain, S., Moshkov, M.: Sequential optimization of binary search trees for multiple cost functions. In: Potanin, A., Viglas T. (eds.) Seventeenth Computing: The Australasian Theory Symposium, CATS 2011, Perth, Australia, January 2011, CRPIT, vol. 119, pp. 41–44. Australian Computer Society (2011)
2. Cormen, T.H., Leiserson, C.E., Rivest, R.L., Stein, C.: Introduction to Algorithms, 2nd edn. MIT Press, Cambridge (2001)

3. Knuth, D.E.: Optimum binary search trees. Acta Inform. **1**(1), 14–25 (1971)
4. Mehlhorn, K.: Data Structures and Algorithms 1: Sorting and Searching. Monographs in Theoretical Computer Science. An EATCS Series, vol. 1. Springer, Heidelberg (1984)

Chapter 18
Global Sequence Alignment

The *sequence alignment* problem is a generalization of a classic problem known as the *edit distance* problem, see Levenshtein [2], which measures the difference between two sequences. The generalized version known as the sequence alignment problem lies at the core of bioinformatics. The solution to this problem helps identify several biological similarities present in long sequences of DNA/RNA and proteins, see Waterman [5] and Setubal and Meidanis [4] for computational and biological aspects of sequence alignment problem in detail.

The biological nature of protein sequences leads to two different kinds of approaches and applications of sequence alignment, namely, the *global alignment* and *local alignment*. Local alignment often aims to align parts of two sequences, which are very long. On the other hand, global alignment techniques try to align the whole sequences. Needleman and Wunsch [3] and Smith and Waterman [6] present general dynamic programming algorithms for global and local alignment problems, respectively.

In this chapter, we concentrate on the study of global alignment problem using an extension of dynamic programming. We provide a method to model the global sequence alignment problem using a directed acyclic graph which allows us to describe all optimal solutions specific to a certain cost function. Furthermore, we can apply a sequence of cost functions to optimize sequence alignments relative to different optimization criteria. In particular, we can begin from minimization of the number of *indels* (the number of gaps in an alignment) and after that continue with minimization of the number of mismatches (the number of matches of different letters in an alignment). The results considered in this chapter were published in [1].

18.1 Preliminaries

Let Σ be a finite alphabet, $X = x_1x_2 \ldots x_m$ and $Y = y_1y_2 \ldots y_n$ be sequences (words) over this alphabet.

H. AbouEisha et al., *Extensions of Dynamic Programming for Combinatorial Optimization and Data Mining*, Intelligent Systems Reference Library 146,
https://doi.org/10.1007/978-3-319-91839-6_18

We consider the problem of *sequence alignment* for X and Y, and denote this problem as $S(m, n)$. Furthermore, we consider subproblems of the initial problem. For $0 \leq i \leq m$ and $0 \leq j \leq n$, we denote by $S(i, j)$ the problem of sequence alignment for the prefixes $x_1 \dots x_i$ and $y_1 \dots y_j$. If $i = 0$ or $j = 0$ we have empty prefixes, denoted as λ.

We describe inductively the set $P(i, j)$ of alignments for $S(i, j)$. Each alignment for $S(i, j)$ is a pair of sequences of equal length obtained from prefixes $x_1 \dots x_i$ and $y_1 \dots y_j$ by insertion of gaps "␣" such that there is no gap in the same position in both sequences. We have $P(0, 0) = \{(\lambda, \lambda)\}$ where λ is the empty sequence, $P(0, j) = \{(\text{␣} \dots \text{␣}, y_1 \dots y_j)\}$, and $P(i, 0) = \{(x_1 \dots x_i, \text{␣} \dots \text{␣})\}$ for any $0 \leq i \leq m$ and $0 \leq j \leq n$. Furthermore, for $0 < i \leq m$ and $0 < j \leq n$,

$$
\begin{aligned}
P(i, j) = & \{(\alpha x_i, \beta \text{␣}) : (\alpha, \beta) \in P(i-1, j)\} \\
& \cup \{(\alpha \text{␣}, \beta y_j) : (\alpha, \beta) \in P(i, j-1)\} \\
& \cup \{(\alpha x_i, \beta y_j) : (\alpha, \beta) \in P(i-1, j-1)\} .
\end{aligned}
$$

We consider now the notion of *cost function* ψ. We fix integral values $w(x_i, y_i)$, $w(x_i, \text{␣})$, and $w(\text{␣}, y_j)$ for $1 \leq i \leq m$ and $1 \leq j \leq n$. Then, for any alignment $(\alpha_1 \dots \alpha_t, \beta_1 \dots \beta_t)$, we have $\psi(\alpha_1 \dots \alpha_t, \beta_1 \dots \beta_t) = \sum_{i=1}^{t} w(\alpha_i, \beta_i)$. We can also define this function inductively: $\psi(\lambda, \lambda) = 0$ and

$$
\psi(\alpha_1 \dots \alpha_t, \beta_1 \dots \beta_t) = \psi(\alpha_1 \dots \alpha_{t-1}, \beta_1 \dots \beta_{t-1}) + w(\alpha_t, \beta_t) .
$$

18.2 Representation of the Set of Alignments

We describe a *directed acyclic graph* (DAG) G_0 which allows us to represent all alignments $P(i, j)$ for each subproblem $S(i, j)$, $0 \leq i \leq m$, $0 \leq j \leq n$. The set of vertices of this graph coincides with the set $\{S(i, j) : 0 \leq i \leq m, 0 \leq j \leq n\}$. If $i = 0$ or $j = 0$ then $S(i, j)$ has no outgoing edges. Let $i > 0$ and $j > 0$, then $S(i, j)$ has exactly three outgoing edges. These edges end in $S(i-1, j)$, $S(i, j-1)$, and $S(i-1, j-1)$ and labeled with indexes 1, 2, and 3, respectively.

Let G be a subgraph of G_0 which is obtained from G_0 by removal of some edges such that, for each vertex $S(i, j)$ with $i > 0$ and $j > 0$, at least one edge outgoing from $S(i, j)$ remains intact. Such a subgraph will be called *proper* subgraph of G_0.

Now, for each vertex $S(i, j)$, we define by induction the set $P_G(i, j)$ of alignments corresponding to $S(i, j)$ in G. We have $P_G(0, 0) = \{(\lambda, \lambda)\}$, $P_G(0, j) = \{(\text{␣} \dots \text{␣}, y_1 \dots y_j)\}$ and $P_G(i, 0) = \{(x_1 \dots x_i, \text{␣} \dots \text{␣})\}$, $1 \leq i \leq m$, $1 \leq j \leq n$.

For $i > 0$ and $j > 0$, let $K_G(i, j)$ be the set of indexes of remaining edges outgoing from $S(i, j)$ in G. Then $P_G(i, j) = \bigcup_{k \in K_G(i,j)} P_G^{(k)}(i, j)$, where

$$P_G^{(1)}(i, j) = \{(\alpha x_i, \beta \sqcup) : (\alpha, \beta) \in P_G(i - 1, j)\} ,$$
$$P_G^{(2)}(i, j) = \{(\alpha \sqcup, \beta y_j) : (\alpha, \beta) \in P_G(i, j - 1)\} ,$$
$$P_G^{(3)}(i, j) = \{(\alpha x_i, \beta y_j) : (\alpha, \beta) \in P_G(i - 1, j - 1)\} .$$

One can show that $P_{G_0}(i, j) = P(i, j), 0 \le i \le m, 0 \le j \le n$.

18.3 Procedure of Optimization

Let ψ be a cost function. We consider the procedure of optimization of alignments corresponding to vertices of G relative to ψ. For each vertex $S(i, j)$ of the graph G, we mark this vertex by a number $\psi_{i,j}$ (later we will prove that $\psi_{i,j}$ is the minimum value of ψ on $P_G(i, j)$) and may remove some edges outgoing from $S(i, j)$, if $i > 0$ and $j > 0$.

We set $\psi_{0,0} = 0$, $\psi_{i,0} = w(x_1, \sqcup) + \cdots + w(x_i, \sqcup)$, and $\psi_{0,j} = w(\sqcup, y_1) + \cdots + w(\sqcup, y_j)$ for $1 \le i \le m$ and $1 \le j \le n$.

Let $1 \le i \le m$ and $1 \le j \le n$. Then $\psi_{i,j} = \min_{k \in K_G(i,j)} \psi_{i,j}^{(k)}$, where

$$\psi_{i,j}^{(1)} = \psi_{i-1,j} + w(x_i, \sqcup) ,$$
$$\psi_{i,j}^{(2)} = \psi_{i,j-1} + w(\sqcup, y_j) ,$$
$$\psi_{i,j}^{(3)} = \psi_{i-1,j-1} + w(x_i, y_j) .$$

Now we remove from G each edge with index $k \in K_G(i, j)$ outgoing from $S(i, j)$ such that $\psi_{i,j}^{(k)} > \psi_{i,j}$.

We denote by G^ψ the resulting subgraph of G. It is clear that, for each vertex $S(i, j), i \ge 1, j \ge 1$, at least one edge outgoing from $S(i, j)$ is left intact. So G^ψ is a proper subgraph of G_0.

For $0 \le i \le m$ and $0 \le j \le n$, we denote by $P_{G,\psi}^{opt}(i, j)$ the set of all alignments from $P_G(i, j)$ which have minimum cost relative to ψ among alignments from $P_G(i, j)$.

The following theorem summarizes the procedure of optimization.

Theorem 18.1 *Let G be a proper subgraph of G_0, and ψ be a cost function. Then, for any i and j such that $0 \le i \le m$ and $0 \le j \le n$, $\psi_{i,j}$ is the minimum cost of an alignment from $P_G(i, j)$, and the set $P_{G^\psi}(i, j)$ coincides with the set of all alignments $(\alpha, \beta) \in P_G(i, j)$ for which $\psi(\alpha, \beta) = \psi_{i,j}$, i.e., $P_{G^\psi}(i, j) = P_{G,\psi}^{opt}(i, j)$.*

Proof The proof is by induction on $i + j$. One can show that the considered statement is true if $i = 0$ or $j = 0$ (in this case $P_G(i, j)$ contains exactly one alignment, and this alignment has the minimum cost relative to ψ among alignments from $P_G(i, j)$). Hence the given statement is true if $i + j \le 1$. Let us assume that, for some $t \ge 1$, the considered statement is true for each i and j such that $i + j \leq t$. Let $i + j = t + 1$. If $i = 0$ or $j = 0$ then the statement holds. Let $i > 0$ and $j > 0$. We know that

$$P_G(i, j) = \bigcup_{k \in K_G(i,j)} P_G^{(k)}(i, j) . \tag{18.1}$$

Since $(i - 1) + j \leq t$, $i + (j - 1) \leq t$ and $(i - 1) + (j - 1) < t$, we have, by the inductive hypothesis, that $\psi_{i-1,j}$, $\psi_{i,j-1}$, and $\psi_{i-1,j-1}$ are the minimum costs of alignments from $P_G(i - 1, j)$, $P_G(i, j - 1)$, and $P_G(i - 1, j - 1)$, respectively. From here it follows that $\psi_{i,j}^{(1)} = \psi_{i-1,j} + w(x_i, \_)$ is the minimum cost of an alignment from $P_G^{(1)}(i, j)$, and similarly $\psi_{i,j}^{(2)} = \psi_{i,j-1} + w(\_, y_j)$ and $\psi_{i,j}^{(3)} = \psi_{i-1,j-1} + w(x_i, y_j)$ are, respectively, the minimum costs for alignments from $P_G^{(2)}(i, j)$ and $P_G^{(3)}(i, j)$. From here and (18.1) it follows that $\psi_{i,j}$ is the minimum cost of an alignment from $P_G(i, j)$.

After applying the procedure of optimization to G relative to ψ, we have

$$P_{G^\psi}(i, j) = \bigcup_{k \in K_{G^\psi}(i,j)} P_{G^\psi}^{(k)}(i, j) , \tag{18.2}$$

where

$$P_{G^\psi}^{(1)}(i, j) = \{(\alpha x_i, \beta \_) : (\alpha, \beta) \in P_{G^\psi}(i - 1, j)\} ,$$
$$P_{G^\psi}^{(2)}(i, j) = \{(\alpha \_, \beta y_j) : (\alpha, \beta) \in P_{G^\psi}(i, j - 1)\} ,$$
$$P_{G^\psi}^{(3)}(i, j) = \{(\alpha x_i, \beta y_j) : (\alpha, \beta) \in P_{G^\psi}(i - 1, j - 1)\}$$

and $K_{G^\psi}(i, j) = \{k \in K_G(i, j) : \psi_{i,j}^{(k)} = \psi_{i,j}\}$.

By the inductive hypothesis, $P_{G^\psi}(i - 1, j)$, $P_{G^\psi}(i, j - 1)$, and $P_{G^\psi}(i - 1, j - 1)$, respectively, coincide with the sets of alignments in $P_G(i - 1, j)$, $P_G(i, j - 1)$, and $P_G(i - 1, j - 1)$ for which the respective costs are $\psi_{i-1,j}$, $\psi_{i,j-1}$, and $\psi_{i-1,j-1}$, i.e., $P_{G^\psi}(i - 1, j) = P_{G,\psi}^{opt}(i - 1, j)$, $P_{G^\psi}(i, j - 1) = P_{G,\psi}^{opt}(i, j - 1)$, and $P_{G^\psi}(i - 1, j - 1) = P_{G,\psi}^{opt}(i - 1, j - 1)$.

From here and (18.2) it follows that the cost of each alignment from $P_{G_\psi}(i, j)$ is equal to $\psi_{i,j}$ and $P_{G^\psi}(i, j) \subseteq P_{G,\psi}^{opt}(i, j)$. Let us show that $P_{G,\psi}^{opt}(i, j) \subseteq P_{G^\psi}(i, j)$.

Let $(\alpha, \beta) \in P_G(i, j)$ and $\psi(\alpha, \beta) = \psi_{i,j}$. From (18.1) it follows that $(\alpha, \beta) \in P_G^{(k)}(i, j)$ for some $k \in K_G(i, j)$. Let $k = 3$. Then $(\alpha, \beta) = (\alpha' x_i, \beta' y_j)$, where $(\alpha', \beta') \in P_G(i - 1, j - 1)$, and $\psi(\alpha, \beta) = \psi(\alpha', \beta') + w(x_i, y_j)$. Since $\psi(\alpha, \beta) = \psi_{i,j}$, we have $\psi(\alpha', \beta') = \psi_{i,j} - w(x_i, y_j)$. It is clear that $\psi_{i-1,j-1} \leq \psi(\alpha', \beta')$. Therefore $\psi_{i,j}^{(3)} = \psi_{i-1,j-1} + w(x_i, y_j) \leq \psi_{i,j} - w(x_i, y_j) + w(x_i, y_j) = \psi_{i,j}$. Hence $3 \in K_{G^\psi}(i, j)$ and $P_{G_\psi}^{(3)}(i, j) \subseteq P_{G^\psi}(i, j)$.

Let us assume that $\psi_{i-1,j-1} < \psi(\alpha', \beta')$. Then $\psi_{i,j}^{(3)} < \psi_{i,j}$ which is impossible. Therefore $\psi_{i-1,j-1} = \psi(\alpha', \beta')$, $(\alpha', \beta') \in P_{G^\psi}(i - 1, j - 1)$, $(\alpha, \beta) \in P_{G^\psi}^{(3)}(i, j)$

and $(\alpha, \beta) \in P_{G^{\psi}}(i, j)$. The cases when $k = 1$ and $k = 2$ can be considered in the same way. Therefore $P^{opt}_{G,\psi}(i, j) \subseteq P_{G^{\psi}}(i, j)$. □

Above theorem proves that during the optimization procedure all optimal solutions and only optimal solutions remain intact.

18.4 Multi-stage Optimization

The optimization procedure presented in previous section allows multi-stage optimization relative to different cost functions.

Initially we get a DAG $G = G_0$. This graph G can be optimized according to a cost function ψ_1. As a result, we obtain the proper subgraph G^{ψ_1} of G_0 which describes the whole set of alignments that are optimal relative to ψ_1. We can apply the procedure of optimization relative to another cost function ψ_2 to G^{ψ_1}. As a result, we obtain the proper subgraph $(G^{\psi_1})^{\psi_2}$ of G_0, which describes the set of all alignments that are optimal regarding ψ_2 among all alignments optimal regarding ψ_1, etc.

18.5 Computational Complexity of Optimization

Initially we get a DAG G_0. After optimization relative to a sequence of cost functions we get a proper subgraph G of G_0, which is obtained from G_0 by removal of some edges. Let us now consider the process of optimization of G relative to a cost function ψ.

The number of subproblems $S(i, j)$ (vertices in G) for a problem of aligning two sequences X and Y of lengths m and n, respectively is exactly $(m + 1)(n + 1)$. It is enough to make three operations of addition and four operations of comparison for the vertex $S(i, j)$ to compute the value $\psi_{i,j}$ and to find all edges for which $\psi^{(k)}_{i,j} > \psi_{i,j}$, if the values $\psi_{i-1,j}$, $\psi_{i,j-1}$ and $\psi_{i-1,j-1}$ are already known. Therefore, the total number of arithmetic operations (comparison is also considered as an arithmetic operation) to obtain G^{ψ} from G is at most $7(m + 1)(n + 1) = O(mn)$.

18.6 Example

We consider a very simple example of two sequences X and Y over $\Sigma = \{\mathtt{A}, \mathtt{T}\}$, where

$$X = \mathtt{AAT}, \quad \text{and} \quad Y = \mathtt{AT}\ .$$

We consider a cost function ψ such that

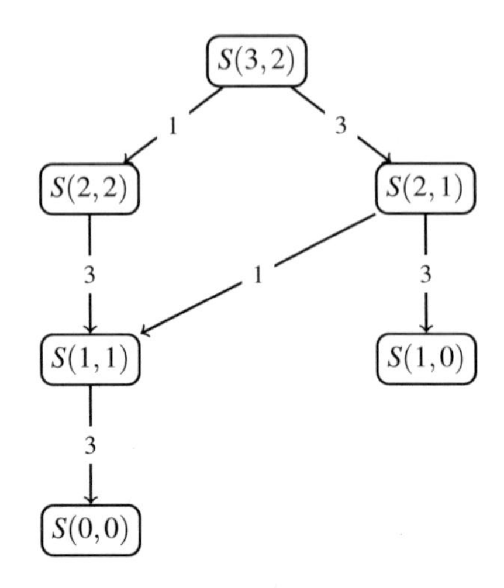

Fig. 18.1 Subgraph of G^{ψ} after optimizing G relative to the cost function ψ

$$w(x_i, \llcorner) = w(\llcorner, y_j) = 1, \quad \text{and} \quad w(x_i, y_j) = 0 \; .$$

That is we consider matching a gap with penalty one and a match/mismatch with penalty zero.

One can construct the initial DAG $G = G_0$ for the considered problems and check that there are exactly 25 possible alignments. However, there are only three different alignments that are optimal relative to ψ, which are

$$\begin{matrix}\texttt{AAT} \\ \texttt{A}\llcorner\texttt{T}\end{matrix}, \quad \begin{matrix}\texttt{AAT} \\ \texttt{AT}\llcorner\end{matrix}, \quad \text{and} \quad \begin{matrix}\texttt{AAT} \\ \llcorner\texttt{AT}\end{matrix},$$

see Fig. 18.1 for the resulting DAG G^{ψ} (for simplicity, we remove from the DAG all vertices which are not reachable from $S(3, 2)$).

We further optimize the resulting DAG G^{ψ} relative to another cost function φ such that

$$w(x_i, \llcorner) = w(\llcorner, y_j) = 0, \quad \text{and} \quad w(x_i, y_j) = \begin{cases} 0, & \text{if } x_i = y_j \; , \\ 1, & \text{otherwise.} \end{cases}$$

That is, this time we consider matching a gap with penalty zero, match with a penalty zero, and a mismatch with penalty one.

It is easy to see that now there are only two different alignments possible which are optimal relative to φ on the DAG G^{ψ}, which are

$$\begin{matrix}\texttt{AAT} \\ \texttt{A}\llcorner\texttt{T}\end{matrix} \quad \text{and} \quad \begin{matrix}\texttt{AAT} \\ \llcorner\texttt{AT}\end{matrix},$$

see Fig. 18.2 for the resulting DAG $(G^{\psi})^{\varphi}$ (again for simplicity, we have removed from this DAG all vertices which are not reachable from $S(3, 2)$).

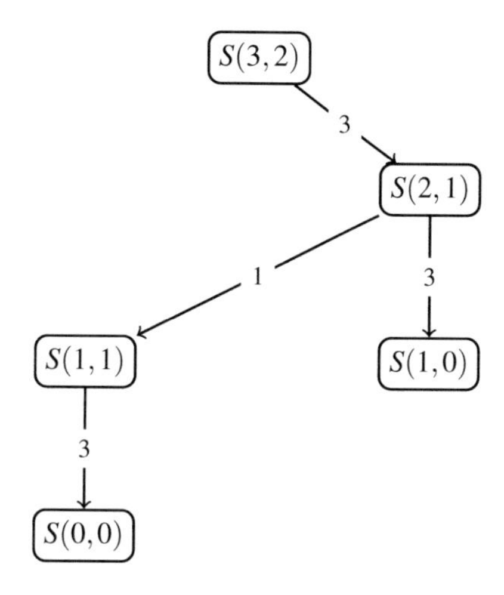

Fig. 18.2 Subgraph of $(G^\psi)^\varphi$ after optimizing G^ψ relative to the cost function φ

References

1. Chikalov, I., Hussain, S., Moshkov, M., Odat, E.: Sequential optimization of global sequence alignments relative to different cost functions. In: ACM International Conference on Convergence and Hybrid Information Technology, ICHIT 2010, Daejeon, Korea, August 26–28, 2010. ACM (2010)
2. Levenshtein, V.I.: Binary codes capable of correcting deletions, insertions and reversals. Soviet Phys. Dokl. **10**, 707–710 (1966)
3. Needleman, S.G., Wunsch, C.D.: A general method applicable to the search for similarities in the amino acid sequence of two proteins. J. Mol. Biol. **48**, 443–453 (1970)
4. Setubal, J., Meidanis, J.: Introduction to Computational Molecular Biology. PWS Publishing, Boston (1997)
5. Waterman, M.S.: Introduction to Computational Biology: Sequences, Maps, and Genomes. Chapman and Hall, London (1995)
6. Waterman, M.S., Smith, T.F.: Identification of common molecular subsequences. J. Mol. Biol. **147**, 195–197 (1981)

Chapter 19
Optimal Paths in Directed Graphs

In this chapter, we present an algorithm for multi-stage optimization of paths in a directed graph G relative to different cost functions. Let w_1, w_2 be weight functions which correspond to each edge nonnegative real weights, and p be a directed path in G. We consider two types of cost functions: ψ_{w_1} where $\psi_{w_1}(p)$ is equal to the sum of weights of edges in p, and φ_{w_2} where $\varphi_{w_1}(p)$ is equal to the minimum weight of some edge in p. We study the problem of minimization for ψ_{w_1} and the problem of maximization for φ_{w_2}.

Let us consider an interpretation of G as a network of roads. In this case, each vertex can be interpreted as an intersection point, and each edge as a road connecting two intersection points. If weight (the value of weight function w) is the length of a road, then $\psi_w(p)$ is the length of the path p. If weight is the fee we need to pay for using the road, then $\psi_w(p)$ is the total fee of the path. In such a way, we can also consider the minimization of the risk, or the expected time. Furthermore, if weight is the constraint on the height, width or weight of a vehicle, then $\varphi_w(p)$ is the strongest constraint on the height, width or weight of vehicle in the path p.

Note that we can consider another interpretation of G as a computer network. The first type of the cost functions ψ_w can be used for minimization of delay or packet loss rate, and the second type of the cost functions φ_w can be used for maximization of bit rate, for example.

The problem of optimization of paths in G relative to ψ_w is very well known. There are algorithms based on *greedy approach* such as Dijkstra's algorithm [4], algorithms based on *dynamic programming* such as Bellman–Ford algorithm [2, 6], and Floyd–Warshal algorithm [5, 7]. Cherkassky et al. discuss these and many other algorithms in [3].

The novelty of results considered in this chapter is connected with the following: we represent all initial paths from G (satisfying conditions (a) and (b) from Sect. 19.1) by a *directed acyclic graph* (DAG) Γ_0 and apply to this graph a procedure of optimization relative to a cost function. As a result, we have a subgraph Γ_1 of

H. AbouEisha et al., *Extensions of Dynamic Programming for Combinatorial Optimization and Data Mining*, Intelligent Systems Reference Library 146,
https://doi.org/10.1007/978-3-319-91839-6_19

Γ_0, which represents all optimal paths from Γ_0. We can apply to Γ_1 the procedure of optimization relative to another cost function, etc. The results presented in this chapter were published in [1].

19.1 Preliminaries

Let $G = (V, E)$, where V is the set of vertices and E is the set of edges, be a directed graph without loops and multiple edges. Let s, t be two vertices in G such that there exists a directed path from s to t. We call s the *source* vertex, and t the *target* vertex.

We consider the problem of finding optimal paths in the graph from s to t . The optimization criteria are defined by means of cost functions.

We consider two types of cost functions. Each *cost function* assigns a *cost* to each directed path in G. Let w be a *weight function* that assigns to each edge (v_i, v_j) of G a weight $w(v_i, v_j)$, which is a nonnegative real number.

We consider a directed path $p = v_1, v_2, \ldots, v_m, v_{m+1}$ in G. A cost function of the first type ψ_w is defined as follows: $\psi_w(p) = \sum_{i=1}^{m} w(v_i, v_{i+1})$. If $m = 0$ then $\psi_w(p) = 0$. We aim to minimize the value of this function. A cost function of the second type φ_w is defined as follows: $\varphi_w(p) = \min\{w(v_1, v_2), ..., w(v_m, v_{m+1})\}$. If $m = 0$ then $\varphi_w(p) = +\infty$. We aim to maximize the value of this function.

Let p be a directed path in G from s to t. One can show that there exists a directed path p' from s to t in G, which can be obtained from p by removal of some vertices and edges such that it satisfies the following conditions:

(a) The length of p' is at most $|V| - 1$,
(b) Both s and t appear exactly once in p',
(c) $\psi_w(p') \leq \psi_w(p)$,
(d) $\varphi_w(p') \geq \varphi_w(p)$.

Therefore, later on we will focus on the study of paths in G from s to t satisfying the conditions (a) and (b). We denote by $P(G)$ the set of all such paths.

19.2 Description of Paths

In this section, we introduce a directed acyclic graph $\Gamma_0 = \Gamma_0(G)$, which represents the set of directed paths $P(G)$ from s to t satisfying conditions (a) and (b).

Let $V = \{v_1, \ldots, v_n\}$, where $v_1 = s$ and $v_n = t$. The set of vertices of Γ_0 is divided into n layers. The first layer contains the only vertex v_1^1. The last layer contains the only vertex v_n^n. For $k = 2, \ldots, n-1$, the k-th layer contains $n-2$ vertices $v_2^k, \ldots, v_{n-1}^k$.

Let $i, j \in \{2, \ldots, n-1\}$, we have an edge from v_1^1 to v_j^2 if and only if $(v_1, v_j) \in E$. For $k = 2, \ldots, n-2$, we have an edge from v_i^k to v_j^{k+1} if and only if $(v_i, v_j) \in E$.

Let $2 \leq j \leq n-1$ and $2 \leq k \leq n-1$. Then, we have an edge from v_j^k to v_n^n if and only if $(v_j, v_n) \in E$. We have an edge from v_1^1 to v_n^n if and only if $(v_1, v_n) \in E$. There are no other edges in Γ_0.

A weight function w on G can be extended to Γ_0 in a natural way: $w(v_i^{k_1}, v_j^{k_2}) = w(v_i, v_j)$ if $(v_i^{k_1}, v_j^{k_2})$ is an edge in Γ_0.

We denote by $P(\Gamma_0)$ the set of all directed paths in Γ_0 from v_1^1 to v_n^n. It is clear that there is one-to-one mapping from the set $P(G)$ to the set $P(\Gamma_0)$. The path $v_1, v_{j_2}, \ldots, v_{j_t}, v_n$ from $P(G)$ corresponds to the path $v_1^1, v_{j_2}^2, \ldots, v_{j_t}^t, v_n^n$ from $P(\Gamma_0)$.

19.3 Procedures of Optimization

Let Γ be a subgraph of Γ_0 which can be obtained from Γ_0 by removal of some vertices and edges such that there is at least one directed path from v_1^1 to v_n^n. We denote by $P(\Gamma)$ the set of all such paths. We consider procedures of optimization of paths in $P(\Gamma)$ relative to ψ_w and φ_w.

19.3.1 Optimization Relative to ψ_w

We remove from Γ all vertices (and their incident edges) such that there is no directed paths from v_1^1 to the considered vertex. We denote by Γ' the obtained subgraph of Γ.

For each vertex v_j^k of Γ' we label it with $\Psi_w(v_j^k)$ which is the minimum cost of a directed path in Γ' from v_1^1 to v_j^k relative to ψ_w and remove some edges incoming to v_j^k. We label v_1^1 with $\Psi(v_1^1) = 0$. Let $v_j^k \neq v_1^1$ and $v_{i_1}^{k_1}, \ldots, v_{i_r}^{k_r}$ be all vertices in Γ' from each of which there is an edge to v_j^k. Then

$$\Psi_w(v_j^k) = \min_{1 \leq t \leq r} \left\{ w(v_{i_t}^{k_t}, v_j^k) + \Psi_w(v_{i_t}^{k_t}) \right\} . \tag{19.1}$$

We remove from Γ' all edges $(v_{i_t}^{k_t}, v_j^k)$ incoming to v_j^k such that

$$w(v_{i_t}^{k_t}, v_j^k) + \Psi_w(v_{i_t}^{k_t}) > \Psi_w(v_j^k) ,$$

Denote the obtained subgraph of Γ_0 by Γ^{ψ_w}. It is clear that there is at least one path in Γ^{ψ_w} from v_1^1 to v_n^n. We denote by $P(\Gamma^{\psi_w})$ all such paths.

Theorem 19.1 *The set of paths $P(\Gamma^{\psi_w})$ coincides with the set of paths from $P(\Gamma)$ that have minimum cost relative to ψ_w.*

Proof We denote by $P(\Gamma')$ the set of all directed paths in Γ' from v_1^1 to v_n^n. It is clear that $P(\Gamma') = P(\Gamma)$. Let us prove by induction on vertices in Γ' that, for each vertex v_j^k of Γ', the number $\Psi_w(v_j^k)$ is the minimum cost of a directed path in Γ' from v_1^1 to v_j^k relative to ψ_w and the set of directed paths from v_1^1 to v_j^k in Γ^{ψ_w} coincides with the set of directed paths from v_1^1 to v_j^k in Γ' that have minimum cost relative to ψ_w. If $v_j^k = v_1^1$ we have $\Psi_w(v_1^1) = 0$. We have exactly one path from v_1^1 to v_1^1 both in Γ' and Γ^{ψ_w}. The length and the cost of this path are equal to zero. So the considered statement holds for v_1^1. For $v_j^k \neq v_1^1$, let $v_{i_1}^{k_1}, \ldots, v_{i_r}^{k_r}$ be all vertices in Γ' from each of which there is an edge to v_j^k . Let us assume that the considered statement holds for $v_{i_1}^{k_1}, \ldots, v_{i_r}^{k_r}$. Since Γ' is a directed acyclic graph, each path from v_1^1 to v_j^k has no v_j^k as an intermediate vertex. From here, (19.1) and inductive hypothesis it follows that $\Psi_w(v_j^k)$ is the minimum cost of a directed path in Γ' from v_1^1 to v_j^k.

Using (19.1), inductive hypothesis, and description of optimization procedure we have that each path in Γ^{ψ_w} from v_1^1 to v_j^k has the cost $\Psi_w(v_j^k)$. So the set of directed paths from v_1^1 to v_j^k in Γ^{ψ_w} is a subset of the set of directed paths from v_1^1 to v_j^k in Γ' that have minimum cost relative to ψ_w.

Let p be a path from v_1^1 to v_j^k in Γ' with minimum cost relative to ψ_w. Then, for some $t \in \{1, \ldots, r\}$, the path p passes through the vertex $v_{i_t}^{k_t}$. Therefore, $\psi_w(p) = \Psi(v_j^k) = \psi_w(p') + w(v_{i_t}^{k_t}, v_j^k)$ where p' is the first part of p from v_1^1 to $v_{i_t}^{k_t}$. It is clear that $\psi_w(p') \geq \Psi_w(v_{i_t}^{k_t})$. From here and the description of optimization procedure it follows that the edge $(v_{i_t}^{k_t}, v_j^k)$ belongs to Γ^{ψ_w}. Let us assume that $\psi_w(p') > \Psi_w(v_{i_t}^{k_t})$. Then $\Psi_w(v_j^k) > \Psi_w(v_{i_t}^{k_t}) + w(v_{i_t}^{k_t}, v_j^k)$ but this is impossible (see (19.1)), so $\psi_w(p') = \Psi_w(v_{i_t}^{k_t})$. According to the inductive hypothesis, p' belongs to the set of paths in Γ^{ψ_w} from v_1^1 to $v_{i_t}^{k_t}$ and therefore p belongs to the set of paths in Γ^{ψ_w} from v_1^1 to v_j^k. Thus the considered statement holds. From this statement (if we set $v_j^k = v_n^n$) it follows the statement of the theorem. □

19.3.2 Optimization Relative to φ_w

We remove from Γ all vertices (and their incident edges) such that there is no directed path from v_1^1 to the considered vertex. We denote by Γ' the obtained subgraph of Γ.

For each vertex v_j^k of Γ', we label it with $\Phi_w(v_j^k)$ which is the maximum cost of a directed path in Γ' from v_1^1 to v_j^k relative to φ_w. We label v_1^1 with $\Phi_w(v_1^1) = +\infty$. Let $v_j^k \neq v_1^1$ and $v_{i_1}^{k_1}, \ldots, v_{i_r}^{k_r}$ be all vertices in Γ' from each of which there is an edge to v_j^k. Then

$$\Phi_w(v_j^k) = \max_{1 \leq t \leq r} \left\{ \min \left\{ w(v_{i_t}^{k_t}, v_j^k), \Phi_w(v_{i_t}^{k_t}) \right\} \right\} . \quad (19.2)$$

Let $c = \Phi_w(v_n^n)$. We denote by Γ^{φ_w} the graph obtained from Γ' by removal of all edges $(v_{j_1}^{k_1}, v_{j_2}^{k_2})$ for which $w(v_{j_1}^{k_1}, v_{j_2}^{k_2}) < c$. It is clear that there is at least one path in Γ^{φ_w} from v_1^1 to v_n^n. We denote by $P(\Gamma^{\varphi_w})$ all such paths.

Theorem 19.2 *The set of paths $P(\Gamma^{\varphi_w})$ coincides with the set of paths from $P(\Gamma)$ that have maximum cost relative to φ_w.*

Proof We denote by $P(\Gamma')$ the set of all directed paths in Γ' from v_1^1 to v_n^n. It is clear that $P(\Gamma') = P(\Gamma)$. Let us prove by induction on vertices in Γ' that, for each vertex v_j^k of Γ', the number $\Phi_w(v_j^k)$ is the maximum cost of a directed path in Γ' from v_1^1 to v_j^k relative to φ_w. If $v_j^k = v_1^1$ we have $\Phi_w(v_1^1) = +\infty$, and the considered statement holds for v_1^1 since there exists a path p from v_1^1 to v_1^1 with length equals to zero and cost $\varphi_w(p)$ equals to $+\infty$. For $v_j^k \neq v_1^1$, let $v_{i_1}^{k_1}, \ldots, v_{i_r}^{k_r}$ be all vertices in Γ' such that from each of which there is an edge to v_j^k. Let us assume that the considered statement holds for $v_{i_1}^{k_1}, \ldots, v_{i_r}^{k_r}$.

Since Γ' is a directed acyclic graph, each path from v_1^1 to v_j^k has no v_j^k as an intermediate vertex. From here, (19.2), and the inductive hypothesis it follows that $\Phi_w(v_j^k)$ is the maximum cost of a directed path in Γ' from v_1^1 to v_j^k.

Let $c = \Phi_w(v_n^n)$ then we know that c is the maximum cost of a path in Γ' from v_1^1 to v_n^n. So each path in Γ' from v_1^1 to v_n^n, that does not contain edges with weight less than c, is an optimal path (a path with maximum cost). Also each optimal path in Γ' from v_1^1 to v_n^n does not contain edges whose weight is less than c. From here it follows that the set $P(\Gamma^{\varphi_w})$ coincides with the set of paths from $P(\Gamma)$ that have maximum cost relative to φ_w. □

19.4 Multi-stage Optimization

The optimization procedures presented in this chapter allow multi-stage optimization relative to different cost functions. Let us consider one of the possible scenarios.

Let w_1 and w_2 be two weight functions for G. Initially, we get the DAG $\Gamma_0 = \Gamma_0(G)$. We apply the procedure of optimization relative to ψ_{w_1} to the graph Γ_0. As a result, we obtain the subgraph $\Gamma = (\Gamma_0)^{\psi_{w_1}}$ of the graph Γ_0. By Theorem 19.1, the set of directed paths in Γ from v_1^1 to v_n^n coincides with the set of directed paths $\mathcal{D}$ in Γ_0 from v_1^1 to v_n^n that have the minimum cost relative to ψ_{w_1}.

We can apply the procedure of optimization relative to φ_{w_2} to the graph Γ. As a result we obtain the subgraph $\Gamma^{\varphi_{w_2}}$ of the graph Γ_0. By Theorem 19.2, the set of directed paths $P(\Gamma^{\varphi_{w_2}})$ in $\Gamma^{\varphi_{w_2}}$ from v_1^1 to v_n^n coincides with the set of directed paths from $\mathcal{D}$ that have the maximum cost relative to φ_{w_2}.

We can continue the procedure of optimization relative to different cost functions.

19.5 Example

We consider a directed graph $G = (V, E)$ (as depicted in Fig. 19.1) where $s = v_1$ and $t = v_4$, and two weight functions w_1 and w_2 for this graph such as

$$w_1(e) = 1 \text{ for all } e \in E ,$$

and

$$\begin{aligned} w_2(s, v_2) = 1, \quad w_2(v_2, t) = 1, \quad w_2(s, v_3) = 2 , \\ w_2(v_3, t) = 2, \; w_2(v_2, v_3) = 2, \; w_2(v_3, v_2) = 2 . \end{aligned}$$

Initially, we get a DAG $\Gamma_0 = \Gamma_0(G)$ (as depicted in Fig. 19.2).

We apply the procedure of optimization relative to ψ_{w_1} to the graph Γ_0 and obtain the graph $\Gamma = (\Gamma_0)^{\psi_{w_1}}$ as shown in Fig. 19.3. The set $P(\Gamma)$ contains exactly two paths.

We apply the procedure of optimization relative to φ_{w_2} to the graph Γ. As a result, we obtain the subgraph $\Gamma^{\varphi_{w_2}}$ of Γ_0, see Fig. 19.4. The set $P(\Gamma^{\varphi_{w_2}})$ contains exactly one path.

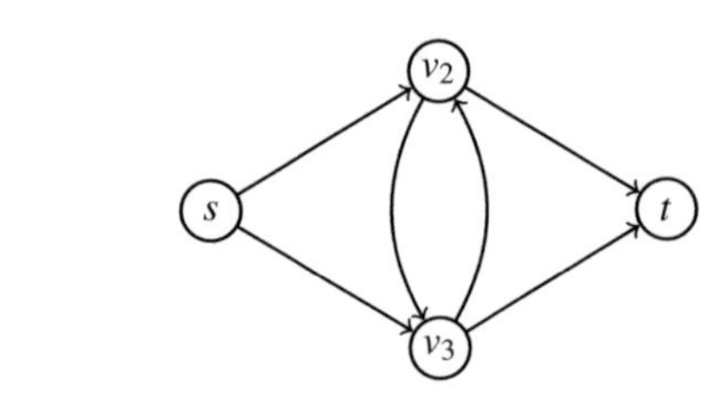

Fig. 19.1 Directed graph G

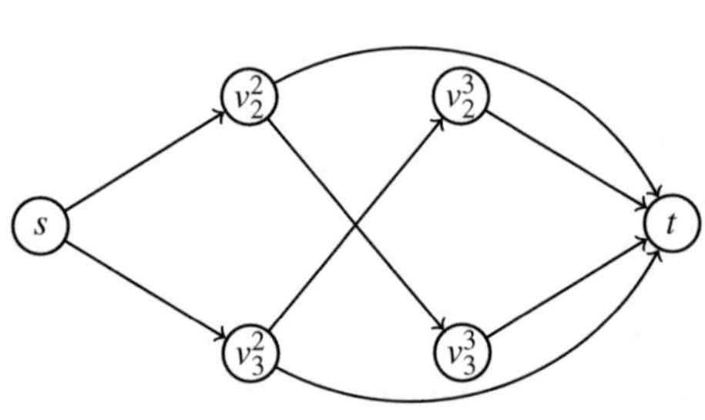

Fig. 19.2 Initial DAG Γ_0

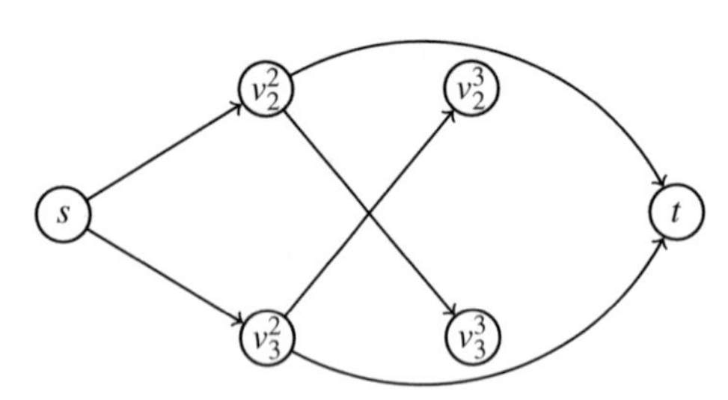

Fig. 19.3 Subgraph $\Gamma = (\Gamma_0)^{\psi_{w_1}}$

Fig. 19.4 Subgraph $\Gamma^{\varphi_{w_2}}$

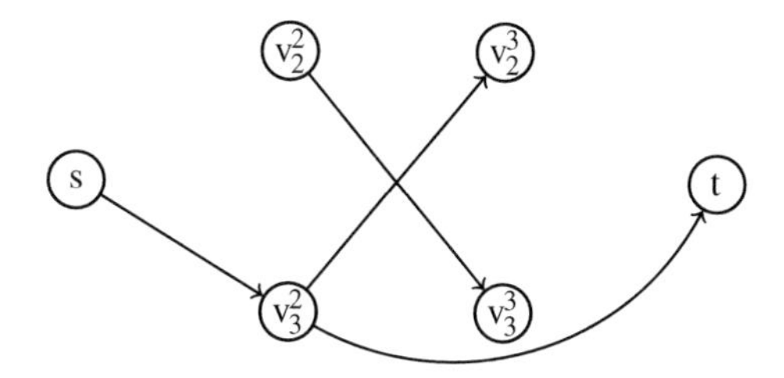

19.6 Computational Complexity

Let $G = (V, E)$ be a directed graph with $|V| = n$ and $|E| = m$, and let $\Gamma_0 = \Gamma_0(G) = (V_0, E_0)$ be the directed acyclic graph corresponding to G. It is clear that $|V_0| = O(n^2)$ and $|E_0| = O(mn)$. Let $\Gamma = (V_1, E_1)$ be a subgraph of the graph Γ_0 obtained from Γ_0 after some number of applications of the optimization procedures. Let us evaluate the time complexity of optimization procedure relative to ψ_w (to φ_w) applied to Γ.

We can find all vertices in Γ such that there is a directed path from v_1^1 to the considered vertex using *breadth-first search* algorithm with the running time $O(|V_1| + |E_1|)$. It is clear that $|V_1| \leq |V_0|$ and $|E_1| \leq |E_0|$. Therefore, the running time is $O(n^2 + mn)$.

After removal from Γ all vertices (and their incident edges), which are not reachable from v_1^1, we obtain a subgraph Γ' of Γ. According to the description of optimization procedure relative to ψ_w we need to make at most two comparisons and one addition per each edge to obtain the graph Γ_{ψ_w} from Γ'. So we need $O(mn)$ operations of comparison and addition to construct the graph Γ^{ψ_w} from Γ'. Similarly, according to the description of optimization procedure relative to φ_w we need to make at most three comparisons per each edge of Γ' to obtain the graph Γ^{φ_w} from Γ'. So we need $O(mn)$ operations of comparison to construct the graph Γ^{φ_w} from Γ'.

References

1. AbuBekr, J., Chikalov, I., Hussain, S., Moshkov, M.: Sequential optimization of paths in directed graphs relative to different cost functions. In: Sato, M., Matsuoka, S., Sloot, P.M.A., van Albada, G.D., Dongarra, J. (eds.) International Conference on Computational Science, ICCS 2011, Nanyang Technological University, Singapore, June 1–3, 2011, Procedia Computer Science, vol. 4, pp. 1272–1277. Elsevier (2011)
2. Bellman, R.: On a routing problem. Q. Appl. Math. **16**(1), 87–90 (1958)
3. Cherkassky, B., Goldberg, A.V., Radzik, T.: Shortest paths algorithms: Theory and experimental evaluation. Math. Programm. **73**, 129–174 (1993)

4. Dijkstra, E.W.: A note on two problems in connexion with graphs. Numerische Mathematik **1**, 269–271 (1959)
5. Floyd, R.W.: Algorithm 97: shortest path. Commun. ACM **5**(6), 345 (1962)
6. Ford, L.R.: Network flow theory. Technical Report P-923, RAND, Santa Monica, CA (1956)
7. Warshall, S.: A theorem on boolean matrices. J. ACM **9**(1), 11–12 (1962)

Appendix A
Dagger Software System

Dagger[1] software system is being developed in KAUST for research and educational purposes. The system implements several algorithms dealing with decision trees, systems of decision rules, and more general class of problems than can be solved by dynamic programming. Dagger has modular structure and can be extended further.

This appendix describes capabilities and architecture of Dagger. Section A.1 describes features of the system, lists third-party libraries used, and states requirements to environment for running Dagger. Section A.2 overviews key design decisions.

A.1 Introduction

Dagger is capable of performing several tasks that can be divided into two classes: general tasks related to dynamic programming and more specific tasks related to building decision trees and systems of decision rules. The first group of tasks operates with abstract objects (subproblem, uncertainty measure, cost function, etc.) that are defined differently for each domain. It contains the following operations:

- building of DAG up to specified uncertainty threshold α;
- computing minimum cost of a solution relative to a given cost function;
- optimizing DAG such that it describes optimal solutions only;
- finding a set of Pareto-optimal points for a pair of criteria (each criterion can be either uncertainty measure or cost function);
- counting the number of optimal solutions.

The second group contains the following operations:

- building one of decision trees (system of decision rules) described by DAG;
- greedy algorithms for building decision trees and systems of decision rules.

[1]The name origins from DAG, directed acyclic graph.

H. AbouEisha et al., *Extensions of Dynamic Programming for Combinatorial Optimization and Data Mining*, Intelligent Systems Reference Library 146,
https://doi.org/10.1007/978-3-319-91839-6

Table A.1 List of main libraries

Library name	Functions
data_set	Generating, loading, storing and other transformations of data sets
partition	Representation of decision table and uncertainty measures for decision table
model	Common operations with classifiers: pruning, error estimation, etc.
tree	Implementation of decision tree object. Greedy algorithms for decision tree construction
rule	Implementation of system of decision rules object. Greedy algorithms for construction of a system of decision rules
dag	Implementation of dynamic programming problem solver. Support functions for parallel computing
dag_table	Implementation of objects common to decision trees and systems of decision rules used by dynamic programming algorithm
dag_tree	Implementation of objects specific to decision trees used by dynamic programming algorithm
dag_rule	Implementation of objects specific to systems of decision rules used by dynamic programming algorithm

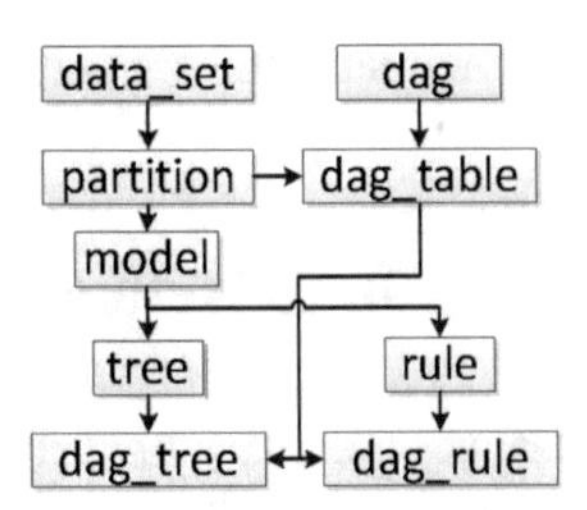

Fig. A.1 Graph of library dependencies

In addition to the key operations listed above, Dagger performs several auxiliary tasks such as loading and preparing decision tables, generating decision tables for Boolean functions and random decision tables, calculating prediction error for models, and others.

Dagger consists of a set of applications and dynamic libraries. Each dynamic library in the system covers certain functional area. Table A.1 contains the list of main libraries and short description of their functions. Library dependencies form an acyclic graph shown on Fig. A.1.

Dagger uses several third-party libraries:

- GetPot [1] for parsing command line arguments and configuration files,
- PThreads [4] for thread management,
- MPI [6] for organizing multiprocessor computations.

The front end of Dagger is a set of console applications taking parameters from command line or configuration files and reporting result to standard output and log files. Decision table can be loaded from a comma-separated values (CSV) file of a special format or ARFF [7] file. The objects that require graphical representation (DAG, decision tree) are saved in GraphML [5] format and can be displayed by third-party applications (for example, yEd [3]). Dagger is capable of drawing plots of functions in LaTeX using PGFPlots package [2].

Most operations allow for parallel execution in a large-scale multiprocess environment. The system uses hybrid parallelization. Computational load is distributed between several processes, and the number of processes can vary in a wide range. Each process contains two threads that provides nonblocking interprocess communication.

The most part of code is platform-independent, however, some thread synchronization primitives and means for console output are Windows-specific.

A.2 Architecture

Building the graph of subproblems differs from a typical problem solved by dynamic programming. First, for some domains, the graph has an irregular structure. For example, consider a graph of subtables of a decision table. The number of children in each node depends on values of conditional attributes for each row and assignment of decisions. Graph structure remains unknown until the graph is constructed. The second peculiarity is absence of data locality. If we consider parents of a particular node, layer of the parents may vary greatly (by layer of a node we mean the minimum length of a path from this node to the root). In any division of the graph into a few parts of equal size, the number of edges crossing the part boundaries is large.

The above-mentioned characteristics influence implementation of the algorithms for a multiprocessor environment. Absence of prior knowledge about graph structure makes it impossible to do explicit partitioning of the graph and assign each part to own process. Absence of data locality implies high amount of interprocess communications, so traditional star-like communication scheme is ineffective as communication with the main process become a bottleneck.

Dagger makes implicit graph partitioning in the following way. A hash function is defined for subproblems and each process owns a sub-range of its values. Thus to assign a subproblem to a process, value of the hash function is calculated and then the problem passed to the process that owns corresponding range of hash values. It is important the hash value to be the same for a subproblem regardless of the order of decompositions of the initial problem. For example, description of a subtable by set of constraints of the form "attribute = value" does not work as different sets of constraints can describe the same subtable. Instead, a subtable in Dagger is described by a vector of row indices relative to the initial decision table. This property allows for effective identification of duplicate subproblems. If the same subproblem appears multiple times, all related requests will be addressed to the same process based on the calculated hash value. The owner process stores solutions for all problems and reuses it when receives the second and further requests for the same subproblem. As a result, each process keeps a part of graph nodes and performs operations on it that makes algorithm scalable and applicable to larger problems.

Computations are organized as a number of worker processes and the master process that synchronizes work of others and interact with the front end modules. Most of the time worker processes communicate with each other directly in order to reduce synchronization overhead and avoid overloading the master process. Communication scheme can be represented as a virtual network whose structure is described by the graph of subproblems. In this network, each node can communicate with its parent and child nodes. Worker processes communicate with each other asynchronously using queue of incoming and outgoing messages.

The whole task is described by working scenario that is a sequence of jobs such as graph building, computing the minimum cost, optimization, etc. Working scenario is provided by application that calls methods of the object representing the master process. Upon calling a particular method, the master process announces to worker processes start of a particular job and then waits for job termination. There are barrier synchronization points between jobs so the next job is started only when all processes are done with the previous job.

For example, consider implementation of the operation that calculates the minimum cost of a solution after the graph of subproblems has been constructed. On start of the operation the master process broadcasts to all worker processes the name of the cost function that will be used. Upon receiving this message, each worker process iterates through own list of graph nodes. For each terminal node, the cost of solution is calculated and passed to all parent nodes. Each intermediate node accumulates downstream messages until messages from all children are received. Then the minimum cost of solution for the node is calculated and passed to all parents. Following these rules, messages are propagated in the graph from the terminal nodes towards the root. The operation is completed when the root node receives messages from all children and calculates minimum cost of solution of the original problem.

References

1. GetPot project page. http://getpot.sourceforge.net/
2. PGFPlots project page. http://pgfplots.sourceforge.net/
3. yEd product page. http://www.yworks.com/en/products_yed_about.html
4. IEEE Standard for Information Technology – Portable Operating System Interface (POSIX). Shell and Utilities. IEEE Std 1003.1, 2004 Edition The Open Group Technical Standard. Base Specifications, Issue 6. Includes IEEE Std 1003.1-2001, IEEE Std 1003.1-2001/Cor 1-2002 and IEEE Std 1003.1-2001/Cor 2-2004. Shell (2004)
5. Brandes, U., Eiglsperger, M., Herman, I., Himsolt, M., Marshall, M.: GraphML progress report, Structural layer proposal. In: Mutzel, P., Junger, M., Leipert S. (eds.) Graph Drawing – 9th International Symposium, GD 2001, Vienna, Austria, September 23–26, 2001. Revised Papers, Lecture Notes in Computer Science, vol. 2265, pp. 501–512. Springer, Berlin (2002)

6. Message Passing Interface Forum: MPI: A message-passing interface standard, version 2.2. Technical report, High Performance Computing Center Stuttgart (HLRS) (2009)
7. Witten, I., Frank, E., Hall, M.: Data Mining: Practical Machine Learning Tools and Techniques. The Morgan Kaufmann Series in Data Management Systems, 3rd edn. Elsevier, Burlington (2011)

Final Remarks

The aim of this book is to extend the framework of dynamic programming and to create two new dynamic programming approaches: multi-stage optimization and bi-criteria optimization. Both approaches are applicable in the cases when there are two or more criteria of optimization on the considered set of objects. The first approach allows us to optimize objects sequentially relative to a number of criteria, to describe the set of optimal objects, and to count its cardinality. The second approach gives us the possibility to construct the set of Pareto optimal points and to study relationships between the considered criteria.

We concentrate on the study of four directions related to decision trees, decision rules and rule systems, element partition trees, and classic combinatorial optimization problems. For each of these directions, we created algorithms for multi-stage optimization, and for the first three directions – for bi-criteria optimization. We proved the correctness of these algorithms and analyzed their time complexity. We described classes of decision tables for which algorithms for decision trees, rules and rule systems have polynomial time complexity depending on the number of conditional attributes in the input tables. The algorithms for element partition trees are polynomial depending on the size of input meshes, and the algorithms for four classic combinatorial optimization problems studied in this book are polynomial depending on the input length.

For each of the considered four directions, we described different applications of the proposed techniques related to studying problem complexity, data mining, knowledge representation, machine learning, PDE solver optimization, and combinatorial optimization.

Future study will be devoted to the extension of the proposed methods to the decision tables with many-valued decisions and to inhibitory trees and rules. Decision tables with many-valued decisions are, sometimes, more adequate models of problems than usual decision tables. For some decision tables, inhibitory trees and rules describe more information contained in the table than decision trees and rules.

H. AbouEisha et al., *Extensions of Dynamic Programming for Combinatorial Optimization and Data Mining*, Intelligent Systems Reference Library 146,
https://doi.org/10.1007/978-3-319-91839-6

Index

H. AbouEisha et al., *Extensions of Dynamic Programming for Combinatorial Optimization and Data Mining*, Intelligent Systems Reference Library 146,
https://doi.org/10.1007/978-3-319-91839-6

Printed in the United States
By Bookmasters